MULTICORE
TECHNOLOGY

ARCHITECTURE, RECONFIGURATION, AND MODELING

Embedded Multi-Core Systems

Series Editors

Fayez Gebali and Haytham El Miligi
University of Victoria
Victoria, British Columbia

MULTICORE TECHNOLOGY

ARCHITECTURE, RECONFIGURATION, AND MODELING

Edited by
Muhammad Yasir Qadri
Stephen J. Sangwine

CRC Press
Taylor & Francis Group
Boca Raton London New York

CRC Press is an imprint of the
Taylor & Francis Group, an **informa** business

CRC Press
Taylor & Francis Group
6000 Broken Sound Parkway NW, Suite 300
Boca Raton, FL 33487-2742

First issued in paperback 2017

© 2014 by Taylor & Francis Group, LLC
CRC Press is an imprint of Taylor & Francis Group, an Informa business

No claim to original U.S. Government works

ISBN-13: 978-1-4398-8063-0 (hbk)
ISBN-13: 978-1-138-07250-3 (pbk)

Library of Congress Cataloging-in-Publication Data

Multicore technology : architecture, reconfiguration, and modeling / editors, Stephen J. Sangwine, Muhammad Yasir Qadri.
      pages cm. -- (Embedded multi-core systems)
   Includes bibliographical references and index.
   ISBN 978-1-4398-8063-0 (hardback)
   1. Multiprocessors. 2. Simultaneous multithreading processors. 3. Computer architecture. I. Samgwine, Stephen J. II. Qadri, Muhammad Yasir.

TK7895.M5M85 2013
621.39--dc23   2013019701

Visit the Taylor & Francis Web site at
http://www.taylorandfrancis.com

and the CRC Press Web site at
http://www.crcpress.com

Contents

vi

List of Figures

List of Tables

Preface

Multicore processor architectures are now mainstream even in applications such as mobile or portable telephones. For decades, computer architecture evolved through increases in the size and complexity of processors, and reductions in their cost and energy consumption. Eventually, however, there came a point where further increases in complexity of a single processor were less desirable than providing multiple cores on the same chip. The advent of the multicore era has altered many concepts relating to almost all of the areas of computer architecture design including core design, memory management, thread scheduling, application support, inter-processor communication, debugging, power management, and many more. This book provides a point of entry into the field of multicore architectures, covering some of the most researched aspects.

What to look for in it...

This book is targeted not only to give readers a holistic overview of the field but also to guide them to further avenues of research by covering the state-of-the-art in this area. The book includes contributions from renowned institutes across the globe with authors from the following institutes contributing to the book (ordered alphabetically):

Barcelona Supercomputing Center, Spain
Bengal Engineering and Science University, Shibpur, India
Boston University, Boston, USA
CEA LIST, Embedded Computing Lab, France
Eindhoven University of Technology, The Netherlands
Google Inc., USA
Jamia Millia Islamia (Central University), New Delhi, India
Laboratoire TIMA, Grenoble, France
Massachusetts Institute of Technology, USA
McGill University, Montreal, Canada
McMaster University, Canada
National University of Singapore, Singapore

University of Glasgow, Glasgow, UK
University of Massachusetts, Lowell, MA, USA
University of Texas at Austin, USA
VMware Inc., USA

The book is divided into five parts: Architecture and Design Flow, Parallelism and Optimization, Memory Systems, Debugging, and Networks-on-Chip.

The contents of each section are discussed in the following.

Architecture and Design Flow

This part contains three chapters.

Chapter 1, MORA: High-Level FPGA Programming Using a Many-Core Framework, presents an overview of the MORA framework, a high-level programmable multicore FPGA system based on a dataflow network of Processors-in-Memory. The MORA framework is targeted to simplify dataflow-based FPGA programming in C++ using a dedicated Application Programmer's Interface (API). The authors demonstrate an image processing application implemented using over a thousand cores.

Chapter 2, Implementing Time-Constrained Applications on a Predictable Multiprocessor System-on-Chip, presents a Synchronous Data Flow (SDF)-based design flow that instantiates different architectures using a template. The proposed design flow can generate an implementation of an application on a MPSoC while providing throughput guarantees to the application. Therefore the platform presented supports fast design space exploration for real-time embedded systems and is also extendable to heterogeneous applications.

Chapter 3, SESAM: A Virtual Prototyping Solution to Design Multicore Architectures for Dynamic Applications, presents an asymmetric MPSoC framework called SESAM. The MPSoC exploration environment can be used for a complete MPSoC design flow. It can help the design and sizing of complex architectures, as well as the exploration of application parallelism on multicore platforms, to optimize the area and performance efficiency of embedded systems. SESAM can integrate various instruction set simulators at the functional or cycle-accurate level, as well as different networks-on-chip, DMA, a memory management unit, caches, memories, and different control solutions to schedule and dispatch tasks. The framework also supports the energy modeling of the MPSoC design.

Parallelism and Optimization

This part contains two chapters.

Chapter 4, Verified Multicore Parallelism using Atomic Verifiable Operations, presents an extension to Atomic Verifiable Operation (AVOp) streams to support loops. An AVOp is the basic instruction in the Domain Specific Language (DSL) proposed by the authors. AVOp streams allow performance to be maximized by introducing an algorithm for scheduling across different threads of execution so as to minimize contention in a synchronous operation. The authors also present a verification algorithm that guarantees hazard avoidance for any possible execution order. This framework enables a programmer to express complex communication patterns and hide communication latencies in an approach similar to software pipelining of loops.

Chapter 5, Accelerating Critical Section Execution with Asymmetric Multi-Core Architectures, presents a mechanism for Accelerated Critical Sections (ACS) to reduce performance degradation due to critical sections. Critical sections are those sections of code that access mutually shared data among the cores. The principle of Mutual Exclusion dictates that threads cannot be allowed to update shared data concurrently; thus, accesses to shared data are encapsulated inside critical sections. This in effect can serialize threads, and reduce performance and scalability. Therefore, in order to avoid this performance loss, the authors propose acceleration of critical sections on a high-performance core of an Asymmetric Chip Multiprocessor (ACMP).

Memory Systems

Chapter 6, TMbox: A Flexible and Reconfigurable Hybrid Transactional Memory System, presents a multicore design space exploration tool called TMbox. This flexible experimental systems platform is based on an FPGA and offers a scalable and high-performance multiprocessor System-on-Chip (SoC) implementation that is configurable for integrating various Instruction Set Architecture (ISA) options and hardware organizations. Furthermore, the proposed platform is capable of executing operating systems and has an extensive support for Hybrid Transactional Memory Systems.

Chapter 7, EM^2: A Scalable Shared Memory Architecture for Large-scale Multicores, presents a technique to provide deadlock-free migration-based coherent shared memory to the Non-Uniform Cache Access (NUCA) family of architectures. Using the proposed Execution Migration Machine (EM^2), the authors claim to achieve performance comparable to directory-based architectures without using directories. Furthermore, the proposed scheme is both energy and area efficient.

Chapter 8, CAFÉ: Cache-Aware Fair and Efficient Scheduling for CMPs, introduces an efficient online technique for generating Miss Ratio Curves (MRCs) and other cache utility curves that uses hardware performance counters available on commodity processors. Based on these monitoring and inference techniques, the authors also introduce methods to improve the fairness and efficiency of CMP scheduling decisions.

Debugging

In Chapter 9, Software Debugging Infrastructure for Multi-Core Systems-on-Chip, the authors present an overview of the existing multithreaded software debugging schemes and discuss challenges that are being faced by the designers of multi-core systems-on-chip. The authors conclude that traditional debugging methods are not suitable for debugging concurrently executing multi-threaded software. Furthermore, the use of trace generation to complement traditional debugging methods is gaining traction and is expected to take an increased role in the debugging of future multi-threaded software. However, for trace generation based schemes, transfer of massive amounts of trace data off-the-chip for analysis is one of the major problems. The authors present an instruction-address trace compression scheme that aims to mitigate this problem.

Network-On-Chip

This section contains four chapters. Chapter 10, On Chip Interconnects for Multi-Core Architectures, presents a detailed study of the state of the art in interconnects used in multi-core architectures. The technologies discussed by the authors include Three Dimensional, Photonic, Wireless, RF Waveguide, and Carbon Nanotubes based Interconnects.

Chapter 11, Routing in Multi-Core NoCs, presents an overview and survey of routing topologies, router architecture, switching techniques, flow control, traffic patterns, routing algorithms, and challenges faced by the existing architectures for on-chip networks.

Chapter 12, Efficient Topologies for 3-D Networks-on-Chip, presents a comparison between mesh- and tree-based NoCs in a 3D SoC. The authors conclude that for 3D SoCs both mesh- and tree-based NoCs are capable of achieving better performance compared to the traditional 2D implementations. However, proposed tree-based topologies show significant performance gains in terms of network diameter and degree and number of nodes, and achieve significant reductions in energy dissipation and area overhead without any change in throughput and latency.

Finally, Chapter 13, Network-on-Chip Performance Evaluation Using an Analytical Method, presents an analytical performance evaluation method for NoCs that permits an architectural exploration of the network layer for a given application. Additionally, for a given network architecture, the method allows examination of the performance of different mappings of the application on the NoC. The proposed method is based on the computation of probabilities and contention delays between packets competing for shared resources, and provides a comprehensive delay analysis of the network layer.

Target Audience

The book may be used either in a graduate-level course as part of the subject of embedded systems, computer architecture, and multi-core systems on-chips, or as a reference book for professionals and researchers. It provides a clear view of the technical challenges ahead and brings more perspective into the discussion of multi-core embedded systems. This book is particularly useful for engineers and professionals in industry for easy understanding of the subject matter and as an aid in both software and hardware development of their products.

Acknowledgments

We thank the team at CRC Press: Nora Konopka, Publisher, for supporting our proposal for this book; Kari Budyk and Michele Dimont, for keeping us on track and for assisting us promptly and courteously with our many questions and queries; and Shashi Kumar, for helping us around our LaTeX difficulties.

We also thank our wives, Dr Nadia N. Qadri and Dr Elizabeth Shirley, who, although they have never met, have shared an experience that we inflicted on them, as we worked long hours editing this book, when we should have been spending time with them. We have often read such thanks (or indeed apologies) in other books, but now we understand why we must acknowledge their contribution to this book. Nadia and Elizabeth, thank you both for your support and patience.

Muhammad Yasir Qadri
Islamabad, Pakistan

Stephen J. Sangwine
Colchester, United Kingdom

December 2012

Editors

Muhammad Yasir Qadri was born in Pakistan in 1979. He graduated from Mehran University of Engineering and Technology in Electronic Engineering. He obtained his PhD in Electronic Systems Engineering from the School of Computer Science and Electronic Engineering, University of Essex, UK. His area of specialization is energy/performance optimization in reconfigurable MPSoC architectures. Before his time at Essex, he was actively involved in the development of high-end embedded systems for commercial applications. He is an Approved PhD Supervisor by the Higher Education Commission of Pakistan, and is currently working as a Visiting Faculty Member at HITEC University, Taxila, Pakistan.

Stephen J. Sangwine was born in London in 1956. He received a BSc degree in Electronic Engineering from the University of Southampton, Southampton, UK, in 1979, and his PhD from the University of Reading, Reading, UK, in 1991, for work on digital circuit fault diagnosis. He was a Lecturer in the Department of Engineering at the University of Reading from 1985 – 2000, and since 2001 has been a Senior Lecturer at the University of Essex, Colchester, UK. His interests include color image processing and vector signal processing using hypercomplex algebras, and digital hardware design and test.

Contributors

Christopher Kumar Anand
Department of Computing and
   Software
McMaster University
Hamilton, Ontario, Canada

Caaliph Andriamisaina
CEA LIST
Gif-sur-Yvette, France

Abdul Quaiyum Ansari
Department of Electrical Engineering
Jamia Millia Islamia
New Delhi, India

Oriol Arcas
Barcelona Supercomputing Center
Universitat Politècnica de Catalunya
Barcelona, Spain

Sai Rahul Chalamalasetti
Department of Electrical and
   Computer Engineering
University of Massachusetts
Lowell, MA, USA

Myong Hyon Cho
Massachusetts Institute of
   Technology
Cambridge, MA, USA

Henk Corporaal
Department of Electrical Engineering
Eindhoven University of Technology
Eindhoven, The Netherlands

Adrián Cristal
Barcelona Supercomputing Center
CSIC — Spanish National Research
   Council
Barcelona, Spain

Tuhin Subhra Das
Department of Information
   Technology
Bengal Engineering and Science
   University
Shibpur, India

Srinivas Devadas
Massachusetts Institute of
   Technology
Cambridge, MA, USA

Michal Dobrogost
Department of Computing and
   Software
McMaster University
Hamilton, Ontario, Canada

Sahar Foroutan
TIMA Laboratory, SLS Team
Grenoble, France

Prasun Ghosal
Department of Information
   Technology
Bengal Engineering and Science
   University
Shibpur, India

Warren J. Gross
Department of Electrical and
    Computer Engineering
McGill University
Montreal, Canada

Alexandre Guerre
Embedded Computing Lab
CEA LIST
Gif-sur-Yvette, France

Roel Jordans
Department of Electrical Engineering
Eindhoven University of Technology
Eindhoven, The Netherlands

Wolfram Kahl
Department of Computing and
    Software
McMaster University
Hamilton, Ontario, Canada

Mohammad Ayoub Khan
Center for Development of Advanced
    Computing
Noida, India

Omer Khan
Department of Electrical and
    Computer Engineering
University of Connecticut
Storrs, CT, USA

Akash Kumar
Department of Electrical and
    Computer Engineering
National University of Singapore
Singapore

Mieszko Lis
Massachusetts Institute of
    Technology
Cambridge, MA, USA

Martin Margala
Department of Electrical and
    Computer Engineering
University of Massachusetts
Lowell, MA, USA

Bojan Mihajlović
Department of Electrical and
    Computer Engineering
McGill University
Montreal, Canada

Onur Mutlu
Department of Electrical and
    Computer Engineering
Carnegie Mellon University
Pittsburgh, PA, USA

Frédéric Pétrot
TIMA Laboratory, SLS Team
Grenoble, France

Soumyajit Poddar
School of VLSI Technology
Bengal Engineering and Science
    University
Shibpur, India

Tanguy Sassolas
Embedded Computing Lab
CEA LIST
Gif-sur-Yvette, France

Hamed Sheibanyrad
TIMA Laboratory, SLS Team
Grenoble, France

Keun Sup Shim
Massachusetts Institute of
    Technology
Cambridge, MA, USA

Satnam Singh
Google, Inc.
Mountain View, CA, USA

Nehir Sonmez
Barcelona Supercomputing Center
Universitat Politècnica de Catalunya
Barcelona, Spain

Sander Stuijk
Department of Electrical Engineering
Eindhoven University of Technology
Eindhoven, The Netherlands

M. Aater Suleman
Department of Electrical and
    Computer Engineering
The University of Texas at Austin
Austin, TX, USA

Osman S. Unsal
Barcelona Supercomputing Center
Universitat Politècnica de Catalunya
Barcelona, Spain

Wim Vanderbauwhede
School of Computing Science
University of Glasgow
Glasgow, Scotland

Nicolas Ventroux
Embedded Computing Lab
CEA LIST
Gif-sur-Yvette, France

Carl A. Waldspurger
(Formerly at) VMware Inc.
Palo Alto, CA, USA

Richard West
Department of Computer Science
Boston University
Boston, MA, USA

Puneet Zaroo
VMware Inc.
Palo Alto, CA, USA

Xiao Zhang
Google, Inc.
Mountain View, CA, USA

Željko Žilić
Department of Electrical and
    Computer Engineering
McGill University
Montreal, Canada

Acronyms

AML	Average Memory Latency	LRU	Least Recently Used
API	Application Programming Interface	LUT	Look-Up Table
		MB	Megabytes
ASIC	Application-Specific Integrated Circuit	MOC	Model of Computation
		MPKC	Misses Per Kilo-Cycle
BRAM	Block RAM	MPKI	Misses Per Kilo-Instruction
CABA	Cycle Accurate Bit Accurate	MPKR	Misses Per Kilo-Reference
CMP	Chip Multi-Processor	MPSoC	Multi-Processor System-on-Chip
CPI	Cycles Per Instruction		
CPU	Central Processing Unit	MRC	Miss-Ratio Curve
DCT	Discrete Cosine Transform	NI	Network Interface
DDR	Double Data Rate	NoC	Network-on-Chip
DMA	Direct Memory Access	NUCA	Non-Uniform Cache Access
DRAM	Dynamic RAM	NUMA	Non-Uniform Memory Access
DSP	Digital Signal Processor		
DWT	Discrete Wavelet Transform	OS	Operating System
ECC	Error-Correcting Code	OSI	Open System Interconnection
FPGA	Field Programmable Gate Array		
		PIM	Processor In Memory
FPU	Floating-Point Unit	QoS	Quality of Service
GALS	Globally Asynchronous Locally Synchronous	RA	Remote Access
		RAM	Random Access Memory
GB	Gigabytes	RISC	Reduced Instruction Set Computer
GHz	Gigahertz		
GPU	Graphics Processing Unit	RTL	Register Transfer Level
HAL	Hardware Abstraction Layer	SDAR	Sampled Data Address Register
HDL	Hardware Description Language		
		SoC	System-on-Chip
HLL	High-Level Language	SPEC	Standard Performance Evaluation Corporation
HPC	High Performance Computing		
		SRAM	Static RAM
I/O	Input/Output	SDRAM	Synchronous Dynamic RAM
IP	Intellectual Property	TDM	Time Division Multiplexing
ISA	Instruction Set Architecture	TLB	Translation Look-Aside Buffer
ISS	Instruction Set Simulator		
ITRS	International Technology Roadmap on Semiconductors	TLM	Transaction Level Modeling
		VC	Virtual Channel
KB	Kilobytes	Vtime	Virtual Time

Part I

Architecture and Design Flow

1

MORA: High-Level FPGA Programming Using a Many-Core Framework

Wim Vanderbauwhede

School of Computing Science, University of Glasgow, Glasgow, UK

Sai Rahul Chalamalasetti and Martin Margala

Department of Electrical and Computer Engineering, University of Massachusetts, Lowell, MA, USA

CONTENTS

This chapter presents an overview of the current state of the MORA framework, a high-level programmable multicore FPGA system based on a dataflow network of Processors-in-Memory. The aim of the MORA framework is to simplify dataflow-based FPGA programming while still delivering excellent performance, by providing a streaming dataflow framework that can be programmed in C++ using a dedicated Application Programmer's Interface (API). Many of the restrictions common to most other C-to-gates tools do not apply to MORA because of the adoption of processors rather than LUTs as the smallest unit of the design. MORA's processors are unique as they are specialised in terms of instruction set, data path width, and memory size for the particular section of the program that runs on them. As a result, we have demonstrated an image processing application implemented using over a thousand cores.

The chapter starts with the background and rationale for this work and the state of the art. The subsequent sections discuss in detail the hardware and software aspects of the MORA framework: architecture, hardware infrastructure, and tool chain for the FPGA; design of the MORA-C++ API, the Intermediate Representation, compiler, and assembler. The final sections present and discuss benchmark results for several streaming data processing algorithms to demonstrate the performance of the current system, and outline avenues for future research.

1.1   Overview of the State of the Art in High-Level FPGA Programming

Media processing architectures and algorithms have come to play a major role in modern consumer electronics, with applications ranging from basic communication devices to high-level processing machines. Therefore architectures and algorithms that provide adaptability and flexibility at a very low cost have become increasingly popular for implementing contemporary multimedia applications. Reconfigurable or adaptable architectures are widely being seen as viable alternatives to extravagantly powerful General Purpose Processors (GPP) as well as tailor-made but costly Application Specific Integrated Circuits (ASICS). Over the last few years, FPGA devices have grown in size and complexity. As a result, many applications that were previously restricted to ASIC implementations can now be deployed on reconfigurable platforms. Reconfigurable devices such as FPGAs offer the potential of very short design cycles and reduced time to market.

However, with the ever increasing size and complexity of modern multimedia processing algorithms, mapping them onto FPGAs using Hardware Description Languages (HDLs) like VHDL or Verilog provided by many FPGA vendors has become increasingly difficult. To overcome this problem several groups in academia as well as industry have engaged in developing high-level language support for FPGA programming. The most common approaches fall into three main categories: *HLL-to-gates*, *system builders*, and *soft processors*.

The *HLL-to-gates* design flow starts from a program written in a High-Level Language (HLL, typically a dialect of C) with additional keywords and/or pragmas, and converts these programs into a Hardware Description Language (HDL) such as Verilog or VHDL. Examples of commercial tools in this category are Handel-C (Sullivan, Wilson, and Chappell 2004), Impulse-C (Santambrogio et al. 2007), Xilinx' AutoESL (Cong 2008), and Maxeler's /MaxCompiler/ (Howes et al. 2006). Academic solutions include Streams-C (Gokhale et al. 2000), Trident (Tripp et al. 2005), and ROCCC (Buyukkurt, Guo, and Najjar 2006). Despite the advantage of a shorter learning curve for programmers to understand these languages, a significant disadvantage of this C-based coding style is that it is customized to suit Von Neumann processor architectures, which cannot fully extract parallelism out of FPGAs.

By *system builders* we mean solutions that will generate complex IP cores from a high-level description, often using a wizard. Examples are Xilinx' CoreGen and Altera's Mega wizard. These tools greatly enhance productivity but are limited to creating designs using parameterized predefined IP cores. Graphical tools such as MATLAB-Simulink and NI LabVIEW also fall into this category.

Finally, *soft processors* have increasingly been seen as strong players in this category. Each FPGA vendor provides its own soft cores such as Microblaze and Picoblaze from Xilinx and Nios from Altera. However, the traditional architectures with shared memory access and mutual memory access are far from ideal to exploit the inherent parallelism inherent in FPGAs for media processing applications. To address this problem, different processor architectures are needed. One such architecture has been proposed and commercialized by Mitrionics: the 'Mitrion Virtual Processor' (MVP) is a massively parallel processor that can be customized for the specific programs that run on it (Kindratenko, Brunner, and Myers 2007). Other alternatives are processor arrays such as proposed by Craven, Patterson, and Athanas (2006), which are based on the OpenFire processor or the MOLEN reconfigurable processor compiler (Panainte, Bertels, and Vassiliadis 2007).

1.2 Introduction to the MORA Framework

In this section we introduce the MORA framework. We discuss the application domain and rationale for the framework and introduce the main concepts and building blocks of the MORA framework, the MORA system abstraction, and the adopted approach to high-level FPGA programming.

1.2.1 MORA Concept

The MORA framework (Vanderbauwhede et al. 2009, 2010) is targeted at the implementation of high performance streaming algorithms. It allows the application developer to write an FPGA application using a C++ API (MORA-C++) which essentially implements a Communicating Sequential Processes (CSP) paradigm (Hoare 1978). The toolchain converts the program into a compile-time generated network (a directed dataflow graph) where every node is implemented on a compile-time configurable Processor-in-Memory (PIM), called the MORA Reconfigurable Cell (RC) (Chalamalasetti et al. 2009). Each RC is tailored (in terms of instruction set and memory size) to the specific code implementing the process. As a result, the MORA framework fundamentally differs from other HLL-to-gates languages as it allows memory-based constructs and algorithms (e.g., stack machines and pointer-based data structures).

1.2.2 MORA Tool Chain

Figure 1.1 shows the MORA tool chain, which consists of the MORA-C++ compiler, the MORA assembler, and the FPGA back-end (currently targeting the SGI RC-100 platform).

From the MORA-C++ source code, the compiler emits MORA Intermediate Representation (IR) language, which is transformed into Verilog and implemented using the Xilinx ISE tools. The assembler can also transform the IR language into a cycle-approximate SystemC model, allowing fast simulation of the design. Combined with the ability to compile the source code using g++, this provides a powerful development system allowing rapid design iterations.

1.3 The MORA Reconfigurable Cell

The MORA architecture consists of a compile-time generated network of Reconfigurable Cells. Although MORA supports access to shared memory, the

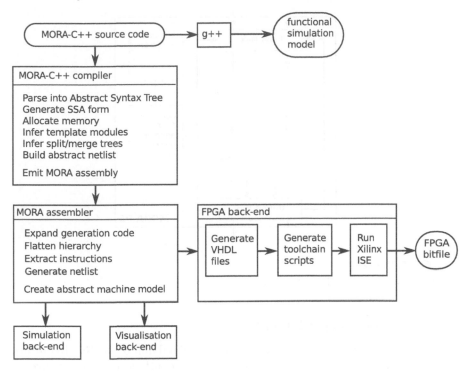

FIGURE 1.1

MORA-C++ tool chain

memory architecture is distributed: storage for data is partitioned among RCs by providing each RC with internal data memory. As each RC is a Processor-in-Memory (PIM), computations can be performed close to memory, thereby reducing memory access time. This architecture results in the high memory access bandwidth needed to efficiently support massively parallel computations required in multimedia applications while maintaining generality and scalability of the system.

The external data exchange is managed by an I/O Controller which can access the internal memory of the input and output RCs through standard memory interface instructions. The internal data flow is controlled in a distributed manner through a handshaking mechanism between the RCs.

Each RC (Figure 1.2) consists of a Processing Element (PE) for arithmetic and logical operations with configurable word size, a small (order of 1 KB) dual-port data memory implemented using FPGA block RAMs, and a Control Unit (CU) with a small instruction memory. The PE forms the main computational unit of the RC; the control unit synchronises the instruction queue within the RC and also controls inter-cell communication, allowing each RC to work with a certain degree of independence. The architectures of the PE and CU are discussed in the following sub-sections.

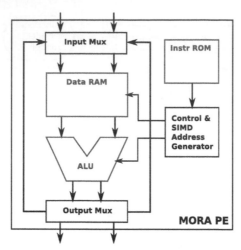

FIGURE 1.2
MORA reconfigurable cell (RC)

1.3.1 Processing Element

The Processing Element is the main computational unit of the RC. In our prior work (Purohit et al. 2008) we optimized the data path design and organization for efficient single-cycle unsigned integer arithmetic operations. As the MORA RC requires additional functionality, we modified the PE to include signed arithmetic, logic, shifting, and comparison operations.

Figure 1.3 shows the organization of the modified PE. It includes the hybrid (signed/unsigned) arithmetic data path along with additional blocks for shifting and comparison operations. The PE is designed by using preprocessor parameters, so that, depending on the instructions assigned to the RC, the required modules of the PE are instantiated. This parameterized approach results in a dramatic improvement in resource utilization. The arithmetic data path is organized to provide single-cycle addition, subtraction, and multiplication operations. The PE also provides two sets of registers at the input and output to enable accumulation-style operations, as often required for media processing applications. Output is available at the registers every clock cycle.

1.3.2 Control Unit and Address Generator

The control unit provides the handshaking signals between memory and data path, and ensures that the two units work in perfect synchronization with each other. The unit consists of a small instruction memory (compile-time configurable, typically 10–100 instructions), three address generators (one for each operand), instruction decoders, and instruction counters. The wide instruction word (the actual size depends on the size of the local memory, e.g.,

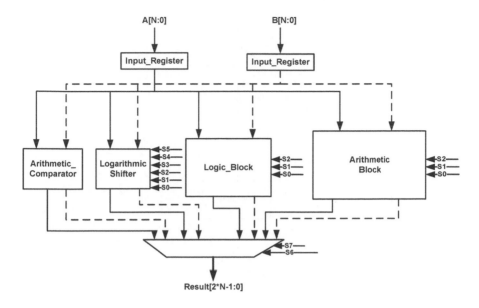

FIGURE 1.3
Block diagram of the MORA PE

92 bits for a 512-word memory) encodes the operation and base addresses for an instruction's operands, and the output data set and address offsets for traversing through memory, as well as the number of times a specific operation is to be performed. The overall flow of the control unit is as shown in Figure 1.4.

The address generator is shown in Figure 1.5. It accepts four data fields: base_address, step, skip, and subset. The base address is initially loaded into the address generator, and, depending on the values of step, skip, and subset, the address of the next memory location to fetch the data is calculated. The three fields allow the controller to move anywhere throughout the available data memory. The address generation algorithm can be written as shown in Algorithm 1.1. The address generator thus generates the range of addresses on which a given instruction is to be performed. This is a key feature for media processing applications which frequently involve operations on matrices and vectors of data.

1.3.3   Asynchronous Handshake

To minimize the impact of communication networks on the power consumption of the array, each RC is equipped with a simple and resource efficient communication technique. As every RC can in principle operate at a different clock speed, an asynchronous handshake mechanism was implemented. As

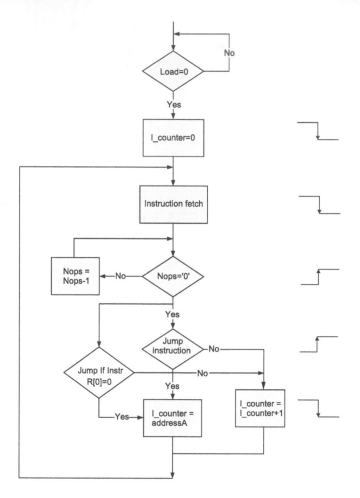

FIGURE 1.4
MORA control unit flow chart

MORA is a streaming architecture, a two-way communication mechanism is required, one to communicate with the upstream RCs and another to communicate with the downstream RCs. Altogether, a total of four communication I/O signals are used by each RC to communicate with the other RCs efficiently in streaming fashion. They are described as follows:

- *rc_rdy_up* is an output signal signifying that the RC is idle and ready to accept data from upstream RCs.

- *rc_rdy_down* is an input signal signifying that the downstream RCs are idle and ready to accept new data.

Algorithm 1.1: Address Generation Algorithm

```
address = base_address
if nops > 1
  tsubset = subset
  for 1 .. nops
    tsubset -= 1
    if tsubset == 0
      address += skip
      tsubset = subset
    else
      address += step
    end
  end
end
```

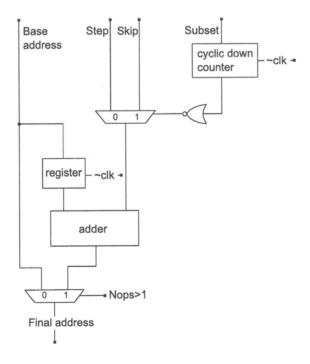

FIGURE 1.5
MORA address generator

- *data_rdy_down* is an output signal asserted when all the data transfers to the downstream RCs are completed.

- *data_rdy_up* is an input signal to the RC corresponding to the *data_rdy_down* signal from upstream RCs.

Each RC can accept inputs either from two output ports of a single RC or from two individual ports of different RCs. The output of each RC can be routed to, at most, four different RCs. In order to support multiple RC connections to a single cell, a two-bit vector is used for *data_rdy_up* (data_rdy_up[1:0]) and a four-bit vector for *rc_rdy_down* (rc_rdy_down[3:0]).

1.3.4   Execution Model

The RC has two operating modes: processing and loading. When the RC is operating in processing mode, it can either write the processed data back into internal memory or write to a downstream RC. For a formal description of MORA's execution model we refer the reader to Vanderbauwhede et al. (2009). Each RC has two execution modes while processing input data. One is sequential execution used for normal instructions (ADD, SUB, etc.) with write back option. The second is pipelined execution for accumulation and instructions with write out option. Instructions with sequential execution take three clock cycles to complete, with each clock cycle corresponding to reading, executing, and writing data to the RAM. A prefetching technique is used for reading instructions from the instruction memory; this involves reading a new instruction word while performing the last operation of the previous instruction. This approach enables the design to save one clock cycle for every new instruction.

For pipelined operations the controller utilizes the pipelining stage between the RAM and PE. This style of implementation allows the accumulation and write out operations to complete in $n + 2$ clock cycles. The latency of 2 clock cycles results from reading and execution of the first set of operands. The single-cycle execution for instruction with write out option makes the RC very efficient for streaming algorithms.

1.4   The MORA Intermediate Representation

The aim of the MORA Intermediate Representation (IR) language is to serve as a compilation target for high-level languages such as MORA-C++ whilst at the same time providing a means of programming the MORA processor array at a low level.

The language consists of three components: a *coordination* component which permits expression of the interconnection of the RCs in a hierarchical fashion, an *expression* component which corresponds to the conventional assembly languages for microprocessors and digital signal processors (DSPs), and a *generation* component which allows compile-time generation of coordination and expression instances.

1.4.1 Expression Language

The MORA expression language is an imperative language with a very regular syntax similar to other assembly languages: every line contains an instruction which consists of an operator followed by a list of operands. The main differences with other assembly languages are

- Typed operators: the type indicates the word size on which the operation is performed, e.g., bit, byte, short.

- Typed operands: operands are tuples indicating not only the address space but also the data type, i.e., word, row, column, or matrix.

- Virtual registers and address banks: MORA has direct memory access and no registers (an alternative view is that every memory location is a register). Operations take the RAM addresses as operands; however, 'virtual' registers indicate where the result of an operation should be directed (RAM bank A/B, output L/R/both)

1.4.1.1 Instruction Structure

An instruction is of the general form:

instr ::= *op nops dest opnd*+

op ::= *mnemonic*:(B|N|C|S|I|L|F|D)?

dest ::= *virtreg addrtup*

opnd ::= *addrtup* | *const*

virtreg ::= MEM|OUTL|OUTR

addrtup ::= (*ram_id*:)?*addr*(:*type*)?

ram_id ::= D|S

addr ::= 0..(MEMSZ-1)

type ::= (W| *num* x *num*)(R|F)?P?

const ::= C:*num*

num ::= −(MEMSZ/2−1)..(MEMSZ/2−1)

We illustrate these characteristics with an example. The instruction for a multiply-accumulate of the first row of an 8×8 matrix of 16-bit integers with the first column of an 8×8 matrix in the data memory reads in full:

```
MULTACC:S 8 OUTL D:0:W D:0:1x8 D:64:8x1
```

The '8' indicates that eight operations need to be performed on a range of addresses. The 'OUTL' is a 'virtual register' indicating that the resulting 2 bytes must be written to the left output port. The ':S' indicates the type of operand of the operation, in this case 'short' (2 bytes). The groups 0:W, etc., are the address tuples. They encode the address and the type of the data stored at the address, in this case a short integer (2 bytes) stored at address 0 of the data memory. The tuple D:64:8x1 encodes an 8 × 1 column of shorts starting at address 64 of the data memory.

1.4.1.2   Address Types

As discussed in Section 1.3.2, the RC supports complex address scan patterns through the use of 4 fields in the instruction word: base_address, step, subset, and skip. The MORA assembler supports a subset of all possible values of step, subset, and skip through its type system. The type component of the address tuple indicates the nature of the data structure referenced by the base address *(ram_id:addr)*:

W: word (single byte)

num   x   *num*: $N \times N$ matrix

The type suffix (F—R) indicates a forward or reverse scan direction. Thus MORA's simple address type system supports the typical vector operations required for $N \times N$ matrix manipulation.

    The final optional suffix P indicates that the address is a pointer rather than a value.

1.4.1.3   Operation Types

The operator of an instruction can be explicitly typed, indicating the length of the word on which the operation should be performed. This information is used to generate the step and the virtual output register. As the MORA RAM is byte-addressable, operation types B (bit) and N (nybble) have no effect on the address generation but result in single-byte output; operations on multiple bytes (types S, I, and L, respectively, 2, 4, and 8 bytes) result in a step of the number of bytes; the assembler generates the individual byte operations that make up the multi-byte operation.

    The operation types F and D stand for Float and Double, meaning IEEE-754 single- and double-precision floating point operations. MORA's floating point support is discussed in Section 1.5.4.

1.4.1.4   Address Spaces

The data memory of the RC stores words of the size of the largest operation type in the program for a given RC. For example, if the program contains only

8-bit instructions, the data memory will be 8 bits wide. However, it is possible that the data memory will be larger than 2^8, in which case the addresses will not fit in the data memory locations. This is only an issue if the program uses pointers for dynamic addressing because for static addressing, the size of the addresses in the instructions is configured at compile time to accommodate the full address space. To store pointers, we create a separate scratchpad memory (denoted with 'S:' in the address tuple) with the word size equal to the address size. As the number of instructions in a typical MORA program is small, the number of pointers will be small too, and the scratchpad can be implemented using synthesized RAM.

1.4.2 Coordination Language

MORA's coordination language is a compositional, hierarchical netlist-based language similar to hardware design languages such as Verilog and VHDL. The language consists of *primitive definitions*, *module definitions*, *module templates*, and *instantiations*.

Primitives describe a MORA RC. They have two input ports and two output ports. Modules are groupings of instantiations, very similar to non-RTL Verilog. Modules can have variable numbers of input and output ports. Instantiations define the connections between different modules or RCs, again very similar to other netlist-based languages.

1.4.2.1 Primitive Definitions

Primitives describe a MORA RC and are defined as `prim_name{...}`, e.g., a primitive to compute a determinant of a 2×2 matrix would be

```
DET2x2 {
MULT:I 1 MEM 64:W 0:W 9:W
MULT:I 1 MEM 65:W 1:W 8:W
ADD:I 1 OUT1 0:W 64:W 65:W
}
```

1.4.2.2 Instantiations

Instances are defined as $(net_{out1}, \ldots) = name(net_{in1}, \ldots)$; unconnected ports are marked with a '_'.

1.4.2.3 Module Definitions

Modules are groupings of instantiations, very similar to non-RTL Verilog. As modules can have variable numbers of input and output ports (but no inout ports), the definition is `module_name(inport1,inport2,...) { ...} (outport1,outport2,...)`. For example, a module to compute 16-bit addition can be built out of 8-bit addition primitives (`ADD8`) as follows:

```
ADD16 (b1,b0,a1,a0) {
(c0,z0) = ADD8 (b0,a0)
(c1,s1) = ADD8 (b1,a1)
(_,z1) = ADD8 (c0,s1)
} (c1,z1,z0)
```

1.4.3   Generation Language

The generation language is an imperative mini-language with a simple and clean syntax inspired mainly by Ruby (Thomas, Hunt, and Fowler 2001). The language acts similarly to the macro mechanism in C, i.e., by string substitution, but is much more expressive.

The generation language allows instructions to be generated in loops or using conditionals. The generation language can also be used to generate module definitions through a very powerful mechanism called module templates. Instantiation of a module template results in generation of a particular module (specialisation) based on the template parameters. This is similar to the way template classes are used in C++.

1.4.3.1   Instruction Generation

Although MORA supports pointer-based access, static addressing is often preferable as it requires fewer cycles. The generation language allows static instructions to be generated in loops or using conditionals. For example, a multiplication of two 8×8 matrices requires 64 multiply-accumulate instructions (every instruction multiplies a row with a column). Using the generation language we can write

```
for i in 0..56 step 8
for j in 0..7
out = i+j
MAC D:out:W D:i:1x8 (j+64):8x1
end
end
```

The variables defined using the generation language syntax are interpolated in the instruction string.

1.4.3.2   Module Templates

The generation language is a complete language with support for string and list manipulation. As a result it can also be used to generate module definitions through a very powerful mechanism called *module templates*.

Instantiation of a module template results in generation of a particular module based on the template parameters. As a trivial example, consider a generator that will generate a primitive with N additions:

```
ADD<n> {
for i in 0..n-1
ADD 1 OUT1 i i (i+64)
end
}
```

On instantiation of this generator for, e.g., N=8, $(z1,z0)$ = ADD<8> $(a1,a0)$, the primitive definition ADD<8> with 8 ADD statements will be generated.

The example module definition from Section 1.4.2.3 can be turned into a template as follows:

```
TMPL16<prim> (b1,b0,a1,a0) {
(c0,z0) = prim (b0,a0)
(c1,s1) = prim (b1,a1)
(_,z1) = prim (c0,s1)
} (c1,z1,z0)
```

The instantiation of this template for PRIM = ADD8 specializes it into a 16-bit addition:

```
(c,z1,z0) = TMPL16<ADD8> (b1,b0,a1,a0)
```

For completeness we note that the number of template parameters is unlimited and that they can be used anywhere in the module definition, including the port statements. The module template system makes it possible to provide a library of configurable standard modules that can be targeted by a high-level language compiler, thus to a large extent abstracting the physical architecture. In particular, this approach is used to implement 16-bit and 32-bit arithmetic.

1.4.4   Assembler

The MORA processor network is configured for a given program by configuring the instruction and data memories for each RC and connecting the RCs as required. The assembler generates the required memory configurations and the interconnect configurations from the Intermediate Representation.

1.4.4.1   Memory Configuration

The data and instruction memory configurations are provided using external files at synthesis time. As every RC can have a different memory configuration, a template-driven generator is used to create multiple instances of the configuration file templates. It then generates and runs a script which builds the actual memories.

1.4.4.2   RC Generation

The RC contains the memories for data and instructions, the PE, and the control unit. Each of these has to be specialized based on the program. The assembler performs an analysis to determine the size of the instruction memory and the required instructions for the PE and control unit. Based on this information the template-driven generator creates specialized instances for every enclosing module.

1.4.4.3   Interconnect and Toplevel Generation

The final step of the generator instantiates all the generated RCs and creates the required interconnections. This is the most complicated step in the process as the generator must infer the control nets as well as wiring up the data nets, and must also correctly strap unused nets to ground or leave them open.

1.4.4.4   FPGA Configuration Generation

Using a template for a Xilinx ISE project in Tcl, the assembler finally builds the complete project including synthesis, place and route, and bitfile generation. All the steps are completely automated, resulting in a truly high-level FPGA programming solution.

1.5   MORA-C++ API

MORA-C++ is a C++ API for high-level programming of FPGAs using the MORA framework. The API is used to describe the connections between RCs or groups of RCs (modules). It essentially implements a netlist language using C++ function templates.

The rationale for this approach (i.e., giving the programmer full control over the RC functionality and interconnections) is based on performance: to write a high-performance program for a given architecture requires in-depth knowledge of the architecture. For example, to write high-performance C++ code one needs a deep understanding of stack and heap memory models, cache, bus access, and I/O performance. Conversely, to write a high-performance program for MORA, one needs to understand the MORA architecture and make the best possible use of it. Therefore we do not attempt to hide the architecture from the developer but we expose it through a high-level API. On the other hand, because of the processor array abstraction, there is no need for the programmer to have in-depth knowledge of the FPGA architecture.

An RC is modeled as a function; input ports are mapped to the function arguments and output ports to a tuple-type return value:

```
tuple<T1,T2,...> rc (T3 in1, T4 in2,...) {
// ...
tuple< T1, T2,... > res(out1,out2,...);
return res;
}
```

The nets are modeled as tuple-type variables, e.g.,

```
tuple<T1,T2> n34= rc1(n1,n2,...);
tuple<T3> n5 = rc2(n34.get<1>);
tuple<T4> n6 = rc3(n34.get<2>);
```

Plain C/C++ is used to describe the functionality of the RC or module. As the MORA RC is a complete Harvard-architecture PIM, there are in principle no restrictions on the language features.

MORA-C++ uses a static type system with operator overloading to determine the compilation route for given expressions and automatic static memory allocation. The most significant feature of MORA-C++ is that the local memory of each RC is separate, requiring the compiler to allocate memory across the network of RCs. For more details on the MORA-C++ language, including benchmark examples, we refer to Vanderbauwhede et al. (2010).

1.5.1 Key Features

The MORA-C++ API relies heavily on the type system to determine the compilation route for given expressions. Because MORA is targeted at vector and matrix operations, these are fundamental types. Operators are overloaded to support powerful matrix and vector expressions. The type system is also used to infer RCs to split and merge signals, so that there is no need for the developer to explicitly instantiate them. Finally, MORA-C++ uses automatic static memory allocation. Apart from the powerful syntactic constructs, this is probably its most significant feature. Static memory allocation is of course standard in C/C++, but essentially the compiler assumes that the memory is infinite. In MORA, the local memory of each RC is very small, requiring the compiler to check whether there is sufficient memory available for a given program. Furthermore, the allocation is distributed across the RCs, as opposed to the common shared-memory programming model.

1.5.2 MORA-C++ by Example

In this section we illustrate the features of MORA-C++ using an implementation of a DWT and a DCT algorithm as examples.

1.5.2.1 Discrete Wavelet Transform

As an example application to illustrate the features of the MORA assembly language we present the implementation of the Discrete Wavelet Transform (DWT) algorithm. An 8-point LeGall wavelet transform is implemented using a pipeline of 4 RCs; each RC computes the following equations:

$$y_i = x_i - (x_{i-1} + x_{i+1})/2$$
$$y_{i-1} = x_{i-1} + (y_i + y_{i-2})/4$$

The MORA-C++ code for the pipeline stages is implemented as a single function template:

```
template <int N,typename TL,typename TR>
UCharPair dwt_stage(TL x,TR y_in) {
UChar y_l;UChar y_r;
if (N==6) {
    y_r = x[1] - x[0]/2;
} else {
    y_r = x[1] - (x[0]+x[2])/2;
}
if (N==0) {
    y_l = x[0] + y_r/4;
} else {
    y_l = x[0] + (y_r+y_in)/4;
}
UCharPair out(y_l,y_r); return out;
}
```

Using this template, the complete DWT algorithm becomes

```
Pair<Row8,Nil> dwt (Row8 inA) {
    vector<UChar> v012 = inA.slice(0,2);
    vector<UChar> v234 = inA.slice(2,4);
    vector<UChar> v456 = inA.slice(4,6);
    vector<UChar> v67  = inA.slice(6,7);
    Row3 x012(v012);
    Row3 x234(v234);
    Row3 x456(v456);
    Row2 x67(v67);
    Row8 ny;
    UCharPair res01 =
dwt_stage<0,Row3,Nil>(x012,_);
    ny[1] = res01.left; ny[0] = res01.right;
    UCharPair res23 =
dwt_stage<2,Row3,UChar>(x234,ny[1]);
    ny[3] = res23.left; ny[2] = res23.right;
    UCharPair res45 =
dwt_stage<4,Row3,UChar>(x456,ny[3]);
    ny[5] = res45.left; ny[4] = res45.right;
    UCharPair res67 =
dwt_stage<6,Row2,UChar>(x67,ny[5]);
    ny[6] = res67.left; ny[7] = res67.right;
    Pair< Row8, Nil > res(ny,_);
    return res;
}
```

The example illustrates several features of MORA-C++:

Data Types

The type UCharPair is a typedef for Pair<UChar,UChar>; a Pair is the fundamental template class used for returning data from an RC (for higher-level modules with more than two output ports, there is a Tuple class). The Pair has accessors left and right for accessing its elements.

MORA-C++ defines a number of signed scalar types, Char, Short, Int, Long, and unsigned versions UChar, etc. These map to 1, 2, 4, 8 bytes, respectively. A special scalar type Nil is also defined and used to indicate unconnected ports. The constant variable '_' is of this type.

The types RowN are typedefs for Row<UChar,N>. Apart from the row vector, MORA-C++ also defines a column vector and a matrix type. All three template classes inherit from the STL vector<> template.

Split and merge

The example also illustrates the use of the slice method for accessing a subset of the data and the use of indexing for accessing scalar data. This is an important abstraction as it relieves the developer of having to create and connect RCs purely for splitting and merging data.

1.5.2.2 Discrete Cosine Transform

To illustrate another key feature of MORA-C++, operator overloading, we present the implementation of the 2-D Discrete Cosine Transform algorithm (DCT) on an 8×8 image block. In its simplest form, the DCT is a multiplication of a pixel matrix A with a fixed coefficient matrix C as follows:

$$M_{DCT} = C.A.C^T$$

The DWT is a pipelined algorithm with little or no parallelism, and as such only illustrates the pipelining feature of the MORA array. The DCT, however, provides scope for parallelism, by computing the matrix multiplication using a parallel divide-and-conquer approach (see Section 1.5.3.3).

The implementation of the DCT in MORA-C++ is extremely simple and straightforward:

```
typedef Matrix<UChar,8,8> Mat;
const UChar ca[8][8] = { ... };
const Mat c((const UChar**)ca);
const Mat ct = c.trans();

Pair<Mat,Nil> dct (Mat a) {
        Mat m = c*a*ct;
    Pair<Mat,Nil> res (m,_);
    return res;
}
```

As the example shows, the multiplication operator (and other arithmetic and logic operators) are overloaded to provide matrix operations. The other classes Row and Col also provide overloaded operations, making integer matrix and vector arithmetic in MORA-C++ very simple.

1.5.3 MORA-C++ Compilation

As a MORA-C++ program is valid C++, it can simply be compiled using a compiler such as gcc. This is extremely useful as it allows for rapid testing and iterations. The API implementing the Domain Specific Language attempts to catch as many architecture-specific issues as possible, so that a MORA-C++ program that works correctly at this stage will usually need no modifications for deployment on the FPGA. Some issues, however, cannot be caught by the C++ compiler; for example, it is not possible to determine the exact number of MORA assembly instructions for any given MORA-C++ program (obviously, as gcc will produce assembly for the host platform, not for the MORA array). If the number of instructions exceeds the maximum instruction size of the RC, this error can only be caught by the actual compilation to MORA assembly language.

To be able to emit MORA Intermediate Representation code, the MORA-C++ compiler (Vanderbauwhede et al. 2010) needs to perform several actions. The most important ones are

- memory allocation

- inferring split and merge trees

- inferring template modules

1.5.3.1 Memory Allocation

The MORA-C++ compiler uses the following model for the MORA RC's memory: the maximum available memory per RC is a fixed value MEMSZ (in bytes). The word size is WORDSZ, which is determined by the compiler based on the types of the RC function's arguments and local variables. The memory depth is thus MEMSZ/WORDSZ. Allocation is performed based on the type, i.e., the dimensions of the matrix. This means that allocation is a 1-D bin packing problem, a well-known problem in complexity theory. A recently proposed solution using constraint programming is Pisinger and Sigurd (2007). The outline of the overall allocation algorithm is shown in Algorithm 1.2.

Static Single Assignment (SSA) (Cytron et al. 1991) is an intermediate representation commonly used in compilers for the purpose of memory/register allocation. Essentially, it consists of assigning every expression to a unique variable (hence the name 'single assignment').

Intermediate expressions are those expressions that are not part of the returned tuple.

Algorithm 1.2: Memory Allocation Algorithm

Determine MEMSZ and WORDSZ;
Allocate space for used function arguments;
Allocate space for constant data;
Convert expressions into SSA;
Identify intermediate expressions;
Allocate space for intermediate expressions;

1.5.3.2   Inferring Split and Merge Trees

In most cases, the input data for a program will have to be distributed over a number of RCs for computation. For example, the DWT algorithm requires an 8-byte vector to be split into three 3-byte vectors and one 2-byte vector. Conversely, to collect the final data for output, usually the results from several RCs have to be merged. In MORA-C++, splitting of a vector into subvectors is achieved via the slice method, merging of subvectors into a single vector via the splice method. To split or merge single elements, indexing is used. The compiler has to infer a corresponding 'split tree' and 'merge tree', a tree of RCs that performs the required operations.

Split algorithm

Because of the definition of *slice*, any intermediate slices can be removed. Let v be a vector of N elements $0 \ldots N-1$ (of some type T):

```
Row<T,N> v;
s1=v.slice(b1,e1);
s2=s1.slice(b2,e2);
```

Obviously $b1, b2 \geq 0$, $e1, e2 < N$, also $b2 \geq b1$ and $e2 \leq e1$ or the slice call will throw an exception. With these restrictions on the bounds of the slice, the following identities hold:

```
s1.slice(b2,e2)≡
v.slice(b1,e1).slice(b2,e2)≡
v.slice(b2,e2)
```

The compiler has to infer the tree of RCs required to slice the divided data into the given slices. Let the total number of slices be N_S. Because every RC has 2 outputs, the tree is a binary tree. The process consists of following steps:

- Determine the minimum required number of RCs, N_{RC}

$$N_{RC} = \begin{cases} N_S/2 & N_S \text{ is even} \\ (N_s + 1)/2 & N_S \text{ is odd} \end{cases}$$

- Compute the number of levels in the tree (closest power of 2) $N_{lev} = \lceil \log_2(N_{RC}) \rceil$

- Optimal grouping of the slices. In many cases, some of the slices will overlap to some degree. The RCs have instructions to move a contiguous range of data in an efficient way (1 cycle per word + 1 cycle overhead); moving a non-contiguous set of data requires one instruction per subset, increasing the overhead. Consequently, it pays to move the smallest contiguous range required for the two slices of the leaf RCs. To determine the optimal grouping, we use a recursive algorithm, as shown in Algorithm 1.3:

Algorithm 1.3: Optimal Grouping of Slices

For every slice, group with all other slices in the set. Let $s_i(b_i, e_i)$ and $s_j(b_j, e_j)$ be the grouped slices. Every grouping receives a weight $w_i j = max(e_i, e_j) - min(b_i, b_j) + 1$, i.e., the size of the combined slice; Remove the group with the lowest weight from the set, assign to the first RC;
Repeat the procedure until the set contains 0 or 1 slices (N_S times if N_S is even, $N_S - 1$ times if N_S is odd);
Using the combined slices, repeat the procedure for the next level of the tree;
Finally, if S_N is odd, prune the tree, i.e., remove any intermediate RCs that return the same slice as they receive;

Merge algorithm

The complement of the split algorithm follows entirely the same pattern: a merge tree can be viewed as an upside-down split tree; the main difference is that ranges to be merged should be non-overlapping.

1.5.3.3 Compilation of Matrix and Vector Arithmetic Using Module Templates

As discussed in Section 1.4, the MORA Intermediate Representation language provides the ability to generate code at compile time and the ability to group instantiations of RCs into hierarchical modules. Module templates combine both features: when a module template is instantiated, it generates a module based on the template parameters. This feature of the assembly language was designed with the express purpose of supporting code generation from overloaded matrix and vector operations. Consider, for example, a matrix multiplication:

```
Matrix<UChar, NR1,NC1> m1;
Matrix<UChar, NC1,NC2> m2;
Matrix<UChar, NR1,NC2> m12;
```

```
m12 = m1{*}m2;
```

The multiplication is computed by splitting the matrices into two submatrices
(NR1/2) × NC1 and NC1 × (NC2/2). The computation of m1.m2 results
in four submatrices of size (NR1/2) × (NC2/2) being computed in parallel
and then combined into m12. Of course it is possible to split either one of the
matrices into more submatrices, but this leads to larger numbers of RCs being
used. In Section 1.7 we present both the smaller and the faster implementation
of the DCT.

In terms of implementation, the multiplication is implemented as a template module which takes the dimensions of both matrices as parameters. Furthermore, if the multiplication is part of a compound arithmetic expression,
the intermediate connections will use the full bandwidth available, rather than
inserting merge and split trees. The current algorithm for deciding whether a
template module can be used and whether split/merge trees should be inferred
is simple:

- Only compound arithmetic expressions on matrices or vectors will be implemented as a template module. This means that expressions with control
  constructs are not implemented this way, nor are expressions that result
  in changing the data type by slicing, splicing, or joining.

- For every such compound expression, a split tree will be inferred for the
  leaf terms and a merge tree for the result.

As the actual syntax of the MORA assembly template modules is out of the
scope for this chapter, we present the module template using the equivalent
MORA-C++ syntax. This also gives a good idea of how much complexity is
handled by the compiler when inferring a template module from an overloaded
matrix multiplication. For conciseness we have omitted the type fields of the
template instances.

```
template <int NR1,int NC1,int NC2>
Pair<Matrix<NR1,NC2>,Nil> mmult4
(Matrix<NR1,NC1> m1, Matrix<NC1,NC2> m2){
Matrix<NR1,NC2> m_res = m1*m2;
Pair<Matrix<NR1,NC2>,Nil> out(m\_res);
return out;
}


template <int NR1, int NC1, int NC2>
Tuple<Matrix<NR1,NC2> > mmult
(Matrix<NR1,NC1> m1,Matrix<NC1,NC2> m2) {

// split. In assembly, this is a split tree
b11 = m1.block(0,0,NR1/2-1,NC1);
b12 = m1.block(NR1/2,0,NR1/2,NC1);
```

```
b21 = m2.block(0,0,NC1,NC2/2-1);
b22 = m2.block(0,NC2/2,NC1,NC2);

// compute partial results
Pair<Matrix<NR1/2,NC2/2>,Nil>
    p11 = mmult4<NR1/2,NC1,NC2/2>(b11,b21);
Pair<Matrix<NR1/2,NC2/2>,Nil>
    p12 = mmult<NR1/2,NC1,NC2/2>(b11,b22);
Pair<Matrix<NR1/2,NC2/2>,Nil>
    p21 = mmult4<NR1/2,NC1,NC2/2>(b12,b21);
Pair<Matrix<NR1/2,NC2/2>,Nil>
    p22 = mmult4<NR1/2,NC1,NC2/2>(b12,b22);

// merge. In assembly, this is a merge tree
Matrix<NR1,NC2/2> m_u =
    p11.left.merge<NR1/2,NC2/2>(p12.left);
Matrix<NR1,NC2/2> m_l =
    p11.left.merge<NR1/2,NC2/2>(p12.left);
Matrix<NR1,NC2> m =
    m_u.merge<NR1,NC2/2>(m_l);
Tuple<Matrix<NR1,NC2> > out(m);
return out;


}
```

The equivalent MORA assembly template is structurally identical to the MORA-C++ version. The above code also serves to illustrate the MORA-C++ merge function. This is a method call implemented using a polymorphic function template which works out how to merge the matrices based on the specified return type, i.e., the dimensions of the returned matrix.

1.5.3.4   Compiler Support for Lanes and Channels

The MORA framework has the ability to automatically generate a design with many parallel instances of an algorithm. The compiler infers the data path width from the input RCs and computes the maximum number of possible 'lanes' (parallel instances) from the system's I/O width. Next, it checks how many lanes can be fitted given the available resources on the FPGA. Typically, the block RAMs (BRAMs) will be the limiting resource; however, because of the template-based generation of the RCs, accurate estimation of the slice count is possible.

When the compiler has determined the number of lanes, it generates a host-side software interface which demultiplexes the original input data stream over the lanes. This interface essentially contains a buffer for each lane and performs a barrier synchronization over all buffers.

In a similar fashion, the compiler can support Direct Memory Access

TABLE 1.1
Utilization Results of Single Precision Floating-Point Core on Virtex 4 LX200

Core Functionality	Slices	Registers	DSP	Maximum Pipeline Depth	Frequency (MHz)
FP + − * acc /	1974	1447	1	10	121
FP * acc /	1625	1276	1	10	116
FP + − /	1189	839	0	10	98
FP * /	1416	1165	1	10	108
FP + − * acc	1052	765	1	7	137
FP + − *	867	677	1	4	127
FP * acc	764	594	1	7	147
FP + −	307	171	0	3	141
FP *	546	506	1	4	128
FP /	1097	1195	0	10	200
Int + − * >> (/)	815	169	0	2	69

(DMA) channels: if the resources of the FPGA allow a multiple of the number of lanes to be instantiated, the compiler will generate multiple instances as well as the DMA control logic; it will also generate a host-side software interface instance for each channel, as well as a host-side software DMA interface.

1.5.4   Floating-Point Compiler (FloPoCo) Integration

In order to support floating-point operations, we integrated INRIA's FloPoCo compiler (Dinechin and Pasca 2011), which generates variable width arithmetic cores for FPGAs. The reasons for choosing the FloPoco compiler compared to the floating-point operations provided by Xilinx' Coregen (Xilinx Inc.) are its better performance per unit area, its portability for FPGAs of different vendors, such as Altera, and its ability to generate fixed-pointed arithmetic operations. Moreover, because of FloPoCo's uniform generation of cores, it can relatively easily be integrated into the MORA tool chain. The inclusion of a floating-point core does not change any architectural features of the RC. However, the control unit is modified to support the variable pipeline depths of the cores generated by FloPoCo. Even though our final goal is to provide variable precision floating point core support, currently we only support IEEE-754 single precision floating-point operations, as this is the most common type in typical HPC applications. FloPoCo generates individual cores for operations such as multiply, add, divide, and accumulate. Thus, to provide the complete PE functionality, all the individual cores generated by FloPoCo are connected as shown in Figure 1.6.

The utilization and performance results of different floating-point core configurations generated by FloPoCo for the MORA RC are shown in Table 1.1.

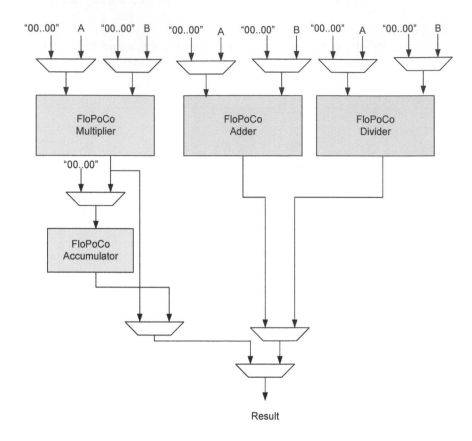

FIGURE 1.6
Reconfigurable cell floating-point core

This table also includes utilization results of regular 32-bit integer PEs used in MORA RC for comparison.

The accumulator stage of the implemented floating-point core from the FloPoCo compiler uses a fast addition and normalization technique (Dinechin et al. 2008) to achieve a higher frequency of operation. Hence, when the multiplier is cascaded with the accumulator stage the performance of the core is not degraded. As a result the operational frequency of the full PE core is higher than 125 MHz, allowing the complete RC to operate at over 100 MHz. Similar to our old PE approach of enabling utilization of required modules, the floating-point core shown in Figure 1.6 is also designed using macros for conditional inclusion, which allows the tool chain to select only the required blocks. Thus, the mapped design achieves a higher resource utilization on FP-GAs. It should be noted that the above proposed results are measured for a complete 32-bit single precision floating-point core. However, if a user program does not intend to use the complete single precision floating-point numbers, the precision and magnitude of the floating-point core can be optimized, allowing a better utilization factor. FloPoCo supports such optimizations. In MORA-C++, the reduced-precision floating-point type can be represented as a template type float<mag,prec>, which makes it very easy for the MORA-C++ compiler to pass the required magnitude and precision on to the FloPoCo compiler.

1.6   Hardware Infrastructure for the MORA Framework

Apart from the core MORA architecture, i.e., the network of Reconfigurable Cells, the MORA Framework requires an infrastructure for I/O and external memory access. This section explains the MORA I/O system and the approach to support shared and external memory access.

1.6.1   Direct Memory Access (DMA) Channel Multiplexing

The DMA data I/O channel width of the SGI RC-100 platform that we used for this work is 128-bit. Therefore, any algorithm which doesn't utilize data I/O more than 64-bits wide could benefit from mapping multiple instances ('lanes') in parallel, thus increasing the throughput. Along with this, the RC-100 platform supports a quad-DMA channel interface allowing multi-operative DMA channels. The additional support of the DMA channel again provides flexibility to map multiple lanes of algorithm in parallel. Apart from the algorithm logic, an auxiliary logic is designed to translate the platform-specific DMA protocol to the RC array handshaking mechanism. The number of DMA channels to be instantiated can be determined by the assembler based on the resource utilisation of a single instance of the algorithm, the available resources on the FPGA, and the maximum number of DMA channels supported by the hardware platform.

1.6.2   Vectorized RC Support

The limited availability and underutilized memory space (for 8- and 16-bit RCs) of BRAMs on the FPGA becomes a bottleneck for implementing algorithms in parallel (i.e., multiple RCs). To overcome this problem we proposed using parallel instances ('lanes') of the algorithm. The core advantage of parallelism in FPGAs is used to map multiple instances (e.g., 4 PEs and 2 PEs for 8- and 16-bit RCs, respectively), illustrated in Figure 1.7, allowing algorithms to achieve higher throughput. A common control unit then executes the same instruction on all the processing cores providing vectorized user application execution.

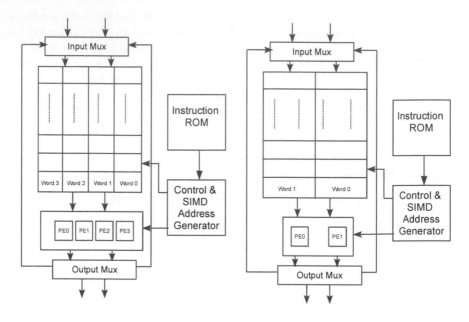

FIGURE 1.7
Vector support for (left) 8-bit and (right) 16-bit RC architecture

This type of vectorization is purely an optimisation so it is not visible in MORA-C++ or the MORA IR. The assembler will decide on vectorization based on program analysis: if the datapath width allows it and the code does not contain conditional instructions, the instances can be vectorised.

1.6.3   Shared Memory Access

The MORA RC is designed to access data from its own local, private dual port memory. However, many applications require some form of global/local shared memory access through which RCs can share their data. Sharing data across RCs with the current setup of distributed memory local to each RC is very inefficient (because RCs can only copy data to downstream neighbors), resulting in a decrease of the overall system throughput for applications that require shared access. Hence, a mechanism for access to a global shared memory by individual RCs is required. The types of memories that can be shared are either external SRAM/SDRAM positioned on the FPGA board or internal BRAMs on the FPGA. The complication involved in accessing a memory shared between multiple RCs is the balanced allocation of memory resources (memory ports) to all the requested RCs. Therefore, we designed an arbitration module using a Round Robin Scheduler (RRS).

The arbitration scheme used is divided into three parts. The purpose of

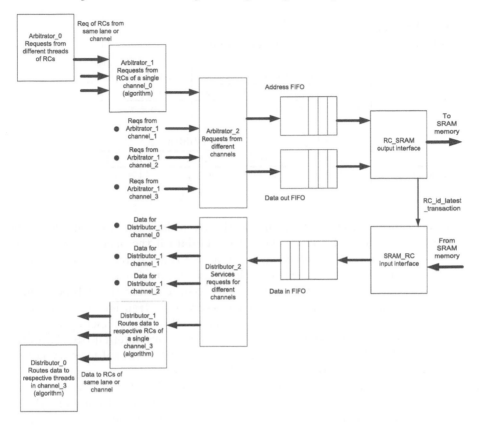

FIGURE 1.8
Shared memory access interface architecture

the first part is to combine requests from RCs from a single algorithm instance. The second and third parts cover lanes and channels, respectively. As each stage of the arbitration has its own RRS, the RC memory access requests from different lanes and channels will be provided fairly, i.e., with equal weight for the shared memory resources. This scheme also eliminates the risk of starvation for the RC requests from multiple lanes and channels. Figure 1.8 shows the architectural description of the implemented shared memory access interface. The inclusion of an additional interface to access shared memory results in a clock cycle penalty, dependent on the number of RCs, lanes, and channels simultaneously connected to the interface. The shared memory access is dependent on the application; we present the latencies induced because of each arbitration stage and other SRAM interface modules in Table 1.2.

In Table 1.2 n stands for the number of extra clock cycles for pipelined data requests from the RC. Hence, for a direct SRAM interface availability

TABLE 1.2
Latency of Shared Memory Interface Modules

Module	Arbitrator			SRAM Output
	RC	Lane	Channel	Interface
Clock Cycles	$3+n$	$2+n$	$2+n$	$6+n$

Modules	SRAM Input	Distributor		
	Interface	RC	Lane	Channel
Clock Cycles	$2+n$	$4+n$	$1+n$	$1+n$

it takes 13 clock cycles to transmit the RC's data transfer request to the shared memory, and 8 clock cycles to read data if the operation is a Read. In addition to regular shared memory access time, it takes 13 clock cycles to write the data and 21 clock cycles to read data from a shared memory module. The latency of this interface may seem high, but this will be reduced if the RC requests a shared-memory block-level data transmission, so that the 21 clock cycles of read latency is only imparted on the memory transaction once. Moreover, if the shared memory is mapped onto internal BRAMs of the FPGA, which have lower access times compared to external memory access, the total memory transaction time will be small.

Shared memory access is implemented in the MORA IR via two special instructions, EXTRD and EXTWR, which, respectively, read from and write to external memory. The next example reads a value from address 0xC0FFEE in the shared memory and returns the result to address 1 at the RC connected to the right output port.

```
EXTWR YL 1:W C:0xC0FFEE
```

This approach is entirely analogous to fetching an address from memory and putting it into a register for an ordinary processor. The MORA RC does not have separate registers but one can consider the data memory and scratchpad as variable-size register files.

In MORA-C++, shared memory is supported very simply by creating a shared static array which is passed as an argument to the RCs which use the shared resource. In this way, several shared memories can be supported simultaneously, although at the moment the RC interface supports only a single shared memory per RC. The assembler decides whether the shared memory will be implemented using BRAMs (if there are enough available on the FPGA) or using the off-chip memory.

1.7 Results

To test the capabilities of our framework, we used the Discrete Cosine Transform (DCT) and Discrete Wavelet Transform (DWT) benchmarks from Vanderbauwhede et al. (2010), adapted for variable data path widths and parallel lanes and channels. The benchmarks were tested on the RC-100 FPGA platform, hosted on an SGI Altix 4700 machine. The RC-100 platform hosts two Virtex-4 LX 200 FPGAs. A direct-I/O DMA transfer over the SGI NUMALink provides the communication between the FPGA platform and the host processor. The NUMALink supports 128-bit wide transfers with a maximum bandwidth of 16 GB/s. The SGI Altix 4700 is a 40-processor dual-core Itanium-2 system at 1.4 GHz.

In terms of resource utilisation, thanks to the custom configuration of the RC for each program, the slice count per algorithm is quite moderate: the largest algorithm, the 'fast DCT', consumes 7,218 slices, which is only 8% of the total available slices on the Virtex-4 LX 200 (89,088). This allows us to deploy many parallel instances of the algorithm, effectively instantiating thousands of cores.

1.7.1 Thousand-Core Implementation

We used two different DCT 'kernels' for this work, DCT_small and DCT_fast, consisting of, respectively, 22 and 44 processors. Figure 1.9 shows the acyclic directed graph (ADG) for the 22-core kernel.

Each kernel consists of a number of concurrent 'threads' of computation (4 for DCT_small, 8 for DCT_fast). For the 8-bit DCT, the I/O width is 16 bits; the I/O width of the FPGA is 128 bits, allowing 8 parallel kernels ('lanes'). Furthermore, based on the allocated memory per processor and the block RAM size, it is possible for a number of kernels (4 in this case) to share common control logic, resulting in a vector processor. Finally, the RC-100 supports DMA channels to demultiplex the high-speed NUMALink I/O. For the DCT_small kernel, the board can support 4 channels and for DCT_fast, 3 channels. Note that this elaborate demultiplexing hierarchy can be completely (and in fact trivially) inferred by the compiler based on the I/O width, memory size, and core count of the application and the system resources.

As discussed in Vanderbauwhede et al. (2011), the DCT_small implementation results in 704 cores, the DCT_fast in 1056 cores. As each of these cores is fully customized for the program it runs, the resource utilization is very efficient. The source code for both implementations is only a few lines of code (see Section 1.5.2.2 for the complete listing). This example illustrates how a few lines of code results in a thousand-core vector processor network.

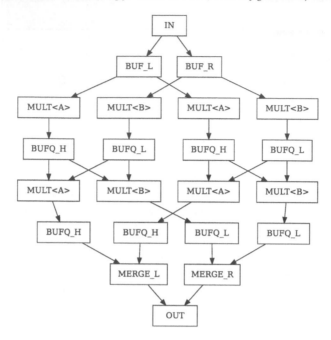

FIGURE 1.9
ADG diagram for the DCT_small algorithm

1.7.2   Results

Figure 1.10 shows the slice count of the RC for different data path widths. As the slice count includes the control unit and instruction memory, it grows less than proportionally with the data path width: e.g., the 32-bit DCT_small kernel uses only twice the number of slices of the 8-bit version. However, as can be seen from Table 1.3, the clock frequency decreases significantly (though again much less than proportionally) with the data path width: the current 32-bit version of the RC only runs at 50 MHz. However, we believe this is largely due to the lack of optimization in the design, as the MORA framework was originally designed with a fixed 8-bit data path.

Figure 1.11 and Tables 1.4 and 1.5 show the efficiency in resource utilization of the vectorized RC when compared to a non-vectorized cell (4-lane → 8-bit, 2-lane → 16-bit). It can be observed that the vectorized RC implementation for the algorithms increases the number of lanes mapped on a single Virtex-4 LX 200 FPGA, resulting in an increase in throughput by a factor of 3 to 4 times when compared to a non-vectorized approach.

In terms of throughput, unsurprisingly, the DWT benchmark performs rather poorly (for reference, a software implementation on a 2.2 GHz Xeon processor, compiled with the most aggressive optimization, has a throughput of 2 GSamples/s). The 1-D DWT algorithm does not contain any parallelism

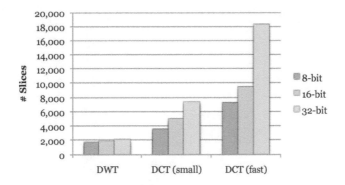

FIGURE 1.10
Slice count for benchmark algorithms with different datapath widths

TABLE 1.3
Benchmark Implementation Results (No Vectorization)

		Slices	BRAMS	Clock frequency (MHz)
DCT (small)	8-bit	3597	22	100
	16-bit	4931	22	66
	32-bit	7393	22	50
DCT (fast)	8-bit	7218	44	100
	16-bit	9482	44	66
	32-bit	18252	44	50
DWT	8-bit	1734	10	100
	16-bit	1810	10	66
	32-bit	2068	10	50

TABLE 1.4
Benchmark Throughput Results for a Single DMA Channel without RC Vectorization (MSamples/s)

Data Path Width	8-bit	16-bit	32-bit	8-bit		16-bit	32-bit
# Lanes	1	1	1	7	8	4	2
DCT (small)	1.6	1.0	0.8	10.9	12.8	4.0	1.6
DCT (fast)	3.1	2.1	1.3	21.5		8.4	2.6
DWT	4.1	2.1	2.1	28.7	32.8	8.4	4.2

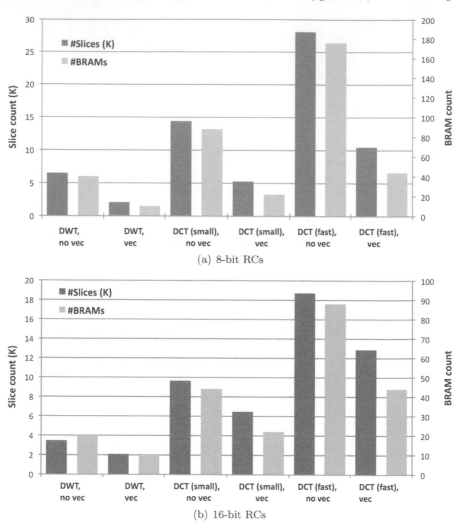

FIGURE 1.11
Effect of vectorization on Slice/BRAM counts for (a) 8-bit and (b) 16-bit RCs

and the different RCs have to synchronize on the last stage. The point of this test bench was not to demonstrate performance but to demonstrate that the RCs automatically synchronize via the handshake protocol. However, the DCT implementation performs very well: the 8-bit version, sample size 64 bytes, has a maximum throughput of 21.5 MSamples/s for a single DMA channel without vectorization and 73.8 MSamples/s for vectorized RCs with 3 DMA channels. We used the Independent JPEG Group's efficient C-code implementation[1]for

---

[1] Available at http://ijg.org/

TABLE 1.5

Benchmark Throughput Results with Multiple DMA Channels and RC
Vectorization (MSamples/s)

	8-bit RC, 8 lanes/channel				16-bit RC, 4 lanes/channel			
# DMA Channels	1	2	3	4	1	2	3	4
DCT (small)	12.5	25.0	37.5	50.0	4.2	8.3	12.6	16.6
DCT (fast)	24.6	49.2	73.8		8.2	16.4	24.6	
DWT	32.8	65.6	98.4	131.2	8.5	16.9	25.4	33.9

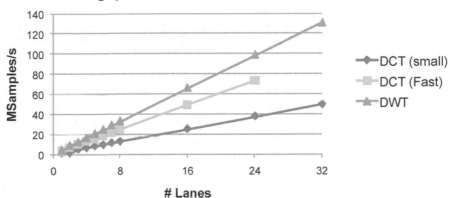

FIGURE 1.12

Throughput versus number of lanes for 8-bit benchmarks

reference, compiled with gcc -O3 (aggressive optimization) and run on a
2.2 GHz Intel Xeon processor, and its throughput is only 3.5 MSamples/s.
So we achieve a speed-up of 6× in absolute terms, or 120× cycle-for-cycle.
The I/O bandwidth of the FPGA is 128 bits/clock cycle, which at 100 MHz
translates to 1.6 GB/s; as the host memory bus on the Altix 4700 operates at
400 MHz, we can multiplex the input stream over several DMA channels (up
to 4 on the RC-100). Thus, our final throughput of 4.7 GB/s is quite close to
the maximum system I/O bandwidth of 6.4 GB/s. If the board contained an
equivalent Virtex-6 device, LXT 240, we could use the 4 DMA channels to
reach the maximum throughput.

Figure 1.12 shows that the throughput increases linearly with the number
of lanes and DMA channels. The first 8 points show the effect of adding lanes,
the 4 last points the effect of adding DMA channels. This illustrates that there
is no degradation in clock speed due to increasing the number of lanes nor in
I/O throughput as a result of demultiplexing the input stream over multiple
DMA channels.

1.7.3   Comparison with Other DCT Implementations

Comparing a new implementation with results reported in the literature is complicated by the very large number of parameters involved in the comparison. In Table 1.6 we list a number of results for various implementations of the Discrete Cosine Transform. However, it is important to annotate these results. The DWARV work (Yankova et al. 2007) is most similar to our work in that it aims to make FPGA programming easier for non-hardware designers. Unfortunately, the results on their DCT implementation are incomplete: the result for the hardware DCT is listed as '3876408 PPC cycles', but it does not detail how many samples were processed in that time. They report a speed-up of almost 10 times against an unspecified software implementation running on a 300 MHz PowerPC processor.

The older results in Loo et al. (2002) compare Handel-C against hand-written VHDL and C code; however, the clock speed of these implementations is very low because they run on a Virtex-E. Still, the ratio of the processor clock speed to the FPGA clock speed is about 20, quite similar to our work; we observe that neither the VHDL nor the Handel-C implementation can outperform their C-code in absolute terms. Finally, we have two native VHDL implementations. The first is a very recent state-of-the art VHDL implementation of a 1-D DCT on a Virtex-5 reported in Alfalou et al. (2009). The authors present throughput results in MSamples/s (a sample for a 1-D DCT is 8 pixels, compared to 64 for a 2-D DCT) and compare these with results from a different implementation done by themselves. Unfortunately, the authors don't specify the clock speed; however, it is reasonable to assume that this implementation runs at least at 100 MHz, and that a 2-D DCT implementation would not be faster than the 1-D DCT. The second (Bukhari, Kuzmanov, and Vassiliadis 2002) is somewhat older, using a Virtex-II Pro. The reported throughput is '2193 CIF frames'; a CIF frame has a resolution of 352×288. To obtain a comparable figure across these implementations, we have normalized the throughput to a 100 MHz clock. We see that our approach has the highest normalized throughput. The normalized throughput is quite instructive in other terms as well: clearly, our IJG JPEG C-code compiled with -O3 is much faster than both other reported C-code results; the discrepancy between the different VHDL implementations is equally significant. These results serve to illustrate how difficult – and even potentially misleading – comparisons with other published results can be. Nevertheless, they also show that the performance of the MORA platform is very respectable.

Based on the LUT counts in Table 1.6, one could argue that MORA trades off performance against resource cost. However, FPGAs used in High-Performance Reconfigurable Computing platforms are typically very large. In other words, the investment in the FPGA platform has already been made and the resource cost is therefore a sunk cost. Using as many as possible of the available LUTs in order to bring the performance as close as possible to the I/O limit is therefore the best strategy. Precisely the fact that MORA

TABLE 1.6
DCT Benchmark Throughput Comparison with Other Implementations

Benchmark	DWARV[a]	C code[b]	VHDL[c]	Handel-C[d]	C code[e]	VHDL[f]	VHDL[g]	MORA	IJG C code
Platform	Virtex II Pro	PowerPC	Virtex-E	Virtex-E	UltraSPARC IIi	Virtex-5	Virtex II Pro	Virtex-4	Intel Xeon
Clock speed (MHz)	100	300	23.7	17.6	440		54	100	2000
Throughput (MB/s)			0.627	0.488	0.639	1648	222	4720	224
Slice count	3307					492	812	62557	
Speed-up (HW/SW)	9.74	1	0.98	0.76	1			21	1
Throughput at 100 MHz			2.7	2.8	0.145	1648	412	4720	11.2

[a](Yankova et al. 2007)
[b](Yankova et al. 2007)
[c](Loo et al. 2002)
[d](Loo et al. 2002)
[e](Loo et al. 2002)
[f](Alfalou et al. 2009)
[g](Bukhari, Kuzmanov, and Vassiliadis 2002)

utilises as much as possible of the available FPGA resources with no extra effort from the developer is therefore a great strength.

1.8 Conclusion and Future Work

In this chapter we have introduced the MORA framework for high-level programming of FPGAs using a many-core network. We have shown how a single C++ program can be compiled to run on thousands of cores simultaneously, thus leveraging the parallelism provided by the FPGA as well as demonstrating the validity of our programming model.

There are many possible avenues for future work. In the short term, the integration of the floating-point compiler and the external memory access needs to be validated. Furthermore, numerous minor improvements need to be considered, such as more aggressive removal of RC functionality based on in-depth program analysis, and replacing the current sequential SIMD engine with a parallel one.

We also want to deploy the MORA framework on different FPGA platforms, notably the Maxwell and Novo-G FPGA supercomputers.

Recently, OpenCL, an open standard for heterogeneous many-core computing, has been gaining momentum and shown great promise for programming of GPUs and multicore CPUs. Altera has announced a proof-of-concept OpenCL framework for FPGA platforms. Our aim is therefore to convert the MORA framework into an OpenCL-compliant FPGA programming solution. The required changes to the compiler and the hardware system will be the main focus for our future work.

2

Implementing Time-Constrained Applications on a Predictable Multiprocessor System-on-Chip

Sander Stuijk

Department of Electrical Engineering, Eindhoven University of Technology, The Netherlands

Akash Kumar

Department of Electrical and Computer Engineering, National University of Singapore, Singapore

Roel Jordans and Henk Corporaal

Department of Electrical Engineering, Eindhoven University of Technology, The Netherlands

CONTENTS

2.1   Introduction

Many embedded systems, such as automotive systems and smart phones, execute time-constrained software applications. Typical examples of these applications are the anti-lock braking system (ABS) in a car or a software-defined radio in a smart phone. Users expect that these applications will exhibit robust behavior and that their performance will be guaranteed (Gangwal et al. 2005). At the same time, especially in battery-powered systems, the energy usage of these applications should be kept as low as possible in order to prolong battery life.

In the architecture domain there is a clear trend to use heterogeneous multi-processor systems-on-chip (MPSoCs) to meet the computational requirements of novel applications at an affordable energy cost (Sangiovanni-Vincentelli and Martin 2001). Programming these systems is a very challenging task, especially since the interaction between all hardware components has to be considered in order to provide timing guarantees to the applications (Martin 2006). Model-based design approaches (see, for example, Bonfietti et al. (2010); Haid et al. (2009); Liu et al. (2008); Moreira et al. (2005); Pimentel (2008); Stuijk et al. (2007)) are being developed to address this challenge, by modeling applications using a dataflow Model-of-Computation (MoC). Several model-based approaches (for example, Haid et al. (2009); Pimentel (2008)) use Kahn Process Networks (KPNs) (Kahn 1974) to model applications. However, relevant properties, such as the minimal storage space needed to avoid deadlock, cannot be determined at design time (Geilen and Basten 2003; Parks 1995). Furthermore, tasks in a KPN cannot be scheduled statically. Hence, a run-time mechanism is needed to detect deadlocks, to schedule tasks, and to reallocate the storage space assigned to the application. This creates a considerable implementation overhead. Moreover, the lack of design-time analysis techniques makes it difficult to use this MoC to design systems that provide timing guarantees to applications. In recent years, several model-based design approaches have been proposed that can provide such timing guarantees (for example, Bonfietti et al. (2010); Liu et al. (2008); Moreira et al. (2005); Stuijk et al. (2007)). These approaches are all based on the (homogeneous) synchronous dataflow ((H)SDF) MoC (Lee and Messerschmitt 1987a; Sriram and Bhattacharyya 2009). The static nature of this model enables design-time analysis (using techniques presented in Ghamarian, Geilen, Stuijk, et al. (2006); Ghamarian, Geilen, Basten, et al. (2006); Ghamarian et al. (2007); Stuijk, Geilen, and Basten (2008); Wiggers et al. (2006)) as well as efficient implementations. The SDF MoC allows, for instance, derivation of exact bounds on the storage requirements of an application and static scheduling of the tasks inside an application. As a result, there is almost no run-time overhead when running an application, modeled with an SDF graph, on an MPSoC.

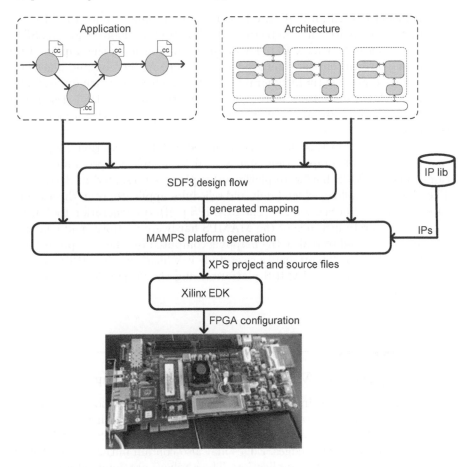

FIGURE 2.1
SDF3/MAMPS design flow

This makes it an attractive MoC to use when developing systems that are running time-constrained applications.

Existing model-based design approaches based on the SDF MoC (such as Bonfietti et al. (2010); Liu et al. (2008); Moreira et al. (2005); Stuijk et al. (2007)) target virtual MPSoC platforms (such as Liu et al. (2008); Moreira et al. (2005); Stuijk et al. (2007)) or MPSoC platforms for which only simulation models are available (an example is Bonfietti et al. (2010)). For any practical system, it is, however, important that a complete trajectory including a physical implementation of the MPSoC platform is provided. This chapter presents a model-based design approach and accompanying hardware platform that provide such a complete design flow from application to implementation. An overview of this design flow is shown in Figure 2.1. The flow takes an application modeled with an SDF graph as input and maps it onto the MAMPS multi-processor platform architecture (A. Kumar et al. 2008) using the SDF3 design flow (Stuijk et al. 2007). This design flow considers the impact of the hardware platform on the timing behavior of the application and allows the design flow to provide timing guarantees to the application when implemented on the MAMPS platform. Hence, the combination of the SDF3 design flow and MAMPS platform makes it possible to build systems with a predictable timing behavior.

The remainder of this chapter is organized as follows. Section 2.2 presents the SDF model-of-computation that is used to model and program applications. The MAMPS hardware platform template is presented in Section 2.3. The SDF3 design flow, which binds and schedules applications on the architecture, is discussed in Section 2.4. The MAMPS platform generation tool, which creates a concrete instance of the MAMPS hardware platform and generates the binaries needed to realize the mapping computed by SDF3, is presented in Section 2.5. Section 2.6 presents a case study in which a motion-JPEG decoder is implemented on the MAMPS platform using the design flow presented in this chapter.

2.2 Application Modeling and Programming

The design flow shown in Figure 2.1 assumes that applications are modeled using a Synchronous Dataflow (SDF) graph. Figure 2.2(a) shows an example SDF graph. The nodes, called *actors*, communicate with *tokens* sent from one actor to another over the edges. Edges may also contain initial tokens that are present at the start of the application. These tokens are depicted with a solid dot in the figure. Actors model application tasks, and the edges model data or control dependencies. An essential property of SDF graphs is that every time an actor *fires* (executes) it consumes the same number of tokens from its input edges and produces the same number of tokens on its output edges. These numbers are called the *rates* (indicated next to the edge ends, while rates of 1 are omitted for clarity). An actor can only fire if sufficient tokens are available on the edges from which it consumes. In our example graph, actor P is the only actor which can initially fire. When it fires, actor P consumes two tokens from the edge between actors R and P, and produces two tokens on the edge from P to Q. The two tokens produced by one firing of actor P enable two firings of actor Q. Tokens thus capture dependencies between actor firings. Such dependencies may originate from data dependencies, but also from dependencies on shared resources. The rates determine how often actors have to fire with respect to each other such that production balances consumption. These rates are constant, which forces an SDF graph to execute in a fixed repetitive pattern, called an *iteration*. An iteration consists of a set

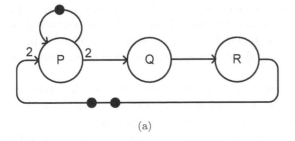

(a)

```
static int local_variable_P;

void actor_P_init(typeRtoP *fromR, typePtoQ *toQ)
{
    local_variable_P = 0;
}

void actor_P(typeRtoP *fromR, typePtoQ *toQ)
{
    toQ[0] = fromR[1] + local_variable_P;
    toQ[1] = fromR[0] + fromR[1] + toQ[0];
    local_variable_P += 1;
}
```

(b)

FIGURE 2.2

(a) Example SDFG and (b) implementation of actor P

of actor firings that have no net effect on the token distribution. These actor firings typically form a coherent collection of computations.[1] An iteration could, for instance, correspond to the processing of a frame in a video stream or an audio sample in an audio stream. The notion of an iteration plays an important role in SDF scheduling techniques. A schedule of an SDF graph that schedules a complete number of iterations of the graph can be repeated indefinitely since the token distribution (and hence the set of enabled actors) is the same every time the schedule is repeated (Sriram and Bhattacharyya 2009). SDF performance metrics (e.g., throughput) rely also on the notion of an iteration. For example, the throughput of an SDF graph is defined as the long-run average number of iterations that can be completed per unit time (Ghamarian, Geilen, Stuijk, et al. 2006). As explained before, an iteration

---

[1] An iteration only exists when the graph is deadlock-free and consistent. Both properties can be verified efficiently (Bhattacharyya, Murthy, and Lee 1996; Lee and Messerschmitt 1987a). SDF graphs that do not have these properties are of no practical use. Therefore, we limit ourselves to consistent and deadlock-free SDF graphs.

of an SDF graph typically corresponds to a meaningful computation in the application (e.g., computing one video frame). Using the aforementioned notation of an iteration, the SDF throughput metric can be translated into a more meaningful throughput metric at the application level (e.g., frames per second).

Most SDF analysis and mapping techniques abstract from the functional behavior of the application. These techniques typically only require information on the worst-case execution time and memory requirements of actors. In addition, some techniques may require information on the worst-case token sizes. Obviously, the final implementation of an application should implement the correct functional behavior. Therefore, the SDF3/MAMPS design flow shown in Figure 2.1 assumes that a set of C functions is attached to each actor in the application SDF graph. These functions implement the functional behavior of the actor and together they implement the complete behavior of the application. Figure 2.2(b) shows an example source code listing of our example actor P. The functional behavior of this actor is described with the function `actor_P`. The function has two parameters, one for the edge from actor R and one for the edge to actor Q. Through the pointer `fromR`, the function has access to an array containing the tokens on the edge from R to P. The size of the array, i.e., the number of tokens that can be accessed by the actor, is equal to the rate on the edge. Note that since a pointer to an array is passed, an actor may perform out of order access on these tokens when it fires. An actor may also access a token multiple times. Similar to the `fromR` pointer, the pointer `toQ` provides access to an array of memory locations where tokens produced by actor P can be stored. Note that the self-edge on actor P is not passed as an explicit argument to the function `actor_P`. Self-edges are handled in a different way, as will be explained below.

Actors in an SDF graph are assumed to be stateless (i.e., no internal actor state is preserved between actor firings). As a result, a system running an SDF graph does not have to store any context after an actor firing has been completed. As a result, the context switching overhead is limited. However, some actors may have a state that must be preserved between firings. For example, an actor implementing a variable length decoder needs to store several decoding tables. These tables must be preserved between subsequent actor firings (i.e., function calls). Any actor state that must be preserved between firings must be modeled explicitly with an initial token on a self-edge. Hence, all the stack, heap, and global variables that must be preserved across actor firings must be explicitly modeled in the graph. In principle, a self-edge could be implemented in the same way as all other edges in the graph (as an array of memory locations that is passed to the function). This will, however, complicate the implementation of an actor since a programmer must then explicitly manage the memory locations where all variables that must be preserved across firings are stored. It would be much more intuitive for a programmer if this process was handled automatically by the design flow and compiler. The programmer could then use global variables and/or static variables inside functions and the design flow should ensure that their content is preserved across actor firings. The SDF3/MAMPS design flow supports such a programming approach. The implementation of actor P (see function actor_P in Figure 2.2(b)) uses a global variable named local_variable_P. This variable can be used throughout the entire function (and any functions called from within it). In line with the C semantics, a programmer may also decide to place this variable inside a function and mark it as static. The SDF3/MAMPS flow will, in both cases, ensure that the state of this variable is preserved across firings while avoiding any requirement for a programmer to explicitly manage the memory locations in which these variables are stored. This greatly simplifies the work of the programmer since he/she does not have to remove static or global variables that are used within the scope of a single actor. As mentioned before, actors communicate with each other by sending tokens over their edges. The use of global variables could potentially open an undesired backdoor to communicate through shared variables. To avoid this misuse of global variables, a programmer should implement each actor in a separate C file and mark all global variables inside each C file as static. This limits the scope of these variables to a single C file and hence a single actor. When this programming constraint is satisfied and the functional implementation of the application produces the correct result when simulated on a workstation (e.g., by embedding the C code input to the SDF3/MAMPS flow in a Y-API process network (Kock et al. 2000)), the SDF3/MAMPS flow guarantees that the functional behavior will also be correct when the application is implemented on an MPSoC.

Initial tokens (both on edges between actors and on self-edges) must be initialized when an application is started. The SDF model abstracts from this initialization behavior since it abstracts from the functional behavior of the application. Obviously, an implementation should support the initialization of initial tokens to ensure a functionally correct execution of the application. The SDF3/MAMPS flow handles the initialization of initial tokens through a special initialization function that must be supplied by the programmer for each actor. Consider again the example source code listing in Figure 2.2(b). The function actor_P_init implements the initialization of the initial tokens on all outgoing edges (including the initial tokens modeling the state of actor P). The arguments passed to this function are identical to the arguments passed to the function that implements a firing of the actor (e.g., the function actor_P).

As mentioned already, the resource allocation and scheduling step (i.e., the SDF3 design flow in Figure 2.1) takes as input an SDF graph that models the application. This graph abstracts from the functional behavior of the application. In order to find a mapping that satisfies the timing constraints of the application, the SDF3 design flow only needs information on the worst-case execution time and memory requirements of actors and the worst-case token sizes of the tokens communicated over the edges. The SDF3 design flow assumes that this information is supplied by the designer. In theory, the worst-case execution times and memory requirements of the actors and edges in an SDF graph could be derived (semi-)automatically from the source code of the actor. At the time of writing, the flow requires, however, that a designer perform this analysis a priori. A designer can perform this analysis using existing worst-case execution time and memory analysis tools such as those presented in Gustafsson (2006); Holsti et al. (2008) and Wilhelm et al. (2008).

2.3   Platform Architecture

The MAMPS platform (A. Kumar et al. 2008) follows the tile-based multiprocessor platform template described in Culler, Singh, and Gupta (1999). In this template, multiple tiles are connected by an interconnection network that provides point-to-point connections between tiles. These connections can, for example, be implemented through a network-on-chip (NoC), using direct hardwired FIFO connections, or by a bus. Figure 2.3 shows an example MAMPS platform instance with four tiles. These tiles form the processing and storage elements of the architecture. All tiles are connected to the interconnect through a standardized network interface (NI). Any tile or interconnect variant that supports this network interface can be directly embedded in the MAMPS platform. This makes it possible to compose easily a concrete platform instance from the elements that are available in the MAMPS architecture. The use of this NI interface also simplifies the extension of the MAMPS architecture. Our example MAMPS platform instance contains four different tile types. Tiles 1 and 2 show simple tile architectures with a processing element (PE) connected to a network interface (NI), local data and instruction memories (dmem and imem), and some optional peripherals (input-output, timers, etc.). Tile 3 shows a similar tile which has been extended with a *communication assist* (CA). This CA handles the sending and receiving of tokens inside the tile. The CA decouples the communication inside a tile from the computation taking place on its processing element. The last tile, Tile 4, shows another configuration in which a hardware IP block (an accelerator) is directly connected to the interconnect using only a network interface. In the remainder of this section, we will discuss the implementation of the most important components in the MAMPS platform architecture (processing elements, network interface, and interconnect) in more detail.

2.3.1   Processing Element

The Xilinx Microblaze soft-core is currently used as a processing element within the tiles. This processing element is used as a bare processor, that is, it runs without an operating system. The MAMPS platform assumes that only a single application is running on the platform. Actors from this application

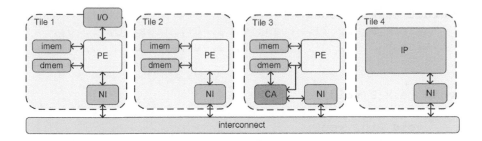

FIGURE 2.3
MAMPS platform architecture

are assumed to be scheduled on the processing element through a static-order schedule. The MAMPS platform generator (explained in detail in Section 2.5) generates C code that loads this scheduler at boot time. It also generates an implementation of the scheduler. A snippet of the code generated for a processing element is shown in Figure 2.4. The `main` function starts by initializing the communication infrastructure and scheduler. Next, the static-order schedule of the actors on this tile is executed using the `while` loop. The next actor that is to be executed is retrieved through the `scheduler_get_next` function. This function uses the `schedule` array to look up this actor. In our example, the array defines a static-order schedule in which actor P is executed first, then actor Q is executed twice, followed by two firings of actor R. After these firings, the schedule is repeated. The `scheduler_get_next` function returns a pointer to a data structure (`actor`) that contains a task handler (i.e., a pointer to the function implementing the actor). For actor P, this handler is implemented through the function `task_P` (see Figure 2.4). The first `while` loop in this function blocks until there are two tokens available on the edge from actor R to actor P. Once these tokens are available, the function `claim_read` returns a pointer to an array containing these two tokens. The second `while` loop blocks until there is space for two tokens on the edge from actor P to actor Q. Hence, once both `while` loops have been executed, the tokens consumed by the actor are guaranteed to be available and the space needed by the actor to store the tokens it produces is also available. This allows the actual actor code to be executed in a non-blocking fashion. As explained in Section 2.2, the function `actor_P` implements the behavior of the actor (see Figure 2.2(b) for the source code of this function). After completion of the function, the `release` functions update the administration of the FIFOs that implement the two edges connected to actor P. The function `release_read` removes two tokens from the edge between actor R and P. The function `release_write` increases the token count on the edge from actor P to actor Q by two (the rate of actor P on this edge).

```c
struct task schedule[] = {
    {task_P, 1, 1}, {task_Q, 1, 0}, {task_Q, 1, 0},
    {task_R, 1, 0}, {task_R, 1, 0}, {NULL,   0, 0}};


void task_P(void)
{
    void *in, *out;
    while( ( in = claim_read (&r2p_fifo, 2)) == NULL );
    while( (out = claim_write(&p2q_fifo, 2)) == NULL );

    actor_P(in, out);

    release_read(&r2p_fifo);
    release_write(&p2q_fifo);
}

void main(void)
{
    struct task *actor;
    init_communication();
    init_scheduler();

    while(NULL != (actor = scheduler_get_next())) {
        actor->handler();
}
```

FIGURE 2.4
MAMPS scheduling code for a processing element

2.3.2   Network Interface

A clear definition of the network interface protocol is crucial to enable a simple and extensible platform architecture. The MAMPS platform defines the Xilinx Fast Simplex Link interface as its network interface protocol. This limits the network interface to communicating 32-bit words. It provides, however, a trivial point-to-point solution for the interconnect since Xilinx Fast Simplex Links (FSL) (Xilinx Inc. n.d.) can always be used to connect the processing elements. To translate arbitrarily sized tokens into one or more 32-bit words and back again requires serialization and de-serialization. These operations can be performed by the processing element of the tile (the PE inside Tile 1 shown in Figure 2.3) or by the addition of some dedicated communication hardware (e.g., the CA block inside Tile 3). The advantage of using the processing element for the serialization and de-serialization of tokens is the simplicity of the generated hardware. However, this simple hardware comes at the cost of extra processing time used in the processing element. This time cannot be used to run the actual actor code. Moreover, it increases the computational load on the processing elements since they need to handle part of the communication. This overhead can be removed by using dedicated communication hardware like the CA described in Shabbir et al. (2010). This CA increases the hardware complexity, but it also relieves the processing element of the serialization and de-serialization of tokens, which improves the actor response time. The MAMPS platform template provides a software based solution to perform serialization and de-serialization of tokens on the Microblaze. It also provides a hardware based solution through an implementation of the CA presented in Shabbir et al. (2010).

2.3.3   Interconnect

The MAMPS platform provides two different options to implement the interconnect. The first option is to use point-to-point connections based on the Xilinx Fast Simplex Links (FSL) (Xilinx Inc. n.d.). The second option is to use a Spatial Division Multiplex (SDM) NoC based on Yang, Kumar, and Ha (2010). Both interconnects comply with the network interface definition, but the NoC interconnect provides more flexibility at the cost of greater logic area and a higher latency while the FSL interconnect relies on the FSL implementation provided by Xilinx. The NoC consists of one router per tile in the design. Each router connects through a set of wires to its neighbors. Each router can also be connected to the network interface of a single tile. The routers are arranged in a 2-dimensional mesh network. The dimensions of this network are based on the number of tiles required in the design and the network is kept as close to square as possible to reduce the maximum distance between two tiles since this distance relates directly to the latency of the network connections. The NoC allows the user to program connections on a point-to-point basis, each connection being assigned a certain bandwidth through the number of wires assigned to that connection, but wires can only be assigned to a single connection at a given time, allowing an efficient usage of network resources. The original NoC presented in Yang, Kumar, and Ha (2010) already complied with the network interface requirements for the MAMPS platform, but it lacked flow control for connections in the network. Flow control is crucial to implement the FIFO based communication between actors of an SDF graph that are running on different tiles. The flow control provides back pressure to the producing actor on an edge. As such, it implements the synchronization between two actors running on different tiles. Flow control was added as part of the integration of the NoC in the MAMPS platform. The changes to the NoC required an increase of approximately 12% in the number of logic slices on the FPGA when compared to the original implementation.

2.4   Design Flow

The design flow shown in Figure 2.1 maps applications onto the MAMPS platform. The first tool in the flow, called SDF3, binds and schedules an application on the platform resources. The output of SDF3 is a mapping that is guaranteed to meet the throughput constraint of the application. In this section, we will explain in detail how SDF3 integrates with the rest of the design flow and how it computes the application mapping.

SDF3 (Stuijk, Geilen, and Basten 2006b) is a tool set for analyzing, transforming, and implementing applications modeled in various dataflow models-of-computation (MoCs). The tool set can, for example, be used to compute the worst-case throughput of an application modeled with an SDF graph. It can also map an application, modeled with the more expressive FSM-based Scenario-Aware Dataflow (FSM-SADF) (Theelen et al. 2006) MoC, onto various MPSoC platforms. Currently, SDF3 supports the mapping of applications onto the MAMPS, CoMPSoC (Akesson et al. 2012), and MPARM platforms (Benini et al. 2005). The tool has been designed in such a way that it is possible to add support for a new platform with only minimal effort. This process will be illustrated using the MAMPS platform as an example. The combined SDF3/MAMPS flow makes use of the FSM-SADF mapping flow (Stuijk, Geilen, and Basten 2010) provided by SDF3. However, the MAMPS platform supports only applications modeled with an SDF graph. Hence, the SDF3 design flow as shown in Figure 2.1 needs to perform an internal conversion from an SDF graph to an FSM-SADF graph prior to running the actual mapping algorithm (see Figure 2.5). Similarly, SDF3 needs to convert the mapping it computes, which is in terms of the FSM-SADF graph, back to a mapping in the format used by the MAMPS platform generator (see Figures 2.1 and 2.5). Since each SDF graph is also an FSM-SADF graph, there exists a straightforward conversion from an SDF graph to an equivalent FSM-SADF graph. Both graphs will have the same set of actors and edges with the same labels (i.e., names). Because of this one-to-one correspondence between actors and edges in both graphs, it is also straightforward to convert a mapping generated by SDF3 based on an FSM-SADF graph into a mapping of the original SDF graph that has been input into the SDF3/MAMPS flow. Some other platforms (e.g., CoMPSoC) which use the Cyclo-Static Dataflow (CSDF) MoC (Bilsen et al. 1996) require more complex conversion steps since no graph transformation exists that can convert a CSDF graph to an FSM-SADF graph while preserving a one-to-one correspondence between the actors and edges in both graphs. A discussion on the graph conversion techniques that should be used in those cases is, however, beyond the scope of this chapter.

As mentioned before, SDF3 supports several MPSoC platforms. The architecture and operation of each of these platforms is different. For example,

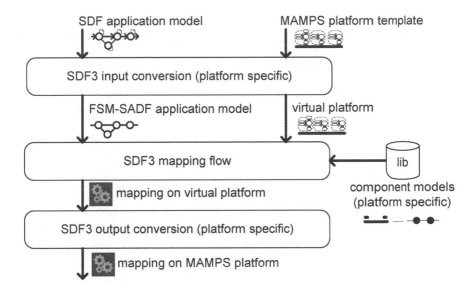

FIGURE 2.5
Overview of the SDF3 mapping flow

MAMPS and CoMPSoC use a different type of NoC as an interconnect, while the MPARM platform assumes a bus-based interconnect. As a result, the delay introduced by an interconnect communication in each of these three platforms is different. In order to provide timing guarantees for a mapping, SDF3 needs to consider the timing behavior of the specific platform that is targeted. In addition to the different hardware components found in the various platforms, these platforms may also support different features. The CoMPSoC platform, for example, allows multiple applications to share the platform resources. The interference from different applications is removed using a Time-Division-Multiplex (TDM) scheduler. The MAMPS platform, on the other hand, does not support the concurrent execution of multiple applications on the same hardware resources. Because these different platforms use a different hardware architecture, SDF3 needs to take different mapping decisions for each of these platforms (e.g., some require the generation of TDM scheduler settings whereas others do not require these settings). In order to support all platforms, the SDF3 mapping flow uses an internal, virtual platform that supports a super-set of the features found in all real platforms it supports. Similar to the application conversion, SDF3 converts the specific platform instance (e.g., from the MAMPS, CoMPSoC, or MPARM architecture) to the internal, virtual platform architecture (see Figure 2.5). Next, it maps the application onto the virtual platform. As mentioned before, each real platform may use different hardware components that each have a different timing behavior. The timing behavior of these components must be considered

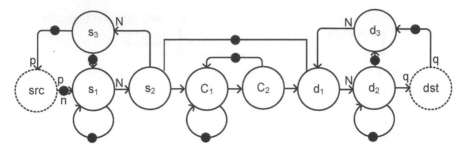

FIGURE 2.6
Dataflow model for interconnect communication in MAMPS

when analyzing the timing behavior of the application when mapped onto the platform. This is ensured by SDF3 through its model-based design approach in combination with its generic virtual platform template. Because of the model-based design approach, SDF3 is able to model all binding and scheduling decisions it takes into the FSM-SADF graph. This requires that each resource allocation decision can be modeled with a dataflow graph. For example, a binding of an edge in the FSM-SADF graph to a connection in the MAMPS architecture can be modeled with the dataflow graph shown in Figure 2.6. (Readers interested in the derivation of this model from the actual hardware behavior are referred to Jordans et al. (2011).) When all resource allocation decisions are modeled in a similar fashion into the application FSM-SADF graph, a new FSM-SADF graph, called a resource-aware FSM-SADF graph (Stuijk 2007), is obtained. This graph extends the application FSM-SADF graph with dataflow models for all resource allocation decisions taken by the mapping flow. The performance analysis techniques provided by SDF3 can then be used on this resource-aware graph to verify whether the throughput constraint of the application is met under the computed resource allocation. This model-based design approach requires that each possible binding of a component (actors and edges) from the application graph to a component in the virtual platform (memory, processing element, connection, network interface, communication assist) is modeled in the dataflow graph by inserting into the application FSM-SADF graph a dataflow model that captures the timing impact of the resource allocation. As mentioned before, this timing impact depends on (among other factors) the specific platform that is targeted. When adding support for a new platform type to SDF3, a library of dataflow graphs must be provided that models the timing behavior of actor and edge allocation to the various resources inside the newly added platform.

After converting the application and platform model input to the SDF3/MAMPS flow, the FSM-SADF application model is mapped onto the virtual platform (see Figure 2.5). This mapping is performed using the design flow presented in Stuijk, Geilen, and Basten (2010). The flow uses the technique from Stuijk, Geilen, and Basten (2006a) to analyze the trade-off between storage-space assigned to the edges of the graph and the throughput of the graph. A Pareto-optimal storage-space assignment with the smallest storage space that satisfies the throughput constraint is then used to constrain the storage requirements of the edges in the graph. In a subsequent step, the flow performs a binding of the actors to the MPSoC resources. Next, static-order schedules are constructed for all processors to which actors are bound. Finally, the flow computes the minimal number of TDM time slices needed on these processors to guarantee that the throughput constraint of the application is met. By minimizing the number of TDM time slices, processor resources are saved for other applications. The output of the flow is a set of Pareto-optimal mappings that provides a trade-off in their resource usage. In some of these mappings, the application could, for example, use many computational resources, but limited storage resources, whereas an opposite situation may be obtained in other mappings. A platform that supports run-time (re)configuration could use this set of mappings to select at run-time the most suitable mapping based on the resource usage of the applications which are already running on the platform (Shojaei et al. 2009; Ykman-Couvreur et al. 2006, 2011).

After completion of the mapping step, the SDF3 mapping flow extracts from this mapping to the virtual platform, a mapping of the application on the real platform (see Figure 2.7). Since the MAMPS platform does not support run-time (re)configuration, the set of mappings computed for the virtual platform must be reduced to a single mapping. Since all mappings computed for the virtual platform are Pareto optimal, it is sufficient to select a random (e.g., the first) mapping. The MAMPS platform does not support multiple applications running concurrently on its resources. Hence, the TDM time slice allocation computed in the SDF3 mapping flow can be discarded. To avoid unnecessarily long run-times because of the TDM time slice computation, the SDF3 input conversion step limits the size of the TDM time wheel inside the virtual platform to a single time slot when targeting the MAMPS platform. This makes the TDM time slice computation trivial and as such it avoids a waste of computational resources when computing the mapping of the application onto the MAMPS platform.

2.5   Platform Generation

After mapping the application onto the MAMPS platform, the flow continues with the platform generation step (see Figure 2.1). This step uses the architecture description input to the flow to generate the hardware platform. The MAMPS platform generation step instantiates and connects the platform tiles. It computes the memory sizes for each tile based on the mapped edges, actors, and the dimensions of the scheduling and communication layer. The

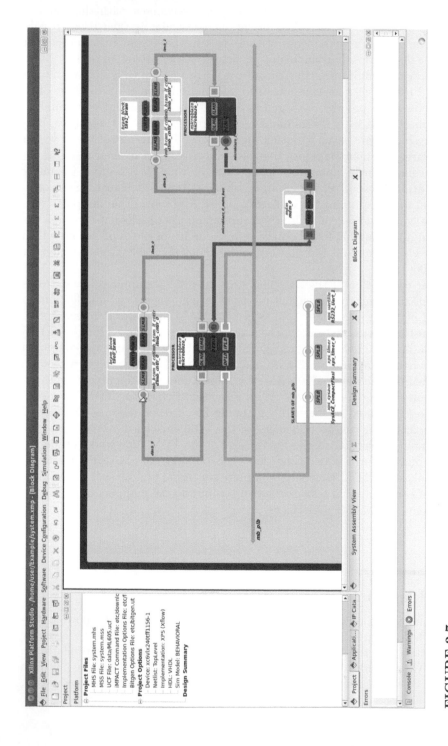

FIGURE 2.7

Two tile MAMPS platform in XPS

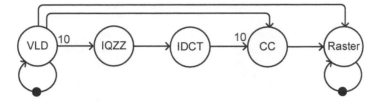

FIGURE 2.8
The SDF graph for the MJPEG decoder

interconnect components are instantiated to match the specified communication architecture. Connections are routed and the VHDL code and peripheral driver connected to the interconnect are also generated. The MAMPS platform generation step also generates the software that must be executed on the processing elements inside the generated platform. This includes generating wrapper code for each actor, translating the static-order schedule provided by SDF3 into C code, and generating initialization code for the communications (see Section 2.3 for an example of this code). The generated code is combined with a template project which already includes an implementation of the scheduling and communication libraries. The Xilinx Platform Studio Tcl script interface is then used to complete the project and to add the required hardware and software targets for the implementation. Using the script interface ensures compatibility over many different versions of XPS and greatly simplifies the generated code.

2.6   Case Study

The application used in the case study is the motion-JPEG (MJPEG) decoder shown in Figure 2.8. The VLD actor parses the input file and decompresses the Minimal Coded Unit (MCU) blocks. MCUs consist of up to 10 blocks of frequency values, depending on the sampling settings used when creating the input file. Each block of frequency values is passed through the inverse quantization and zig-zag reordering (IQZZ) and IDCT actors which transform the frequency values into color components. The color conversion (CC) actor translates the color component blocks of one MCU to pixel values and the rasterization (Raster) actor puts the pixel values at the correct location in the output buffer. The edges from the VLD actor to the CC and Raster actors forward information from the file header (frame size and color composition) to the CC and Raster actors. One graph iteration of the MJPEG decoder decodes a single MCU. Hence, the throughput of the application is defined in MCUs per clock cycle of the generated platform. A method based on Gheorghita et

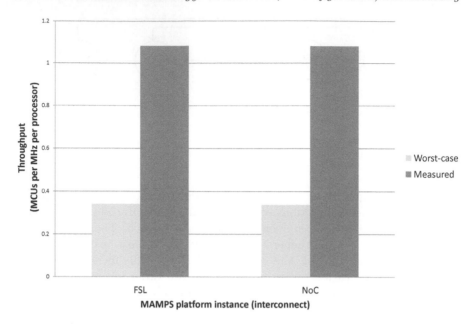

FIGURE 2.9
Measured and guaranteed worst-case throughput

al. (2005) combined with execution time measurement was used to determine the worst-case execution time of the actors in this case study.

2.6.1   Throughput Guarantees

The SDF3/MAMPS flow is able to implement an application, modeled with an SDF graph, on the MAMPS platform while providing throughput guarantees to the application. As explained in Section 2.4, the SDF3 mapping flow tries to minimize the resource allocation of an application while ensuring that even under the worst-case input, the throughput constraint is met. Hence, when running the application (e.g., an MJPEG decoder) with an arbitrary input, the throughput of the system should always be above the throughput computed by SDF3. In our experiment, we used the SDF3/MAMPS flow to implement the MJPEG decoder on two variants of the MAMPS platform. Both variants contained two tiles with MicroBlaze processing elements. One variant used an FSL to connect the two tiles while the other variant used a NoC as interconnect. Using a series of five test images, we measured the throughput with which these images were processed in our system (when implemented on a Xilinx Virtex6 FPGA). Figure 2.9 shows the measured throughput and the worst-case throughput as predicted and guaranteed by SDF3. The results show that the throughput guaranteed by SDF3 is indeed conservative for the supplied images. The figure also shows that there is a substantial difference

TABLE 2.1
Designer Effort

Manual steps	Time spent
Parallelizing the MJPEG code	< 3 days
Creating the SDF graph	5 minutes
Gathering required actor metrics	1 day
Creating application model	1 hour
Automated steps	
Generating architecture model	1 second
Mapping the design (SDF3)	1 minute
Generating Xilinx project (MAMPS)	16 seconds
Synthesis of the system	17 minutes
	Total ∼ 4 days

between the guaranteed worst-case throughput and the measured throughput. This is due to the fact that the execution times of the actors when decoding the series of test images is much shorter than their potential worst-case execution time. However, when an application requires throughput guarantees it may not always be possible to avoid the over-dimensioning of the average case.

2.6.2   Designer Effort

Table 2.1 lists the required designer effort in creating and mapping the MJPEG decoder, as measured by the authors of this chapter. This implies a working understanding of the application as well as previous experience in writing applications for the design flow and platform. The upper part of the table represents manual labor performed by the designer and the lower part is automated by the SDF3/MAMPS design flow. Manually implementing the overall system would cost at least another 2–5 days, depending on the complexity of the hardware (number of tiles) and the number of application mappings tried.

2.6.3   Overhead

The overhead of the generated system when compared to a manually developed system can be split into two categories: modeling and implementation overhead. The primary source of modeling overhead is the fixed output rates of the SDF actors. This can be seen in the MJPEG example as the output rate of the VLD actor which produces up to 10 frequency blocks per MCU, depending on the format of the input stream. In our model, we had to fix this production rate at the worst-case rate (i.e., 10). Another source of modeling overhead can be found in communicating the initialization values on the edges from the VLD actor to the CC and Raster actors. A manual implementation of the algorithm could communicate these values once, during an initialization phase. It is not possible, however, to model such an initialization phase in an SDF graph. As a result, the initialization values have to be passed on each iteration of the graph. Fortunately, these initialization tokens are relatively small and constitute only 1% of the communication data. This limits the modeling overhead of the SDF graph. The implementation overhead of the SDF graph is also very small. Scheduling on the MAMPS platform is done using a static order schedule which reduces the scheduler to a lookup table. A manual implementation would be likely to implement the same schedule in its main loop and this would be similar in efficiency. Communication would also be solved in a similar way and therefore it does not influence the implementation overhead. Overall, it holds that the scheduling overhead will be similar for other applications. The modeling and communication overhead may, however, vary depending on the nature of the application.

2.7   Conclusions

In this chapter, we have presented an automated design flow that is capable of generating an implementation of an application on an MPSoC while providing throughput guarantees. The design flow provides a method for automatically instantiating different architectures using a template-based architecture model. This template-based architecture is easy to extend and allows the automated selection of the correct implementation when heterogeneous systems are designed. This allows the designers to perform a very fast design space exploration for real-time embedded systems. The mapping and platform generation flow presented is publicly available at `http://www.es.ele.tue.nl/sdf3`.

Acknowledgments

The work described in this chapter was supported in part by COMMIT-NL as part of the SenSafety project (CMP 50004079) and in part by Agentschap NL as part of the EUREKA/CATRENE/COBRA project (CA 104).

3

SESAM: A Virtual Prototyping Solution to Design Multicore Architectures for Dynamic Applications

Nicolas Ventroux, Tanguy Sassolas, Alexandre Guerre, and Caaliph Andriamisaina

Embedded Computing Lab, CEA LIST, Gif-sur-Yvette, France

CONTENTS

3.1   Introduction

The emergence of new embedded applications for telecom, automotive, digital television, and multimedia applications has fueled the demand for architectures with higher performance, greater chip area, and improved power efficiency. These applications are usually computation-intensive, which prevents them from being executed by general-purpose processors. Architectures must be able to simultaneously process concurrent information flows, and they must all be efficiently dispatched and processed. This is only feasible in a multithreaded execution environment. Designers are thus showing interest in a System-on-Chip (SoC) paradigm composed of multiple computation resources and a network that is highly efficient in terms of latency and bandwidth. The resulting new trend in architectural design is exemplified by the MultiProcessor SoC (MPSoC) (Jerraya and Wolf 2005).

Another important feature of future embedded computation-intensive applications is the dynamism. Algorithms become highly data-dependent and irregular. Their execution times depend on input data or external interactions with the system. Consequently, on a multiprocessor platform, optimal static partitioning cannot exist. Bertogna, Cirinei, and Lipari (2008) show that the solution consists in dynamically allocating tasks according to the availability of computing resources. Global scheduling maintains the system load balance and supports workload variations that cannot be known off-line. Moreover, the preemption and migration of tasks balance the computational power between concurrent real-time processes. If a task has a higher priority level than another, it must preempt the current task to guarantee its deadline. Besides, the preempted task must be able to migrate to another free computing resource to increase the performance of the architecture. Only an asymmetrical approach can implement global scheduling and efficiently manage dynamic applications.

An asymmetric MPSoC architecture consists of one (sometimes several) centralized or hierarchized control cores, and several homogeneous or heterogeneous cores for computing tasks. The control core handles task scheduling. In addition, it performs load balancing through task migrations between the computing cores when they are homogeneous. The asymmetric architectures usually have an optimized architecture for control. This distinction between control and computing cores makes the asymmetric architecture more transistor/energy efficient than the symmetric architectures.

Designing an MPSoC architecture requires the evaluation of many different features (effective performance, bandwidth used, system overheads, etc.), and the architect needs to explore different solutions in order to find the best trade-off. In addition, he needs to validate specific synthesized components to tackle technological barriers. For these reasons, the whole burden lies on the MPSoC simulators, which should be parameterizable, fast and accurate, easily modifiable, support wide ranges of application specific IPs, and integrate new ones easily.

In this context, we developed the SESAM tool to help the design of new MPSoC architectures. The novelty of SESAM is its support for asymmetrical MPSoC architectures, which includes a centralized controller that manages the tasks for different types of computing resources. The heterogeneity can be used to accelerate specific processing, but the task migration between heterogeneous resources is not supported. The best trade-off between homogeneity, which provides the flexibility to execute dynamic applications, and heterogeneity, which can speed up execution, can be defined in SESAM. Moreover, this tool enables the design of MPSoCs based on different execution models, which can be mixed, to find the best suitable architecture according to the application. In addition, SESAM can support simultaneous different and/or identical applications and mix different abstraction levels; therefore it can take part in a complete MPSoC design flow.

This chapter is organized as follows: Section 3.2 covers related work on MPSoC simulators from both the industrial and academic worlds. Then, Section 3.3 gives an overview of SESAM and introduces the following sections depicting the SESAM framework. Section 3.4 describes SESAM's components and focuses on its infrastructure. Section 3.5 focuses on programming and execution models supported by the SESAM environment. Section 3.6 describes its specific debugging strategy, which is necessary in a dynamic environment. Section 3.7 outlines all the implemented solutions developed to model and manage the energy consumption. Section 3.8 describes how to use SESAM in order to explore the MPSoC design space. Section 3.9 describes a practical use case by modeling a complete asymmetric MPSoC architecture named SESAM. Section 3.10 illustrates the performance results obtained by running this architecture on different real case embedded system applications. Finally, Section 3.11 concludes the chapter by discussing the presented work.

3.2 Existing Work

A number of research papers have been published on single-processor, multiprocessor and full-system simulators (Yi and Lilja 2006; Cong et al. 2008). Some of them focus on the exploration of specific resources. For instance, Flash (Gibson et al. 2000) eases the exploration of different memory hierarchies; SICOSYS (Puente, Gregorio, and Beivide 2002) studies only different Network-on-Chips (NoCs). Taken separately, these tools are very interesting but a complete MPSoC exploration environment is needed in order to analyze all architectural aspects under a real application processing case.

Among complete MPSoC simulators, MC-Sim (Cong et al. 2008) uses a variety of processors, memory hierarchies, or NoC configurations but remains slow due to cycle accuracy. On the contrary, simulators like STARSoC (Boukhechem and Bouernnane 2008) offer a rapid design space exploration but only consider functional level communications. To study network contentions and the impact of communication latencies, a timed simulation is necessary. Others, like ReSP (Beltrame et al. 2008), use generic processors and cannot take into account instruction set specificities. This does not allow sizing and validation of MPSoC architectures. On the contrary, some simulators, like MPARM (Benini et al. 2005), are processor specific and do not allow the exploration of different memory system architectures or different processors, and hence lack flexibility.

Some of the simulators benefit from the genericity of a very high description level, like Sesame (Pimentel, Erbas, and Polstra 2006) or CH-MPSoC (Shen, Gerin, and Pétrot 2008). They use a gradual refinement Y-Chart methodology to explore the MPSoC design space. However, even if they remain very promising tools, they cannot support complex IPs or MPSoC structures with advanced networking solutions. Generated architectures remain very constrained. Less generic projects exist, like SoCLib (Viaud, Pêcheux, and Greiner 2006), but their scope is too limited to fulfil MPSoC exploration and in particular they cannot support automatic MPSoC generation to analyze the impact of its parameters.

Some interesting projects (Wieferink et al. 2004; Paulin, Pilkington, and Bensoudane 2002; August et al. 2007) make a model of a large set of MPSoC platforms. Nonetheless, these solutions do not propose a rich set of Network-on-Chips (NoCs), and it is not possible to easily integrate a centralized element to dynamically allocate tasks to resources. The programming model statically allocates threads onto processors and does not allow the design of architectures optimized for dynamic applications.

To the authors' knowledge, there is no existing work on a simulator that supports asymmetric MPSoC exploration to help the design of MPSoC architectures for dynamic applications. For this reason, we developed a specific

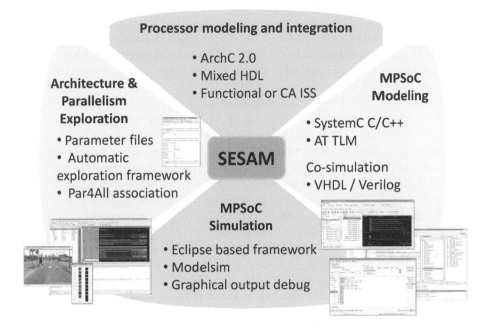

FIGURE 3.1
SESAM overview. The SESAM framework uses ArchC instruction set simulators in a SystemC/TLM environment. It has been specifically designed for fast design space exploration of asymmetric MPSoCs.

MPSoC environment named SESAM, which can easily be integrated into the design flow. The next section will give an overview of this framework.

3.3 SESAM Overview

SESAM is a tool that has been specifically built to ease the design and the exploration of asymmetric multiprocessor architectures (Ventroux, Guerre, et al. 2010; Ventroux, Sassolas, David, et al. 2010) (Figure 3.1). SESAM is based on ArchC (S. Rigo et al. 2004) and can integrate various instruction set simulators at the functional or cycle-accurate level, as well as different networks-on-chips, a DMA controller, a memory management unit, caches, memories, and different control solutions to schedule and dispatch tasks (Section 3.4). All blocks can be timed. This framework is described with the SystemC/TLM description language

SESAM allows MPSoC exploration at the TLM level with fast and cycle-accurate simulations. It reaches 3 MISPS (Million Instruction Simulated Per Second) when modeling an MPSoC of 8 processing elements. SESAM can also be used to analyze and optimize the application parallelism, as well as control management policies.

Besides, SESAM uses approximate-timed TLM with explicit time to provide a fast and accurate simulation of highly complex architectures (more details in Section 3.4.1). Simulating a whole MPSoC platform needs to find an adequate trade-off between simulation speed and timing accuracy. This solution allows the exploration of MPSoCs while reflecting the accurate final design. Regarding communications, we point out a 90% accuracy compared to a fully cycle-accurate simulator. Time information is necessary to evaluate performance and to study communication needs and bottlenecks.

It supports co-simulation within the ModelSim environment (ModelSim) and takes part in the MPSoC design flow, since all the components are described at different hardware abstraction levels. It can be used in a co-simulation environment based on the VHDL or Verilog languages. Co-simulation is essential to refine the architecture and to design specific components into such a complex environment.

The graphical interface of SESAM provides many services to trace and debug the application. It includes a waveform viewer that can display a set of internal signals and different standard output terminals for each processor composing the architecture. In addition, the interface can be integrated in an Eclipse environment to ease the debug of applications.

3.4   SESAM Infrastructure

As depicted in Figure 3.2, SESAM is structured as an asymmetrical MPSoC. It is based on a centralized Control Manager that manages the execution of tasks on processing elements. SESAM proposes the use of different components to design new MPSoCs. Other SystemC IPs can be designed and integrated into SESAM if they have a compatible TLM interface. The main elements are the Memory Management Unit (MMU), the Code Loading Unit (CLU), Memories, a set of Instruction Set Simulators (ISS), traffic generators (TG), a Direct Memory Access (DMA) unit, a Control Manager, and Network-on-Chips (NoC).

3.4.1   Approximate-Timed TLM

Different modeling levels can be chosen depending on the purpose of the study (Guerre et al. 2009). In the case of multiprocessor exploration, the TLM

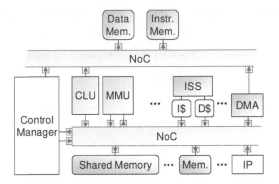

FIGURE 3.2

SESAM infrastructure (Ventroux, Guerre, et al. 2010). SESAM offers a rich set of IPs that can be composed to model complex asymmetric MPSoCs.

or Transaction-Level Modeling is the most adequate level. This modeling type has two implementations which allow for more or less precision.

Approximate-timed TLM and timed TLM are, respectively, based on *event-driven* and *time-driven* techniques (Dally and Towles 2004) which are sketched in Figure 3.3. It is important to mention that the TLM standard library can only be used in *SystemC* threads and not methods.

The *time-driven* approach (Figure 3.3(a)) uses a global clock that synchronizes the whole platform. In this approach, each communication and thread is synchronized by the clock. Each time a thread is woken up by a clock event, a context switch occurs to process the thread computation. In each clock step, all the threads in the simulator start. But the simulation speed of *SystemC* depends on the number of woken-up threads and context switches (Charest et al. 2002). Therefore, the timed TLM approach leads to a huge simulation time overhead due to useless thread awakening.

On the contrary, the *event-driven* approach (Figure 3.3(b)) wakes up only the useful part of the platform when an event occurs. Besides, each element estimates the time spent for its processing. Depending on these estimations, this method limits the number of threads awakened and gives a better trade-off between speed and accuracy than the *time-driven* approach.

In addition, with approximate-timed TLM simulation, time can be *explicit* or *implicit* in the TLM interface. *Implicit time* integrates the *wait* function in the TLM interface. So, according to OSCI (Open SystemC Initiative) (Transaction-Level Modeling Working Group), it obtains a better simulation speed. But the *implicit time* method loses the information about the exact moment when a message leaves or returns from each part of the platform. This moment is important to correctly estimate the time spent by other communications. For example, the time penalty due to memory arbitration is estimated with the last message end time. Therefore, if we do not know when

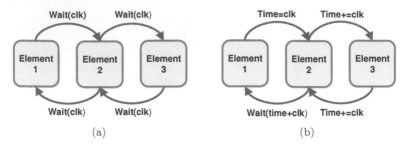

FIGURE 3.3

Two approaches for TLM abstraction level: (a) timed TLM or time-driven, and (b) approximate-timed TLM or event-driven (Guerre et al. 2009)

the last message leaves, it is hard to correctly estimate the communication length. To the contrary, *explicit time* implements the *wait* function in the *SystemC* module description. Thus this module has a better time notion and can approach or enable cycle accuracy to simulate communication. For these reasons, our method uses *explicit time event-driven* TLM communications, whereas OSCI uses *implicit time event-driven* TLM communications. As a result, our approach allows for better time handling and fast simulation that is adapted to MPSoC simulation and evaluation. We will see the benefit of explicit time management with the case of the fast interconnection simulation in the next section.

3.4.2 Interconnections

Using approximate-timed TLM with explicit timing, interconnections allow non-blocking and deterministic data exchanges, regardless of the network load or the execution constraints. The network behavior can be described as follows. First of all, a request is sent to the network and is stored in a list of pending requests. When a request is sent, an entry point ID is given to the request to know the position of the initiator. If this request is the first one in the list, it sets off an event and wakes up the main thread of the network. This thread processes pending requests, calculates the path taken by the request in the network, and computes a penalty in case of contention. Then, the request is sent to the destination. When the response comes back, a wait function is launched with the computed communication time as argument. Finally, the response is sent back to the initiator. In the response, the initiator has some information like the number of routers crossed or the time spent in the different modules of the MPSoC platform.

To be able to calculate contention penalties, depending on the routing type and the ID, the network builds a list of virtual routers and links

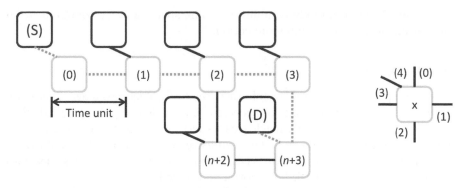

FIGURE 3.4
Routing example for a mesh network. Example of a packet path in a mesh network using XY routing. The router link notation is shown on the right.

(node/link). These couples (node/link) correspond to the router and the link used by the packet during its journey. In Figure 3.4, the packet has a path: $(0, 1) - (1, 1) - (2, 1) - (3, 2) - (n + 3, 4)$ in a Mesh network. Each couple also contains a time reference when the communication enters and leaves the router. This information and the network description are used by the network thread to compute the latency in the network. In order to do this calculation all communication paths are compared with the communication already inside the network to find all contentions. This permits re-adjustment of the latency inside the list of (node/link) pairs. When the latency is known, the communication waits this time duration before delivering the packet.

Many NoC topologies are supported and already implemented in SESAM: a multibus, a mesh, a torus, a multistage, and a ring network. These networks are detailed in Guerre et al. (2009). The multibus can connect all masters to all slaves, but does not allow master to master communications. In the mesh or the torus network, one master and several slaves are linked with a router. An XY routing and a wormhole technique are implemented. The multistage is an indirect fully connected network. It is divided into different stages composed of 4 input-output routers, and linked with a butterfly topology. All masters are on one side and all slaves are on the other side. It also uses a wormhole technique to transfer packets. Finally, in a ring network, a message has to cross each router when it goes through a ring. A parameter can change the number of rings. But each master can connect itself to only one ring. A ring is bi-directional.

In all networks, in order to accept simultaneous requests at its output, two arbiters can be used: a FIFO or a fair round-robin policy. To add a new interconnection, the function which generates the (node/link) list must be described. To summarize, all communications are done at the transactional level and we can accurately estimate the time spent in every communication depending on the chosen interconnection.

3.4.3   ArchC

The main part of an MPSoC simulator is the processing resources. Therefore, having access to a wide range of processor models is compulsory for MPSoC architecture exploration. To reach this variety of resources we use an architecture description language (ADL), which can easily generate an ISS in a specific level of abstraction. The ADLs' modelization levels are classified into three categories: structural, behavioral, and mixed.

Structural or cycle-accurate ADLs describe the processor at a low abstraction level (RTL) with a detailed description of the hardware blocks and their interconnection. These tools, such as MIMOLA (Leupers and Marwedel 1998), are mainly targeted for synthesis and not for design space exploration due to their slow simulation speed and lack of flexibility.

On the contrary, behavioral or functional ADLs abstract the microarchitectural details of the processor and provide a model at the instruction set level. Their low accuracy is compensated by fast simulation speed. Many languages exist, such as nML (Fauth, Van Praet, and Freericks 1995) and ISDL (Hadjiyiannis, Hanono, and Devadas 1997).

Therefore, mixed ADLs provide a compromise solution and combine the advantages of both the structural (accuracy) and behavioral (simulation speed) ADLs. It is the best abstraction layer for design space exploration. EXPRESSION (Halambi et al. 1999), MADL (Qin, Rajagopalan, and Malik 2004), LISA (Pees et al. 1999), and ArchC (Azevedo et al. 2005) are examples of mixed ADLs. The last two will be discussed in this literature review.

LISA, which stands for Language for Instruction Set Architecture, was developed at the University of RWTH Aachen and is currently used in commercial tools for ARM and CoWare (LISATek). Processor models can be described in two main parts: resource and operation declarations (ISA). Depending on the abstraction level, the operations can be defined either as a complete instruction or as a part of an instruction. For example, if the processor resources are modeled at the structural level (pipeline stages), then the instructions' behavior in each of the pipeline stages should be declared. Hardware synthesis is possible for structural processor models.

A recent type of processor description language called ArchC (S. Rigo et al. 2004) is gaining special attention from the research communities (Beltrame et al. 2008; Schultz et al. 2007; Kavvadias and Nikolaidis 2008). ArchC 2.0 is an open-source Architecture Description Language (ADL), developed at the University of Campinas in Brazil. It generates from architecture resource and ISA description files, a functional or cycle-accurate ISS in SystemC. The ISS is ready to be integrated with no effort in a complete SoC design based on the SystemC *Open SystemC Initiative (Open SystemC Initiative)*. In addition, the ISS can be easily deployed in a multiprocessor environment thanks to the interruption mechanism based on TLM, which allows the preemption and migration of tasks between the cores. The main distinction of ArchC is its ability to generate a cycle-accurate ISS with little development time. Only the behavior description of the ISA requires accurate description. As for the microarchitectural details, they are generated automatically according to the architecture resource description file. There exists also a graphical framework, called PDesigner (Araujo et al. 2005), based on Eclipse and ArchC processor models, which allows the development and simulation of MPSoCs in SystemC in a friendly manner. Since ArchC is an open-source language, we can modify the simulator generator to produce a processor with customized microarchitectural enhancements, which makes it a great tool for computer architecture research (Sandro Rigo et al. 2004). However, the processor model cannot be synthesized because it is not supported by ArchC.

3.4.4 Instruction Set Simulators

Thus, we use processors designed with the ArchC language as processing resources with data and instruction cache memories, which are optional. The ArchC tool (Azevedo et al. 2005) generates a functional or cycle-accurate ISS in SystemC with a TLM interface (Bechara, Ventroux, and Etiemble 2010). A new processor is designed in approximately 2 man-weeks, but it depends on the instruction set complexity. Its simulation speed can reach tens of MISPS. Different models are available (Mips, PowerPC, SPARC, ARM, etc. (*ArchC — The Architecture Description Language*)), as well as a complete Mips32 processor (with an FPU) at the functional level. We have also designed an AntX ISS in order to provide a small RISC processor dedicated to control processing (Bechara et al. 2011). Moreover, specific modifications have been carried out to support blocked and interleaved multithreaded computing processors (Bechara, Ventroux, and Etiemble 2011), and even variable pipeline lengths (Bechara, Ventroux, and Etiemble 2010). Preemption and migration of tasks is made possible through the use of an interruption mechanism, which allows us to switch the context of the processing unit, to save it, and to restore the context code from the executed task memory space.

3.4.5 Traffic Generators

SESAM offers the possibility to integrate two kinds of traffic into our traffic generators, in order to represent different application behaviors. The first one is a uniform traffic that matches with an unpredictable load in the network. The second one is a normal distribution traffic that mainly represents local communications. The normal distribution variance gives the main distance that can be reached by messages, whereas the mean represents the closest memory from the traffic generator. Both traffics are representative of a dynamic multi-domain application. The message sending frequency can be dynamically modified. The frequency controls the network load during simulation. When the message creation frequency is higher than the sending rate, messages are buffered in an infinite FIFO memory.

3.4.6   MMU and TLBs

The MMU is optional and can bring advanced capabilities to manage all the shared memory space, which is divided into pages. The whole page handler unit is physically distributed between the MMU and the local Translation Lookaside Buffers (TLB) for each processing core. All the memory functions are available through the SESAM HAL (described in Section 3.5.3). In the SESAM framework, the memory space can be implemented with several banks or with a single memory. Some memory segments are protected and reserved for the Control Manager. The Memory Management Unit (MMU) manages the memory space and shares it between all active tasks. It is possible to dynamically allocate or deallocate buffers. There is one allocated buffer per data block. An identifier is used for each data block to address them through the MMU, but it is still possible to use physical addresses. Different memory allocation strategies are available and can be implemented. Dynamic rights management of pages is also possible to enable a data-flow execution through local synchronizations. For instance, a task can wait for a data block produced by another task in a streaming execution model. Each task can also have write exclusive access. A data access request can be a blocking demand; then another task reads the data when the owner task has released its permission right.

3.4.7   CLU

The CLU dynamically loads task codes from the external memory through a DMA access when it receives a configuration command from the Control Manager. Then, in a dynamic memory management context, it also has to update the MMU to provide the corresponding virtual to physical address translations. A context and a stack are automatically included for each task.

3.4.8   Memory

Different memory elements can be instantiated. The memory space can be implemented as multiple banks or as a single memory. The former is logi- cally private or shared, while the latter is only shared between the processors. Multiple readers are possible and all the requests are managed by the NoC. Instruction and data cache memories can also be activated for each processor. They are fully configurable. We can define the associativity, the block size, or, for instance, different writing policies (write through, write back + write allocate, etc.).

3.4.9   DMA

A DMA controller is necessary to transfer data between the external data memory and the internal memory space. A DMA controller is a standard processing resource and takes part in the heterogeneity of the architecture. It is a fully programmable unit that executes a cross-compiled task for its architecture. A 3-dimensional DMA controller is available. Transfer parameters can afterwards be dynamically modified by other tasks, to specify source and target addresses defined at run-time. Finally, it dynamically allocates the required memory space for the transfer.

3.4.10   Control Manager

The Control Manager can be either a fully programmable ISS, a hardware component, or a mix of both. With the ISS, different algorithms can be implemented. Thanks to the SESAM HAL and an interrupt management unit, the tasks are dynamically or statically executed on heterogeneous computing resources. In addition, multi-application execution is supported by this HAL. A set of scheduling and allocating services in hardware (HW accelerator coupled with the ISS) or software can be easily integrated, modified, and mixed. Besides, a complete hardware real-time operating system is available, named Operating System accelerator on Chip (OSoC) (Ventroux and David 2010). The OSoC supports dynamic and parallel migration, as well as preemption of tasks on multiple heterogeneous resources, under real-time and energy consumption constraints.

As we have seen, SESAM allows the modeling of various asymmetric MP-SoC architectures thanks to many available components. In addition, these elements are modeled at the TLM level to allow fast but accurate simulation compatible with MPSoC exploration. The next section shows how applications can be programmed and executed within the SESAM framework.

3.5   SESAM Programming and Execution Models

SESAM offers the possibility to model many different asymmetric MPSoCs with various execution models. In this section, we will present the SESAM programming model, the supported execution models, and the SESAM Hardware Abstraction Layer (HAL).

3.5.1   Programming Model

The programming model of SESAM is specifically adapted to dynamic applications and global scheduling methods. Obviously, it is inconceivable to carry

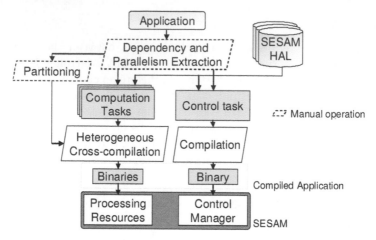

FIGURE 3.5
SESAM programming model (Ventroux, Sassolas, David, et al. 2010)

out a generic programming model for all asymmetrical MPSoCs. Nonetheless, it is possible to add new programming models. The programming model is based on the explicit separation of the control and the computation parts. As depicted in Figure 3.5, each application must be manually parallelized and divided into different tasks. Mainly, tasks are divided according to loop nesting and dependencies must be explicitly expressed.

To help this manual operation for dataflow applications, designers can use the Par4All (HPC Project n.d.) generation tool, which semi-automatically generates the control task and all computation task source code based on the SESAM HAL (see Section 3.5.3). Depending on the execution results, it is then possible to change the kernel tasks by modifying the pragmas used in the input application, and run it again through Par4All, to better adapt the different stage lengths in the application pipeline. More details on Par4All and its interaction with SESAM can be found in Ventroux et al. (2012). Only homogeneous computing resources are supported for the moment.

The control task is a Control Data Flow Graph (CDFG) extracted from the application, which represents all control and data dependencies. The control task handles the computation of task scheduling and other control functionalities, like synchronizations and shared resource management. It must be written in a dedicated and simple assembly language. Each control task, for each different application, needs to define the number of computation tasks, the binary file names corresponding to these tasks, and their necessary stack memory size. Then, we must specify which are the first and last tasks of the application. Finally, for real-time task scheduling, the deadline of the application, as well as the worst-case execution time of each task, must be defined. The processor type of each task is also specified and this information is used during the allocation process. A specific compilation tool is used for binary generation.

Each embedded application can be divided into a set of independent computation tasks. A computation task is a standalone program. The greater the number of extracted independent and parallel tasks, the more the application can be accelerated at runtime. This acceleration comes at the expense of the control overhead. Then, a manual partitioning must be carried out in the case of heterogeneous MPSoCs. Heterogeneous resource management takes place before the task compilation. Finally, all tasks are compiled for the processing resources or the control manager. Depending on the hardware platform to be explored, the designer can use the SESAM HAL or explicit physical addresses without memory virtualization.

3.5.2 Execution Models

Within the SESAM framework, two execution models are supported. They can be mixed or used independently. The first one is called *control-oriented*. With this model, because all data and control dependencies have been extracted, all tasks are executed without interruption until completion. During its execution, a task cannot access data not selected at the extraction step. It cannot use work-in-progress data, and it must follow the execution order established by the control unit. The execution of non-blocking tasks starts as soon as all input operands are available. Consequently, the model eliminates data coherency problems without the need for specific coherency mechanisms. This constitutes an important feature for embedded systems, since their architecture is accordingly simplified.

We also have the possibility to share data with other concurrent tasks via dynamic buffer allocation. In this model, a task can wait for data produced by another task in the dataflow pipeline. Thus, the second execution model is based on streaming execution. To support this execution model, the most important feature is the management of shared buffers between the pipeline stages. A consumer must wait for the shared data to be written before reading it, in order to keep data coherent. This requires the implementation of a specific protocol. Then, to maximize the parallelism in the pipeline and ensure sufficient concurrent executions, the granularity of data synchronizations must be well-sized. A fine-grain synchronization level generates an important hardware and control overhead to implement all semaphores used to store the access status information. For this reason, we decided to synchronize all shared data accesses at the page level. This is a simple and practical implementation of streaming support, but it remains different from direct handshaking work that has been done in Henriksson and van der Wolf (2006). Our synchronization mechanisms are less intrusive and not blocked by acknowledgments. Besides, all this extra control will generate a task execution overhead that will have to be evaluated, in order to validate the efficiency of this execution model.

With streaming execution, the stages of the application pipeline must communicate through these synchronization primitives to access their shared buffers.

3.5.3   Hardware Abstraction Layer

A computation task is a standalone program, which can use the SESAM HAL to manage shared memory allocations and explicit synchronizations. This HAL is summarized in Table 3.1. It can be extended to explore other memory management strategies.

This HAL provides memory allocation, read/write shared memory access, debugging, and page synchronization functions. Each item of data is defined by a data identifier, which is used to communicate between the memory management unit and the computation tasks. For instance, the function call *sesam_data_assignation(10,4,2)* allocates *4* pages for the data ID *10* with *2* consumers for this data. The function call *sesam_write_data(10,&c,4)* writes the word *c* starting from the 4th byte of the data ID *10*. The *sesam_wait_page* function is a blocking wait method. The task waits for the availability of a page in only read or write mode. When all consumers have sent a write availability, the *sesam_send_page* function is used to inform the memory management unit that the content of the page is ready to be read, or that its content has become useless for the consumer task. The memory management unit can then release the page access rights and accept future writes. This handshake protocol is a semaphore-like process and guarantees data consistency.

TABLE 3.1
Hardware Abstraction Layer of SESAM

	HAL functions	Description
Memory allocation functions	sesam_reserve_data()	reserve pages
	sesam_data_assignation()	allocate the data
	sesam_free_data()	deallocate the data
	sesam_chown_data()	change the data owner
Data access functions	sesam_read()	read data
	sesam_write()	write data
	sesam_read_burst()	read a finite number of bytes
	sesam_write_burst()	write a finite number of bytes
	sesam_read_byte()	read a byte
	sesam_write_byte()	write a byte
Debug function	sesam_printf()	display debug
Page synchronization functions	sesam_wait_page()	wait for a page
	sesam_send_page()	page is ready

When a *sesam_send_page* is sent to the MMU, the status of the page is updated. If the page was in a *write* mode, the consumer number is checked and updated. To distinguish multiple requests of a single task from multiple consumers' requests, a consumer list is maintained for each page. When all consumers have read the page, the page status changes and it becomes possible to write into it again. When a *sesam_wait_page* is sent to the MMU, the request is pushed into a *wait_dispo list request* and the information is sent to the controller. As soon as the page becomes available, the MMU sends to the processor an answer that unlocks the waiting *sesam_wait_page* function. Because a task can dynamically be preempted by the controller and migrated to another processing element, the MMU must be able to address the processor executing the waiting task. Thus, a *sesam_wait_page* is sent again when the task is resumed on the new processor in order to update the processing element address. This protocol is explained in more detail in Ventroux and David (2010).

3.6 SESAM Debug

Virtual prototyping solutions are used to allow an early design of applications for not-yet-existing architectures. To perform an efficient development of embedded applications a debugging environment is compulsory. Thus we developed for SESAM a debugging solution based on the GDB tool chain.

In this section we will first survey the basic element composing our debugging solution: the GDB stub. Then we will present the specificities of our solution to allow an efficient debugging for MPSoC architecture allowing dynamic task scheduling with migrations. Lastly we will briefly present how the GDB interface was enhanced to facilitate the use of the newly developed functionalities.

3.6.1 GDB Stub

The GNU Project Debugger (GDB) is a common debugging tool part of the GNU toolchain. It is classically used to debug programs compiled with GCC that embed debugging information in DWARF format. GDB is particularly used for making a program stop on specified conditions and examining what happened when your program stopped. It is then possible to modify the program and test the effects of one bug and go on to learn about another. In addition, GDB supports all the architectures supported by GCC. Only the GCC back-end for a newly built architecture is needed to develop GDB support. This is done through the definition of a .md file (.md stands for machine description). Thus, GDB was the perfect debugging component for the

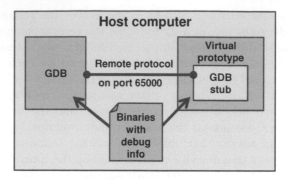

FIGURE 3.6
Structure of the debugging solution implemented in SESAM

SESAM environment since it already supports many processors and allows the development of debug solutions for new processor architectures.

GDB is most often used to debug applications running on the native platform. But it also includes a unique ability to connect to remote processors to debug the code they execute. The connection is established using a port and a standardized protocol, known simply as the GDB remote protocol. To be able to work properly in remote mode, a processor must implement a GDB stub, which is basically anything that can answer properly to basic (compulsory) GDB remote protocol commands. It can therefore be either hard-wired or application based. The GDB end user does not need to know the protocol to debug a program as GDB interprets the usual commands and converts them to remote protocol ones.

When considering debugging on a virtual prototype architecture, the prototype itself can be considered as a remote target. Thus, to be able to debug the architecture we developed a GDB stub for each virtual prototype processor. The debugging infrastructure is summarized in Figure 3.6.

A GDB stub can be seen as a daemon that answers to the request of the GDB instance connected to it. The remote protocol defines a wide range of commands to which the stub must answer. Nonetheless, not all commands are compulsory; only a limited subset is. Indeed a GDB stub must at least answer the requests summarized in Table 3.2. Though handling of some of these requests is not compulsory, they are necessary to reach an efficient execution speed. Among these, we identified from our common GDB usage the commands found in Table 3.3.

Even though ArchC processors embed a debug stub, it is not fully functional yet. As a result we developed our own stub for the SESAM framework. Still, ArchC provided an interesting interface to define standard memory and register accesses. This way any processor designer could provide a unified memory/register view from the processor that our generic GDB stub can use

TABLE 3.2
Basic Remote Protocol Support Commands

command code	command signification
g	read general registers
G	write general registers
m	read at address
M	write at address
c	continue/continue at specified address
s	advance single step
others	send back unsupported request answer

TABLE 3.3
Additional Remote Protocol Commands for Fast Debugging

command code		command signification
p		read single general register
P		write single general register
	Z0	memory breakpoint (not implemented)
	Z1	hardware breakpoint
Z	Z2	write watchpoint
	Z3	read watchpoint
	Z4	access watchpoint

to implement the required remote protocol commands. Our stub communicates with the GDB instance through a specified Linux port.

3.6.2 Debug and Task Migration

In the case of MPSoC architectures modeled within SESAM, the simplest solution consists in instantiating a GDB stub per processor and to connect a GDB instance to every one of them. However, handling at the same time a multiple number of GDB instances is cumbersome for the application developer, especially when the number of processors can reach dozens.

As a result, we chose to develop a solution that would only use a single remote stub interface and would abstract the complexity of the architecture from the application programmer. We also wanted our remote stub to allow the debugging of tasks in the context of dynamic application scheduling. This imposed a requirement to be able to follow the execution of a task no matter which processor executes it.

Then we implemented a single GDB stub for all the processors with the same architecture. (In the case of heterogeneous architectures modeled with SESAM, one GDB stub must be instantiated per processor kind). This common stub keeps all the breakpoints and watchpoints set by the GDB user and shares them with all the ArchC processors.

As an architecture often uses virtual memory with the help of TLB (Translation Lookaside Buffers), these virtual addresses must be cleverly handled. Indeed, two different tasks can have the same virtual addresses when executed on two processors, especially when two different instances of the same application are concurrently executed. Therefore, one must be able to differentiate the tasks executed from one another with information additional to their addresses. As a result, we extended the information transferred from the GDB end-user through the remote protocol so that it could include task identification. In the case of the SCMP architecture later used as an example (Section 3.10), this task ID is composed of an application number, an instance number, and finally a task number. To transfer this information through the remote protocol without modifying it, we extended the remote architecture address size. As a result, in the SCMP architecture case, breakpoint/watchpoint addresses can be set on 64 bits; the first 32 bits are the virtual address and the other 32 include the task identification.

On the stub side, breakpoints/watchpoints are set on the virtual address with the task ID information. A breakpoint is detected only if the task currently executed on the processor is the one in the breakpoint information. This allows a correct debug of a task throughout its potential migration. Note that when a user sets a breakpoint, the breakpoint is set at the virtual address only. To maintain a correct task debug, the stub assumes that the breakpoint is set for the current executing task on the stopped processor.

While this address extension turns out to be very helpful to manipulate a wide range of architectures, it is less natural for the application developer. As a result we needed to hide this mechanism as much as possible from the end-user. This was performed by providing GDB user-defined commands. We will present in the next section the new commands that were implemented for the case of the SCMP architecture.

3.6.3   Specialising GDB's Functionalities

Enhanced GDB commands were defined to keep the GDB debug as simple as possible, while accurately debugging migrating tasks. User-defined commands are described in a .gdbinit file. The .gdbinit file is first searched in the current user's home directory then in the folder from which GDB was launched. The following syntax is used to define user GDB commands:

```
1   define <command>
2   <code>
3   end
4   document <command>
5   <help text>
6   end
```

As a result, we defined the following commands to ease the debug:

- **sesam_connect** connects to the simulator platform while it is waiting for a remote connection.

- **sesam_file** *file_path filename* loads filename as the reference code for GDB.

- **sesam_break** *ida nin idt address32* breaks at the address *address32* of task *idt* of application *ida* instance *nin*. It allows a breakpoint to be set on a task no matter which processor executes it.

- **sesam_proc_status** displays the task executed by the currently inspected processor.
  sesam_proc_status *-1* displays the tasks executed by every processor.
  sesam_proc_status *id_proc* displays the task executed by processor *id_proc*.

- **sesam_select_proc** *proc* sets the current debugged processor to *proc*.

In the case of the SCMP architecture the shared data are accessed through HAL functions. Therefore, to offer debugging that is adapted to the prototyped SCMP architecture, we developed watchpoint functionalities adapted to its programming model. **sesam_watch** *id_data offset* sets a write watchpoint on the shared data *id_data* at *offset*. The scope of such a watchpoint is the whole application of the current task. We also defined **sesam_rwatch** and **sesam_awatch** for read only and write only watchpoints, respectively.

3.7 SESAM Energy Modeling

Energy consumption is a very important parameter to be considered at each step of the design process. Different solutions have been implemented in the SESAM framework to allow the exploration of different energy saving strategies. The first estimates the processor's energy consumption at the instruction level and is fully integrated into the ArchC language. It has been named *PowerArchC* (Gupta et al. 2010). The second is based on power state machines and can model DPM (*Dynamic Power Management*) and DVFS (*Dynamic Voltage Frequency Scaling*) modes. Both solutions are detailed in the following sections. Then we will present an example of an energy saving strategy that relies on the power consumption models.

3.7.1 PowerArchC

PowerArchC is built on top of the ArchC methodology (Gupta et al. 2010). ArchC provides an efficient framework for describing a processor architecture

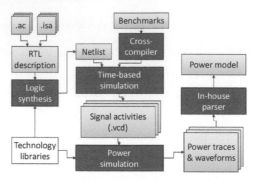

FIGURE 3.7
PowerArchC: Power model generation flow

and ISA at the behavioral or cycle accurate level. In addition, it automatically generates an ISS with a short development time, but it does not have any way to estimate its power consumption. We have extended the ArchC methodology with power capabilities. Using a power model from Instruction Level Power Characterization (ILPC) into PowerArchC we can provide total used energy and average power consumption for a given benchmark, design implementation, and technology node. The PowerArchC methodology is based on two steps.

The first step of PowerArchC consists in creating static and dynamic power models and is depicted in Figure 3.7. A Processor designed at Instruction level or RTL is a lot less detailed about the design. Hence, to characterize power, the gate-level design corresponding to the functional description is required to provide sufficient detail. Consequently, the RTL description of the processor is performed manually relative to the architecture description made in ArchC. From that, an RTL synthesis can be done to obtain a gate-level netlist of the processor. Synthesis is performed with the Synopsys tool Design Compiler (Synopsys Inc. n.d.).

Then we use the gate-level power simulation tool Primetime (Synopsys Inc. n.d.) to provide accurate power consumption of each block, at each clock cycle. Since power simulation tools only provide power consumption of hardware blocks, we designed a parser tool that outputs the average power consumption of each instruction from power and program traces provided by the simulation tool. The parser tracks the power value in the different pipeline stages crossed by the instruction. Based on the data path in each stage of the pipeline, it gathers the power conusmption caused by the execution of one instruction and stores it in a model. Each instruction with the same name will only have one entry in the previously defined model, and their different power values are averaged for each stage. A characterization study has to evaluate all of the instructions for different operand and instruction interrelations. Since it is not

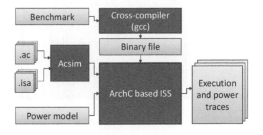

FIGURE 3.8
PowerArchC: Power-aware ISS architecture generation

possible to cover all the possible cases in a reasonable time, only a subset of possibilities is usually considered. In our case, ILPC outputs a power model that provides an average total power value (static and dynamic) for each instruction.

The second step of PowerArchC uses the power models to estimate the energy consumption at the instruction level at runtime. The behavioral description of the instruction now contains a variable that points to the corresponding instruction in the power model. The ISS is now able to output both instruction and power traces and total consumed power of the executed program. The basic and modified architecture of ISS is illustrated in Figure 3.8. From two input description files (.ac, .isa), the tool acsim generates SystemC-based ISS source code that can be easily integrated in a complex MPSoC model. Given that and an input power model, the compiler for the host machine generates the modified ISS with power capabilities. A new ISS output provides power traces at the end of simulation of a benchmark.

From that, it is possible to explore the power consumption at a high level in a quick fashion. Since we experiment with performance accuracy at the instruction level, the resulting power accuracy will obviously be lower than the one obtained after synthesis. This is mainly due to the interrelations between the instructions during execution and the value of the operands that both affect the bit toggling activity of the processor microarchitecture and hence the dynamic power consumption. The maximum error observed is 6.4% compared to Primetime, but with a performance gain of 100. One should note that the ILPC can be performed automatically whenever hardware implementation of the microarchitecture or the technology library changes.

3.7.2 DVFS and DPM

Having the possibility to model the power dissipation of the whole platform is very challenging and very important for the MPSoC designer. Nevertheless, Dynamic Voltage and Frequency Scaling (DVFS) and Dynamic Power Management (DPM) modes must also be integrated into MPSoC to manage

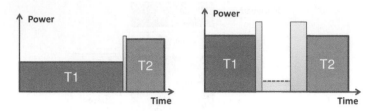

FIGURE 3.9
DPM (left) and DVFS (right) technique timing issues: the deeper the sleep mode is, the less energy it consumes but the longer its wake-up time and energy penalties become.

their energy consumption and adapt it to the user and application needs. It is then mandatory for an MPSoC exploration framework to integrate these functionalities in order to better estimate the global energy consumption but also to develop adapted energy saving policies.

The dissipated power in a CMOS design can be attributed to two major sources: the dynamic power consumption and the static. The dynamic consumption is mainly due to transistor state switching and it can be drastically reduced by lowering the supply voltage. As the transistor delay is a function of the supply voltage, lowering the supply voltage imposes an adapted frequency reduction. This technique is called DVFS. The static consumption is due to various current leakages in the transistors. The DVFS technique has some impact on the static power consumption thanks to the supply voltage reduction. Nonetheless this is not sufficient to drastically reduce static power consumption. To cut down static power consumption the only viable solution consists in switching off unused parts of a circuit. This technique is called DPM. Contrary to the DVFS technique, the resource is made unavailable.

The main drawback of these two techniques lies in the mode switching timing and consumption penalties. If the timing penalties for the DVFS are rather constrained, it is not the same for the DPM, where wake-up time can reach a hundred milliseconds (136 ms for the PXA270 (Intel Corporation 2005)). Therefore, for a processor implementing both techniques, the issue is to discover when reducing the voltage and frequency couple is more energy efficient than running at full speed then switching off the processor. This matter is summarized in Figure 3.9. For a given technological process, the issue is thus to evaluate the duration of future inactivity periods of the resource. This is why, to allow the benchmarking of power consumption control solutions, we have integrated complete power state machines which take into account energy saving and temporal activation overheads.

3.7.3 Scheduling Example

To illustrate what can be performed by using some of SESAM energy consumption features, we propose a scheduling algorithm example, which tackles power consumption issues through an efficient use of Dynamic Voltage and Frequency Scaling (DVFS) and Dynamic Power Management (DPM) modes of the processing resources. This algorithm was first presented in Sassolas et al. (2011).

Our scheduling algorithm has been written to handle streaming applications. A streaming application is a set of tasks with consumer/producer relationships. Data is transferred from a producer task to a consumer task through a circular buffer. Only one task can write on a buffer, while it can be read by multiple consumer tasks. This creates a divergence in the data flow. A consumer task can also read multiple input buffers, creating a convergence in the data flow. This allows the description of parallelism in the processing flow of a given data stream.

Given the previously described application model, one can make a few observations. The throughput of a streaming application is constrained by the duration of its slowest stage. As a result, other pipeline stages can be slowed down to meet the same output rate as the slowest stage. This can be performed by using a slower DVFS mode for the resources with an excessive output rate. Besides, tasks that are further in the pipeline stream than the slowest task are to be blocked waiting for data. These tasks should be preempted if other tasks can execute instead, or the resource should be shut down if not. This implies the use of DPM functionalities. Given these observations, our algorithm used DVFS to balance the pipeline stage length and DPM to shut down unused resources. Our objective was to maintain the same data throughput as if the task were executing at full speed while making substantial energy savings.

To be able to balance an application pipeline, we needed additional information on the dynamic output rate of a task. Thus we introduced monitors on every communication buffer. For every buffer we specified how many datasets it could contain. We also specified two thresholds. When the higher threshold is reached we assume that the producer is executing too fast. When the lower threshold is reached we assume that the producer is not executing fast enough. A specific event is sent to the scheduler when a threshold is crossed. It contains the writing task identifier. An event is also sent when a task is blocked reading an empty buffer, as well as when a task is blocked writing a full buffer. The buffer monitors are summarized in Figure 3.10. One objective of balancing pipeline stage length was to prevent buffers from getting full, which would block the producer. Another was never to reach an empty buffer, which would block the consumer and could result in an increase of the data processing length. Thus, as depicted in Figure 3.10, when a buffer reaches a full threshold the writer task is blocked and can be preempted. When a buffer reaches an empty threshold the reader task is blocked and can be preempted. To avoid these cases we define 2 thresholds that will make the writer enter

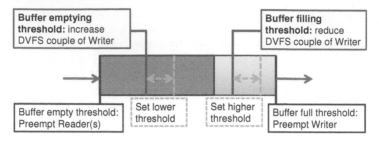

FIGURE 3.10

Summary of buffer monitors and scheduling implications (Sassolas et al. 2011)

or leave half-speed DVFS mode. The positions of these thresholds can be set through their instantiation with the HAL.

To keep our scheduling algorithm as simple as possible the task priorities were made of a static and a dynamic part. We will list the different priority parts by level of importance. First we check the blocked task status, as we do not want to give priority to a blocked task. Then the application priority is taken into account. After that, we study pipeline position priority. Every task is given a priority depending on its position in the streaming pipeline. This allows us to give the priority to tasks handling older datasets, i.e., the ones that are deeper in the pipeline. Finally, for tasks that have the same pipeline position priority, we give the priority to the task with the emptier buffer. This power-aware scheduling solution was benchmarked thanks to SESAM with the case of a WCDMA application executing on the SCMP architecture virtual prototype. The power reduction achieved will be shown in Section 3.10.

3.8   SESAM Exploration

A high level architectural model is not only used for virtual prototyping but also for architecture exploration. Indeed in the early architecture design stages, various architectural options must be compared to reach required criteria for given application domains. Such criteria can be very different for a wide range of target applications. Some domains target raw performance while others consider power consumption, performance per surface unit, or respect for real-time constraints. Thus, finding an appropriate sizing is compulsory to develop cost-effective architectures adapted to their target use-case. These ideas were at the root of the conception of the SESAM simulator. As a result, the SESAM simulator has been built so as to allow an easy architecture sizing. In this section we will present how the SESAM simulator uses dynamic parameters to efficiently reconfigure the simulated architecture. Then we will present two tools that could be joined to SESAM which allow the exploration of application parallelism for both control-oriented and streaming execution models. These parallel applications can help in sizing the architecture and also in finding the best trade-off between parallelism and area/performance efficiency. Finally, we will present how SESAM simulations can be easily launched in parallel to obtain sizing results as early as possible, thus reducing the time-to-market.

3.8.1 Dynamic Parameters

To ease the exploration of MPSoC architectures, SESAM was made very flexible and all the components and system parameters are set at run-time without platform recompilation. Indeed the SESAM binary takes as an argument a parameter file. SESAM analyzes this file and updates the values of the various parameters of the simulator. For all missing parameters, SESAM uses default values. SESAM knows more than 160 different parameters: it is possible to define the memory map, the name of the applications that must be loaded, the number of processors and their type, the number of local memories and their size, the parameters of the instruction and data caches, memory latencies, network types and latencies, etc.

SESAM comes with a Python script that allows an easy generation of parameter files from a range of values for various parameters. This allows a more thorough exploration of the architectural design space.

To further improve the exploration analysis, SESAM natively monitors more than 250 platform statistics. For instance, SESAM collects the miss rate of the caches, the memory allocation history, the processor occupation rate, the number of task preemptions, the time spent to load or save the task contexts, the effective used bandwidth of each network, etc. A complementary macro allows easy import of these statistics to Microsoft Excel, thus allowing efficient analysis to be automatically conducted. This macro drastically reduces the comparison time between various configurations of an architecture.

The complete exploration flow described here is summarized in Figure 3.11. As we have seen, SESAM was conceived for the exploration of MPSoC architecture. We will now focus on additional tools for SESAM that speed up the generation of test applications for the purpose of exploration, and also help in defining the parallelism level that gives the best area and performance efficiency.

3.8.2 Application Parallelism Exploration

Porting an application to an MPSoC platform requires it to be parallelized and to find the best match between the architecture and the final application. Parallelizing an application induces additional control cost, code overheads, or data dependencies, and exploring different possibilities becomes mandatory to determine the best trade-off. However, parallelizing an application takes a

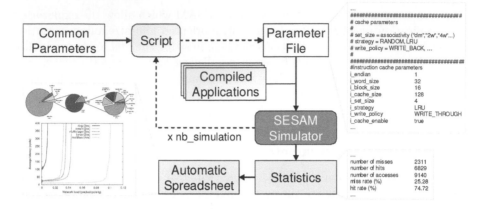

FIGURE 3.11

SESAM exploration tool and environment (Ventroux, Guerre, et al. 2010). A script is used to execute a set of SESAM simulations with different input parameters. Once all simulations are finished, all statistics are imported and automatically shaped to ease the result analysis.

lot of time. This is why we can associate to SESAM specific tools to design an efficient system by exploring both applications and architectures. The first one, named SESAM AGP, can semi-automatically generate control-oriented applications with a variable parallelism. The second one derives from a collaboration with an HPC project and concerns streaming applications. Both solutions have been developed in the SCALOPES project.

3.8.2.1   SESAM AGP

To facilitate the design of parallel applications with explicit control for exploration purposes, we developed an automated tool: SESAM AGP. The tool does not perform an automated parallelization of sequential code. It relies on a manually parallelized version of the application where some parameters depend on the desired degree of parallelism. The user needs to modify initial source code to prepare the parallelism extraction. Source code that will be generated by the tool must be generic, whereas some independent source codes remain static. The user also needs to define a generic control graph that can be easily extended. Then, for every application, explicit data and control dependencies must be manually extracted and expressed in a control graph. The tool makes it easy to express how the parameters should evolve according to the desired degree of parallelism. With this knowledge, the tool performs a source to source generation of parallel applications.

The SESAM AGP engine is based on Python and does not depend on the application. It takes as inputs modifiable and static source codes of the application, as well as a generic control graph with parallelism options.

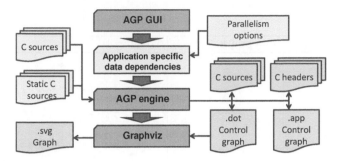

FIGURE 3.12
SESAM AGP toolchain

Options are set in the GTK graphical interface, which is the entry point of the tool. As shown in Figure 3.12, the SESAM AGP engine specializes the control graph according to the options and generates all source code corresponding to the new parallel application. Then, a graphical interface allows the display of a graphical representation of the control graph.

The SESAM AGP tool can generate multiple instances of an application with a different level of parallelism (see Figure 3.13). After using our tool, we only need to compile the control and the computation tasks to execute the new application for the architecture modeled in the SESAM framework.

3.8.2.2 SESAM/Par4All

In the case of dataflow applications, an automatic parallelization tool is also compulsory for both architecture and application exploration. Usually the best execution for a pipelined application is reached when the pipeline is well balanced. In the case of asymmetric MPSoCs, irregular processing length can be partially hidden thanks to the preemption and migration of tasks that balance the computing resource load between homogeneous resources. Thus, the optimal pipelining is obtained when the number of available tasks for execution is high enough so that all resources are always executing at full speed. But, as control latencies and memory usage overheads also need to be taken into account in overall performance, a task size limit exists where additional splitting of tasks is counter-productive. Thus, finding the optimal task size is an open problem that depends on various parameters such as the amount of shared data, the control latencies, and the available memory. Only comparative simulation with various pipeline splitting choices can allow efficient trade-offs to be found. As a result, the issue lies in the development cost of all possible pipelining of the sequential code.

To deal with this issue, we jointly developed with an HPC project a specialized version of the Par4All tool dedicated to the SESAM dataflow execution model (Ventroux et al. 2012). This tool takes as input the sequential code of an application annotated with pragmas identifying execution kernel limits.

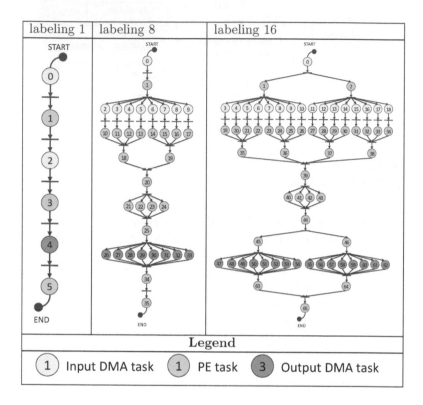

FIGURE 3.13
Example of automatic parallelization with a connected component labeling application. The figure represents the generated CDFG for a sequential version and two parallellized versions with a maximum parallelism rate of 8 and 16.

Based on data dependency analysis, Par4All can identify the data that will be shared between the execution kernels and automatically generate the additional data access code. Thus, the tool fully generates a pipelined version of the sequential code with implicit data dependencies. As this specialization of the tool is based on an SIMD specialization where data processing is forked and then joined, the output pipeline integrates data synchronization tasks between computation stages. Further tool improvements include the removal of unnecessary synchronization tasks and the development of an automated iterative search of an optimally balanced application pipeline.

As a consequence, the SESAM environment comes with two application parallelization tools, SESAM AGP and Par4All, for, respectively, control-oriented and data-flow applications.

3.8.3 Distributed Simulations

Because the exploration of many parameters can take a lot of simulation time, SESAM offers the possibility to automatically dispatch all the simulations to different host PCs. A tool named S^3 (SESAM Simulation Server) was developed to provide an easy way for users to automatically dispatch simulations and share computing power. With S^3, 400 simulations can be carried out with 12 hosts (40 slots) in less than one hour and a half for the execution of a connected component labeling application on an asymmetric MPSoC composed of 8 computing resources.

S^3 is composed of 3 programs which implement a centralized simulation queue. *S^3_client* is the program through which a user can ask for the execution of a simulation. When a user configures a simulation, its execution demand is forwarded to the *S^3_dispatcher*. *S^3_host* is a daemon executing on computers willing to share computing resources. A graphical interface allows the sharer to specify how many resources he wishes to share with S^3 clients. The available resources status is sent to the *S^3_dispatcher*. As a result, the *S^3_dispatcher* daemon implements the global simulation queue. It is responsible for dispatching all simulation requests to available simulation cores according to a simple load balancing algorithm: when there are less simulations than usable slots, the dispatcher chooses the host with the lowest load first. Host load is defined by:

$$load = \begin{cases} 1 & number\ of\ usable\ slots = 0 \\ \dfrac{busy\ slots\ number}{usable\ slots\ number} & number\ of\ usable\ slots > 0 \end{cases}$$

If two hosts have the same load, the dispatcher uses the one with the highest usable slot number.

In terms of implementation, S^3 is structured around a dispatcher and a Network File System (NFS) server. The execution flow of a simulation with the help of S^3 is summarized in Figure 3.14.

As a result, the SESAM tool is a very complete architecture exploration framework. It allows us at the same time to easily set various parameters of a virtual platform and to retrieve statistical data. It also comes with two application parallelization tools, SESAM AGP and Par4All, that enable fast application parallelization exploration. Lastly it can dispatch multiple simulations to distant hosts, thus reducing the simulation time. We will present in Section 3.10 what was achieved thanks to this exploration framework.

3.9 Use Case

To demonstrate SESAM's capabilities to model asymmetric multiprocessors, we have used this framework to model the SCMP architecture (Ventroux and David 2010). The SCMP architecture is a computation-intensive MPSoC

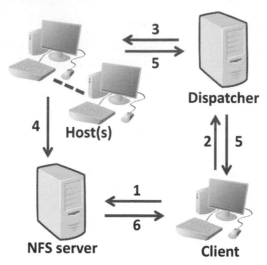

FIGURE 3.14

Parallelization of SESAM simulations. The client copies the simulator and its libraries on an NFS share folder (1). Then the client sends to the dispatcher the list of simulations to be executed with their parameter files (2). The dispatcher looks for an available slot to execute each simulation (3). Then the host changes its working directory to the NFS share folder and executes the simulation. At the end of each simulation, the result is written into the share folder (4), and the host informs the client through the dispatcher (5). Finally, the client gets back all the results (6).

that is seen by the CPU as a coprocessor. This architecture is characterized by the centralized control manager named CCP, that dynamically executes tasks on homogeneous processing elements. The SCMP architecture supports a constrained-task and a streaming execution model, through a logically shared and physically distributed memory. The SESAM environment was used to size the various components of the architecture and resulted in the VHDL implementation of the selected configuration.

3.9.1   SCMP Overview

As depicted in Figure 3.15, the architecture includes three internal NoCs. The system NoC interconnects the external CPU, the external memories, and the TLB dedicated to the application with the core of the architecture. The system NoC is a 10-cycle-latency 32-bit simple bus with a round-robin arbiter. The CPU represents a host interface that allows the user to send on-line new commands to the MPSoC. For instance, it is possible to ask for the execution of new applications. The TLB Application is used to store all the

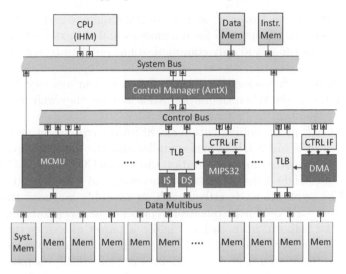

FIGURE 3.15
SCMP architecture

pointers to every task of every application in the external instruction memory. External memories have a latency of 4 cycles. When the simulator starts, it automatically loads all the selected applications into this memory and updates the TLB Application.

The control NoC is used to connect the Control Manager and all processing resources through a control interface. It is similar to the system NoC. In addition, processing resources can communicate with each other, and with the Memory Configuration and Management Unit (MCMU). The MCMU aggregates the MMU and the CLU presented before. The data NoC is only used for communication between the processing resources and the local memories. It is a 64-bit multibus network that connects all PEs and I/O controllers to all shared and banked memory resources. The data network and memory latencies are 3 cycles. A specific memory, named system memory, is accessible through this network and stores all the system code used by the computing resources. The CCP is based on the AntX processor (Bechara et al. 2011). It is used to prefetch a task's code before its execution and manages all the dependencies between tasks. It determines the list of eligible tasks to be executed, based on control and data dependencies. It also manages exclusive accesses to shared resources and non-deterministic processes. Then task allocation follows online global scheduling, which selects real-time tasks according to their dynamic priority and minimizes overall execution time for non-real-time tasks. An interrupt controller is used to communicate with the computing resources. We have deliberately used a homogeneous architecture composed of MIPS32 processors to highlight our parallelism management approach rather than overall system performance. However, we use two DMA units to carry out input image transfers between the internal local memories and the external data memory. All devices are timed and only communications are approximate-timed transactions.

ISSs boot on a read-only memory, named system memory, that contains all the system code. When the initialization is done, they wait for the Control Manager requests.

The SCMP programming within the SESAM environment necessitates the integration of a MIPS32R2 cross-compiler and the AntX compiler. Only C language is supported for computation tasks. The CCP needs a Control Data Flow Graph (CDFG) that represents all control and data dependencies of the parallelized application, described with an assembly language that can easily represent CDFGs. Each transition represents an execution constraint that imposes the task execution order. Finally, a parser generates the binary for the control manager from the CDFG file. Besides, the complete system code needed by the platform is described by the MIPS32R2 assembly language. The SCMP architecture offers a very high degree of parallelism. Thanks to dedicated hardware scheduling and fast reactivity, it also enhances resource utilization. To measure SCMP's performance for dynamic embedded applications, we implemented a radio spectrum sensing application, developed by Thales Communications France (TCF) within the SCALOPES project and a Wide-band Code Division Multiple Access (WCDMA) application.

3.9.2　Implemented Applications

3.9.2.1　Radio Sensing Application

Spectrum Sensing is one of the main functions constituting a cognitive radio. A cognitive radio is a system characterized by the ability of a terminal to interact with its environment. It means that this terminal will have skills to sense its surrounding environment (sensing), to decide (cognitive manager), and to reconfigure (software radio) itself. For instance, it will be able to detect the available frequencies and use them.

The application case that we study is more precisely the spectrum sensing step (ITU 2005) which aims to detect unused spectrum and to share it without interference with other users. In other words, the spectrum already used is detected in order to identify spectrum holes. This application is used in spectrum monitoring devices and electronic warfare devices. The sensing function faces a number of challenges in terms of miniaturization, power consumption, and timing response. These constraints are even more severe for mobile terminals. Three main sensing techniques can be used within the scope of spectrum monitoring and sensing: Cooperative Context (data-aided techniques), Blind Context, and Semi-Blind Context. In this use case, we limit the use case to a Global System for Mobile Communications (GSM) sensing application (cooperative context). The GSM sensing is composed of 3 main steps:

1. digital signal pre-processing consisting of signal wideband filtering and in baseband transposition as well as in the signal channelization.

2. processing of each channelized digital signal consisting of parameter estimation and the measurement of signal amplitude, bandwidth, and modulation parameters.

3. processing of each symbol consisting in the demodulation of the signal and in the measurement of the symbol stream code parameter.

For this application, each processor has an 8 KB data and 4 KB instruction cache, exclusive memory and implements a write-back and write-allocate protocol. It also has access to a 4 MB shared memory organized in 32 memory banks of 128 KB each. The memory space is divided into 2 KB pages.

3.9.2.2 Wide-Band Code Division Multiple Access (WCDMA)

The second implemented application is a complete Wide-Band Code Division Multiple Access (WCDMA) encoder and decoder (Richardson 2006). This communication technology is based on the use of Orthogonal Variable Spreading Factor (OVSF) to allow several transmitters to send information simultaneously over a single communication channel. This application uses a rake receiver with a data aided channel estimation method.

Known pilot symbols are transmitted among data. The channel estimation algorithm operates on the received signal along with its stored symbols to generate an estimate of the transmission channel. The processing of pilot frames generates a dynamic behavior of the application, since this induces a variable execution length.

The application is pipelined into 13 different tasks. To maximize the concurrency between pipelined tasks, a double buffer is used between each task. Thus, tasks can independently execute the next frame from the results of the previous pipelined stage. For this application, each processor has 2 KB data and instruction cache exclusive memory and implements a write-back and write-allocate protocol. It also has access to a 4 MB shared memory organized in 64 memory banks of 16 KB each. The memory space is divided into 256-byte pages.

3.10   Validation

To evaluate the performance of SESAM, we integrated the SCMP model into SESAM and executed the radio sensing and WCDMA applications on it. An RTL-level description of this model has also been developed and implemented on an FPGA-based emulation board provided by EVE (Emulation and Verification Engineering 2010). First of all, by using the radio sensing as the use case, we will evaluate SESAM accuracy compared to a real execution on the hardware prototype. In addition, we will compare the SESAM simulation speed to an RTL simulation and a hardware emulation with the EVE board. As an example, we will evaluate the performance of different Networks-on-Chip and therefore demonstrate the potential of SESAM to help MPSoC design. Finally, based on the WCDMA application, we will evaluate the SESAM performance and the power consumption saving brought by the power optimisation policy presented in Section 3.7.

3.10.1   SESAM Accuracy

We ran the radio sensing application on SESAM and on the hardware prototype by varying the number of cores from 1 processing element (sequential implementation) to 4 processing elements. Figure 3.16 depicts the deviation between SESAM simulation and the FPGA-based emulation. It is shown that the maximum deviation remains below 6%.

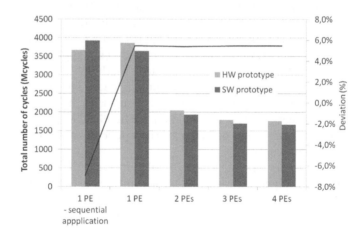

FIGURE 3.16
Evaluation of SESAM accuracy

3.10.2   SESAM Simulation Speed

In our development flow, we have used SESAM to create a virtual prototype of the SCMP architecture and to explore the design space. Then we have designed the complete MPSoC architecture at the RTL, and finally emulated the platform on a hardware emulation board to accelerate our RTL simulations. We measured the necessary simulation time with these three methods and, as depicted in Figure 3.17, we measured a 250× acceleration with SESAM com-

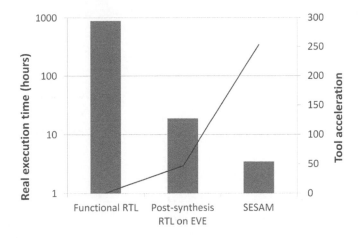

FIGURE 3.17
Comparison of simulation duration between SESAM, functional RTL simulation, and FPGA-based emulation.

pared to a functional RTL simulation, as well as a 5.4× acceleration compared to hardware emulation.

3.10.3 SESAM Sizing Example: NoC Study

To evaluate NoC performances, we replaced ISSs with traffic generators. We consider only shared-memory architectures. Thus, each communication is a request to a memory. Access times and access conflict timings are computed into memories. Traffic generators send only blocking read requests with a fixed data size. For the evaluation, 5 different NoC topologies are studied. This includes the basic interconnection topologies multibus, ring, multistage, mesh, and torus. These networks have been chosen because they are representative of the network design space (Salminen, Kulmala, and Hamalainen 2007). The architecture of these networks can be found in Guerre et al. (2010).

In this study, we will determine the best performance network under a uniform traffic. We consider the global latency of a message as the time between the message creation and the answer reception in the traffic generator. All considered latencies in the next figures are the mean of 20 simulations with a minimum of 5000 requests by traffic generators. One thousand warm-up requests are generated at the beginning of each simulation. The simulation time depends on the packet injection rate. All routing algorithms are deadlock free.

As already explained, the use of uniform traffic allows us to simulate an unpredictable network load. Figure 3.18 shows the NoC average latency as a function of the network load. Because the multibus offers a direct and uniform link to each memory, it has the best performance. Actually, the more

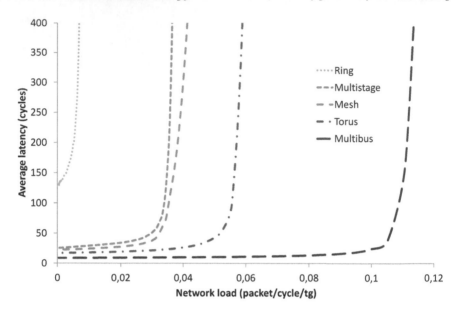

FIGURE 3.18
Average latency of all networks as a function of network load with a uniform
distribution (extracted from Guerre et al. (2010)).

the distance with all memories is heterogeneous and important, the less the
topology is efficient with a uniform distribution. Thus, for instance, the torus
is better than the mesh.

This NoC evaluation shows that SESAM can be used to perform specific studies, even for generic applications, thanks to traffic generators. In the
SCMP architecture, such architectural explorations lead to a multibus data
network.

3.10.4   Performance Study in SESAM

To study the acceleration and the benefits that could be obtained with streaming execution, we compare in Figure 3.19 the execution of the non-parallelized
WCDMA application on a standalone PE, with a streaming implementation
on several processors.

As shown in Figure 3.19(a) and (b), the parallelism obtained with streaming execution can be important. We get an acceleration of 4.5 and an occupation rate beyond 80% with 8 PEs. The acceleration is constrained by the
pipeline length, which is limited to 13 tasks. However, the task execution on
a standalone processing element is penalized by more cache misses, since all
the applications share the same caches. These results depend on the control

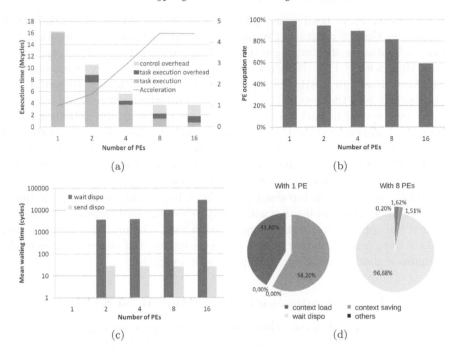

FIGURE 3.19

WCDMA encoder/decoder implementation results on a MPSoC platform with a streaming execution (except with one PE): (a) total execution time, task execution overhead, and control overhead depending on the processing element (PE) number (MIPS32 processors); (b) utilization rate of PEs; (c) mean waiting time for the management of shared buffers; and (d) details of task execution overheads that take part in the task execution time with 1 and 8 PEs (with 1 PE the application is not pipelined) (Ventroux, Sassolas, David, et al. 2010).

overhead, which must be minimized. In our architecture, we use a microkernel especially developed to optimize the reactivity of the control. However, when the number of PEs increases (up to 16 PEs), the task execution and control overhead become predominant. This is mainly due to the control complexity, which increases with the number of tasks and PEs to manage, and also because of the limited parallelism level of the application.

The streaming description of the application and the use of the streaming protocol to access shared data also has a non-negligible cost. Many accesses to a central device, such as the MMU, to get the authorization to write or read each page of a buffer could have a very negative impact on performance. Only simulations help the evaluation of the potential benefit. In Figure 3.19(c) and (d), the task execution overhead is represented and helps us understand the penalty induced by a streaming execution model. Without parallelization, the task execution overhead is insignificant and is only due to the task loading or saving. On the contrary, the task execution overhead must be taken into account within a streaming execution. It represents 22% of the total execution time with 8 PEs. It is mainly due to page availability waits and this increases drastically with the number of processing elements. Regarding the contentions into the network, since we use a multibus network, all PEs have their own access to each memory bank, and therefore only FIFOs on each memory bank can induce an additional latency. For instance, with 8 PEs the increased latency to access a memory bank is about 0.43% for data accesses and 0.12% for instruction accesses. These results are due to a smart allocation of data and instructions between memory banks by the MMU.

These results show interesting benefits when using streaming execution. The parallelization is effective with many processors and turns out to be scalable if the application pipeline length is sufficient. The protocol used to access shared data has a negative impact on performance, but this execution model remains a good trade-off for multiprocessing execution.

3.10.5   Power Management Evaluation in SESAM

To study the impact of our scheduling algorithm, presented in Section 3.7, we chose to compare it to two simpler versions of the algorithm. The first version does not handle power issues. It simply schedules tasks, relying on pipeline stage position and blocked states, and DVFS is not exploited. It is referred to as no energy handling scheduling. The second version is DPM only scheduling. This corresponds to a naive power-aware approach. Here, unused resources and resources executing blocked tasks are put into deep idle mode. Finally, our proposed algorithm will be referred to as DPM + DVFS scheduling.

The results presented in Figure 3.20 were obtained with the same WCDMA application sending 256 frames. The communication buffers were 8 frames long and had a higher threshold identical to the lower one and set to 2 frames. As shown in Figure 3.20(a), the total execution time of the WCDMA application is not affected by our scheduling algorithm no matter how many processing resources there are. The variation in execution time is always maintained below 1.2%. In addition, our algorithm allowed a good acceleration of the processing for streaming applications.

While we managed to maintain the execution time of the scheduling without energy awareness, Figure 3.20(b) shows that substantial energy savings were made. As soon as processor effective occupancy drops, it is directly compensated by our power saving method. With 13 processors, we reduced the power consumption by 45%. In addition, our method obtains better results than the DPM-only scheduling, which only reaches 37% energy saving in that case.

Figure 3.20(c) illustrates how our scheduling algorithm uses the DPM mode in a real application case. This figure shows that when processors spend little time waiting for data or in an unused state (below 17%), the Deep Idle

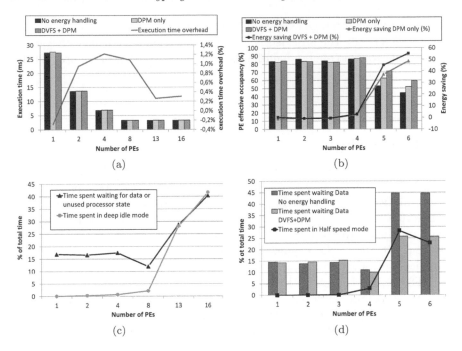

FIGURE 3.20

Power management evaluation: (a) Total execution time for the WCDMA application as a function of the number of processing resources and the scheduling algorithm used; execution time overhead of our solution compared to the *no energy handling* algorithm. (b) Total processor effective occupancy and energy saving as a function of the number of processing resources and the scheduling algorithm used. (c) Average time spent in deep idle mode compared to the time spent in unused state or waiting for data for a processor when using our proposed algorithm. (d) Comparison of the average time a processor spends waiting for data in the case of the no power saving algorithm and of our solution (DPM+DVFS) (Sassolas et al. 2011).

mode is seldom used. When the wasted time increases, the DPM usage curve follows the unused or blocked processor curve as planned. In fact, when the number of processing elements is small, there is often another task ready to be executed immediately. For low PE numbers, the wasted time corresponds to the control overhead. The controller lacks reactivity to reach higher computing performance or power saving.

Finally, Figure 3.20(d) studies the impact of DVFS mode usage on the application execution. We compare the execution of our algorithm to the no energy handling scheduling. The analysis shows that when DVFS modes are used they drastically reduce the amount of time spent in blocking states (42% reduction for 13 processors). Thus, our algorithm succeeds in balancing the streaming pipeline stage execution length efficiently when the processor usage drops. As a result, the processor load is increased with our algorithm compared to the one with no energy handling scheduling, as shown in Figure 3.20(b).

3.11   Conclusion

In this chapter, we have presented an asymmetric MPSoC framework named SESAM. This MPSoC exploration environment can easily take part in a complete MPSoC design flow. It can help the design and sizing of complex architectures, as well as the exploration of application parallelism on multicore platforms, to optimize the area and performance efficiency of embedded systems. It allows the exploration of different approaches for a given set of applications, and the design of new architecture paradigms.

The SESAM framework has been evaluated through the simulation of a complete MPSoC architecture: SCMP. WCDMA and radio sensing applications were implemented on this platform. Results showed that simulations reach a high simulation speed while maintaining good accuracy. SESAM offers the possibility to study many architectural parameters and helps the designer to obtain interesting results easily. With the examples presented in this chapter, thanks to SESAM, it has been demonstrated that good performance can be obtained in SCMP in spite of control and synchronization overheads induced by an asymmetric approach.

We have also presented different energy consumption features that come with SESAM. As a study example, a power-aware scheduling algorithm for pipelined applications was proposed. With the SESAM environment, we were able to analyze the impact of our strategy on the energy consumption and performance. We showed that the use of DVFS and DPM did not impact the application execution speed, while significantly reducing the energy consumption.

Only simulations could demonstrate that our global scheduling did not impact the execution time in spite of the use of DPM and DVFS modes.

Using a virtual prototyping solution such as SESAM is mandatory today when designing complex MPSoC architectures. Making an intensive use of simulation early in the design process drastically reduces development costs and improves final products.

Acknowledgments

Part of the research leading to these results has received funding from the ARTEMIS Joint Undertaking under grant agreement No. 100029. The authors would like to thank Raphaël David, Guillaume Blanc, Nassima Boudouani, and Larbi Moutaoukil for their helpful contributions to this work. We would also like to thank Florian Broekaert from Thales Communications France for providing the radio-sensing application, and Beatrice Creusillet, Ronan Keryel, and Pierre Villalon from the HPC Project for their work on the Par4All compilation tool.

Part II

Parallelism and Optimization

4

Verified Multicore Parallelism Using Atomic Verifiable Operations

Michal Dobrogost, Christopher Kumar Anand, and Wolfram Kahl

Department of Computing and Software, McMaster University, Hamilton, Ontario, Canada

CONTENTS

4.1 Introduction

Parallel computer architectures have now become almost universal, with even mobile phones containing dual core processors. Taking full advantage of the performance available in a computer system requires taking full advantage of the parallelism offered by that system. Opposed to the need for high performance is the need to verify the code's requirements. Programming such architectures has long been recognized as a difficult task. Both requirements for performance and verifiability are difficult to attain.

In a traditional locking architecture, performance can be degraded by thread contention over the same lock. It is fundamental to locks that they cause blocking and it is the programmer's job to attempt to structure the programming in order to minimize this. This is a difficult task requiring careful structuring and ordering of acquiring and releasing the locks. The programmer is also burdened with the job of balancing between the overhead of an additional lock versus contention for the same lock.

Verifying parallel code using locks is difficult. There are a plethora of things that may go wrong, including deadlocks, livelocks, priority inversion, and race conditions. Since the underlying environment is non-deterministic, it is difficult to create test suites or even debug specific problems.

While locks synchronize by letting only one thread touch a specific piece of data, we synchronize by a pair of instructions: SendSignal (non-blocking) and WaitSignal (blocking). The specific coupling of the core sending the signal and the one waiting allows us to keep track of the relative ordering of instructions across cores. In locking, however, any core may acquire a lock, thereby blocking out any other core. In fact, given knowledge about which instruction completed on a specific core, we can keep track of exactly which instructions are known to have finished on all other cores.

In general, locks require two-way messages between two cores to acquire and release the lock since the lock must exist on some core, or otherwise be in shared memory. This means that locking is a fundamentally synchronous mode of communication; however, synchronous communication scales poorly as more cores are added. Our asynchronous approach avoids this problem and exposes opportunities for out-of-order execution. The fact that this approach scales up to many cores makes loops all the more necessary to prevent two possible bottlenecks: first, the language interpreter itself; and second, the transportation of large program texts.

Our approach is especially well suited to architectures such as the Cell processor (IBM/Sony/Toshiba 2008).

4.1.1   Novelty

In this chapter we address these issues by extending Atomic Verifiable Operation (AVOp) streams to allow loops. An AVOp is the basic instruction in our Domain Specific Language (DSL). It is analogous to a processor's instruction.

AVOp streams (introduced by Anand and Kahl (2008)) allow performance to be maximized by introducing an algorithm for scheduling across different threads of execution so as to minimize contention. In this context, contention means the need to stop execution while in a synchronous operation, blocked waiting on another thread. At the same time, a verification algorithm guarantees that no possible execution order will produce a hazard.

AVOp streams form a programming language. The key contribution of this chapter is making this programming language more expressive by supporting repetition via a looping construct. At the same time, the verification algorithm is extended to efficiently verify in the presence of looping constructs. Other language constructs had to be modified to accommodate these changes.

4.1.2   Impact

The AVOp language presented in Anand and Kahl (2008) is meant to be highly scalable and limits itself to asynchronous communication in order to do so. However, because the program text is presented in a linear fashion, the text length is proportional to run time length in the Loopless AVOp language. Adding a looping construct is important for two reasons. First, it significantly extends the set of complicated communication patterns that can be expressed in the language while still being able to reason about them. Second, it is vital in compressing the AVOp stream which minimizes the overhead of feeding AVOps to a large number of cores. This is a key property required for the scalability of the AVOp language to super computers or large clusters.

4.1.3 Chapter Organization

We first introduce notation that may be unfamiliar to the reader. A restricted form of the AVOp language is presented after which simple loops are introduced. The fast verification algorithm is presented, and a proof of correctness is shown. The language is then extended to handle loops with permutation rewrites and the proof is extended to handle this new construct. We incorporate all the AVOps of the language. Finally, some future avenues for research are covered.

4.2 Notation

4.2.1 Map Modification Notation

We make heavy use of maps to maintain the minimal required state for verifying our programs. A map is an association of an element of a domain to an element of a range. Maps are called 'functions' in algebra and theoretical computer science, but we use the term 'map' since conflicting definitions for 'functions' exist in applied mathematics and programming languages. We define a map f taking an element of the domain (of type D) and associating it with an element of the range (of type R) by:

$$f : D \to R$$

Once we have the type of the elements defined, we also want to define which specific element of the range is associated with each element of the domain. We could define a map floor : $\mathbb{R} \to \mathbb{Z}$, of limited practical interest due to its finite range, with the following notation:

$$\text{floor} : \mathbb{R} \to \mathbb{Z}$$

$$\text{floor} = \begin{pmatrix} 3.14 \mapsto 3 \\ 2.71 \mapsto 2 \\ 1.11 \mapsto 1 \end{pmatrix}$$

We denote the element of the range which corresponds to a particular element of the domain using familiar function notation. For example,

$$\text{floor}(3.14) = 3$$

Now that we have a fully defined map, it is also useful to be able to modify it. In particular, we are interested in updating a specific entry or adding a new one, both of which are handled by the same notation. Given a map $f : D \to R$, and a pair (d, r) with $d \in D$ and $r \in R$ which corresponds to the modification we want to make, the updated map g is defined by:

$$g = ([d \mapsto r](f))(d') = \begin{cases} r & \text{if } d = d' \\ f(d') & \text{otherwise} \end{cases} \tag{4.1}$$

As an example of this notation, let us take the floor map defined above and add an additional entry into it. The resulting map, written as $[0.01 \mapsto 0](\text{floor})$, is explicitly visualized as follows:

$$[0.01 \mapsto 0](\text{floor}) = \begin{pmatrix} 3.14 \mapsto 3 \\ 2.71 \mapsto 2 \\ 1.11 \mapsto 1 \\ 0.01 \mapsto 0 \end{pmatrix}$$

Multiple modifications are denoted using common function syntax; in this case we modify the entry for the domain element 3.14 twice.

$$[3.14 \mapsto 0]([3.14 \mapsto 99](\text{floor})) = \begin{pmatrix} 3.14 \mapsto 0 \\ 2.71 \mapsto 2 \\ 1.11 \mapsto 1 \end{pmatrix}$$

4.2.2 Groups

Among all maps, bijective maps which have a unique image point for every domain point are special because they naturally come with the structure of a mathematical group. Bijective maps that form a finite set in themselves are elements of the permutation group: the set of all shuffling operations. The simplest groups are cyclic groups that cycle the elements of the underlying set. For example, modulo addition by a constant is a map, which is a permutation, and the set of arbitrary compositions of this permutation with itself is a group that can be identified by the constants in the set of integers modulo some base, where the permutations are given by addition of the constant modulo the base.

What does this have to do with programs? Many loops are governed by an induction variable, i, which is incremented ($i := i + 1$) on each iteration. When a loop is expressed in this way, it is difficult to reason about it, because it is open. If, however, we consider the group action $+1$ acting on the integers modulo n, for some positive n, we have a closed system. All possible states are determined by the elements of the group, which is finite. This makes it much easier to reason about the loop and calculate the state of the induction variables and even the state of the verification mechanism at an arbitrary point in the future, using a fixed number of operations. This is the key to understanding the mechanisms of this chapter. All else are details, which, although technically dense, work because of this simple principle.

4.2.3   Disjoint Unions

Disjoint unions, denoted by $\oplus$, are used to represent many of the data types used by the verifier algorithm. Disjoint unions allow us to treat multiple distinct types as a single type. For example, we could specify the type for a non-empty singly linked list of integers as

$$\text{IntList} : \begin{pmatrix} \mathbb{Z} \\ \oplus \\ \mathbb{Z} \times \text{IntList} \end{pmatrix}$$

Notice how the data type could be in either of the cases separated by $\oplus$. First, a list can consist of a single integer. Second, an integer can be prepended to a list to form another, longer, list. In this case the type is recursive, but there is no such requirement in general. This can be represented as an Algebraic Datatype in the programming language Haskell (see Peyton Jones (2003)) as:

```
data IntList = Base Integer
             | Recr Integer IntList
```

In object oriented languages this would typically be modeled using inheritance. The key property here is that both Base and Recr are List objects but any List object can be either a Base or a Recr (not both). We show a translation into the Java programming language:

```
public abstract class IntList {
}

public class Base extends IntList {
    public int x;
}

public class Recr extends IntList {
    public int x;
    public IntList rest;
}
```

4.3   Background and Previous Work

Material from Anand and Kahl (2008) serves as a good introduction to verifying AVOp streams and the benefits of verification. We extend it with a new notation required in the context of the new looping construct which slightly changes the presentation of all AVOps. An AVOp stream is a list of instructions (AVOps), where each AVOp is bound to (and executed on) a particular core.

In the example below, we see that core one (c_1) loads a buffer from main memory, performs a computation on it, and then signals to c_2 that the computation has been completed. At some other point in time, when c_2 requires that the computation on c_1 has been completed, it can block waiting for the signal sent by c_1. Despite the WaitSignal occurring later in the stream, it may begin execution earlier since it is bound to a different core. However, the WaitSignal cannot finish execution before the corresponding SendSignal completes.

$$\underline{\qquad\qquad \text{AVOp} \qquad\qquad}$$

$\qquad\qquad\qquad \cdots$

$\qquad$ LoadMemory c_1

$\qquad\qquad\qquad \cdots$

$\qquad$ RunComputation c_1
$\qquad$ SendSignal c_1 $s \rightarrow c_2$

$\qquad\qquad\qquad \cdots$

$\qquad$ WaitSignal c_2 s

$\qquad\qquad\qquad \cdots$

The bulk of verification is in checking that synchronization AVOps, such as SendSignal and WaitSignal, are used correctly. We limit ourselves to just these two AVOps in presenting material, as other synchronization AVOps are handled in a similar manner. The full specification of these AVOps is

- SendSignal c_1 c_2 s — core c_1 sets the boolean signal s on core c_2. If the signal was already set, it remains set (this is a hazard).

- WaitSignal c s — halts execution on core c until signal s is set. Clears the signal s and resumes execution once this condition has been met.

Below is a brief overview of other AVOps whose full coverage is delayed until Section 4.7.

- SendData, WaitData — transfer data between cores (and synchronize).

- LoadMemory, StoreMemory, WaitDma — transfer data to and from main memory.

- RunComputation — run a computation kernel.

4.3.1 Concurrency Verification

Verification needs to satisfy two use cases: first, developers are interested in using it to understand the source of defects in their code. Second, users (including regulators) are interested in guaranteeing that safety-critical software is correct. Concurrent errors are inherently non-local, yet the type of error and one of the instructions involved in creating an unsound program are identified by this verification algorithm.

We envisage two verification steps which make sense in the context of multi-core parallelism: reduction of the parallel program to a pure data-flow graph, the harder of the two, and matching the data-flow graph to a specification. For problems such as structured linear algebra, it is easy to generate single- and multi-level pure data-flow graphs and convert them to an appropriate normal form. This makes the second step easy.

The ideas influencing this approach are those of the superscalar, out-of-order execution model of a CPU. Instructions may be executed out of order, but have to be presented in a linear order. The out-of-order semantics must match the sequential program semantics. This is the case for our Domain Specific Language (DSL) programs that pass verification. We will use the following terminology:

- A *program* is a list of AVOps. The order of AVOps in this list is referred to as the *presentation order*.

  The AVOps originate in this order and are transmitted to cores in this order. The ordering of execution of AVOps on a single core corresponds to the restriction of the program to that core. However, each core buffers its instructions and executes them without additional synchronization. Thus, the original program order is not necessarily the same as the resulting execution.

- A program is *order independent* if, given the same input in main memory, all possible execution orders terminate and produce the same output in main memory. Intermediate values computed and temporarily stored in private or global memory may differ.

- A program is *locally sequential* if every (SendSignal c_1 c_2 i) is followed by a corresponding (WaitSignal c_2 i).

  Note that it is easy to construct order-independent programs without this property, simply by ordering the instructions by core.

- A program is *safely sequential* if it is locally sequential and order independent.

The programmer wants to know if programs are order independent. Checking this is very expensive. The programmer understands sequential programs, and with some difficulty can understand parallel programs 'close to' sequential programs. Locally sequential programs are our attempt to formalize this concept. The key to usability is the fact that we can efficiently identify the safely sequential programs within the class of locally sequential programs. To reiterate these definitions using relational language:

1. The presentation order is a total order.

2. AVOps on a single core are totally ordered.

3. AVOps on different cores have a partial order derived from the use of synchronizing primitives. This partial order is the intersection of all possible execution orders.

4. We will show how to identify the partial order (3) as a sub-relation of the presentation order (1) if the presentation order is safely sequential. Doing this will require memory in $\mathcal{O}(|\text{cores}|^2 \cdot |\text{signals}|)$.

4.3.2 Motivating Example

Before explaining the efficient method of verification, it is important to understand that locally sequential programs are not necessarily sound. We start with the following program:

index	AVOp
1	long computation on c_1
2	SendSignal c_3 $s \to c_1$
3	WaitSignal c_1 s
4	computation on c_1
5	SendSignal c_2 $s \to c1$
6	WaitSignal c_1 s

To be able to better visualize the meaning of this program we expand it by listing each core's AVOps separately.

index	core 1	core 2	core 3
1	long computation		
2			SendSignal $s \to c1$
3	WaitSignal s		
4	computation		
5		SendSignal $s \to c1$	
6	WaitSignal s		

Remember that each core executes independently of the other cores, except for blocking AVOps, which halt execution until some kind of communication (signal, change in data tag, DMA) is confirmed to have been completed. Therefore, in this case the most likely instruction completion order has core 2 executing the SendSignal as soon as it is queued, allowing the signal to be sent before core 1 has received core 1's signal and cleared the signal hardware:

index	core 1	core 2	core 3
2			SendSignal $s \to c1$
5		SendSignal $s \to c1$	
	second signal overlaps the first, only one registered		
1	long computation		
3	WaitSignal s		
4	computation		
	no signal is sent, so the next WaitSignal *blocks*		
6	WaitSignal s		

To be precise, completion of the SendSignal means that the signal has been initiated by the sender and reception may be delayed, so the signal from core 2 could arrive before the signal from core 3. In either case, neither signal will arrive after the first WaitSignal, so the second WaitSignal will wait forever and this program execution will not terminate.

The cause of this problem is that there are no signals or data transmissions enforcing completion of instruction 5 to follow completion of instruction 3.

This example, when considered as part of a longer program, also demonstrates a possible safety violation with the valid completion order:

index	core 1	core 2	core 3
1	long computation		
5		SendSignal $s \to c1$	
3	WaitSignal s		
4	computation **using wrong assumptions**		
2			SendSignal $s \to c1$
6	WaitSignal s		

In this case, computation 4 might rely on assumptions that are only available once the SendSignal 1 completes — for example, it might initiate a DMA to core 3, which could arrive before some earlier instructions on core 3 have finished writing to the DMA's target area, thus invalidating the DMA transfer.

4.3.3   Strictly Forward Inspection of Partial Order

We have seen that locally sequential programs are not always sound. The advantage that locally sequential programs have over general programs is that order-inducing instructions must occur before instructions whose well-definedness depends on them. Using simple inference to maintain a list of known ordering relations between instructions on different cores, we can inspect the program in order (even though it can execute out of order).

Referring to Figure 4.1, we see that after instruction 3, it is known that instructions on core 1 will complete after instruction 2 on core 3 (and any

Good Program

index	core 1	core 2	core 3
1	long computation		
2		SendSignal $s \to c1$	
3	WaitSignal s		
4	computation		
5	SendSignal $s \to c3$		
6			WaitSignal s
7			SendSignal $s \to c2$
8		WaitSignal s	
9		SendSignal $s \to c1$	
10	WaitSignal s		

FIGURE 4.1
Locally sequential program

previous instructions). We know nothing about the relative execution of instructions other than this until instruction 6. At that point we know that instructions after and including 6 on core 3 must execute after any instruction before instruction 5 on core 1. After instruction 8, we have the relation that this and further instructions on core 2 execute after instructions up to instruction 7 on core 3, but since instruction 7 on core 3 executes after instruction 5 on core 1, we also know that instruction 8 on core 2 executes after instruction 5 on core 1.

Contrast this with the example in Section 4.3.2, where we never have any relations between instructions on core 2 and other cores. Keeping track of these relations is the key to efficient verification of correctness.

There is one other class of (bad) example to keep in mind: the situation where, in presentation order, two cores send the same signal to a third core without a wait on the third core to separate them. This case is visually apparent, because the sends overlap in program order, but we have to safeguard against it nonetheless.

Clearly, the same considerations apply to data transfers. With data transfers, we must additionally check the values of the data tags, because if the existing tag matches the tag being transferred in, the WaitData instruction will never block execution, even if the data is not available.

4.3.4   The Follows Map (Φ)

Recall that a locally sequential program is a sequence of AVOps (communication primitives) with a totally ordered presentation. The order does not fully determine the execution order, but it does define the analysis order for the verification algorithm. Without this order, there would be no linear-time verification algorithm. Given the order, it makes sense to talk about the state maintained by the verification algorithm. This state encodes the information about the possible execution states required to verify the soundness of the program in this way. We introduce the following notation:

- I — is the set of AVOps (Instructions).

- C — is the set of Core IDs that cores are referenced by.

- S — is the set of Signal IDs that may be signaled by SendSignal.

- P — is the fixed program we are verifying.

- $N = \{0, \ldots, \text{length}(P) - 1\} \subseteq \mathbb{N}$ — is the set of indices into P considered as a list. This allows us to consider the program as a total function $P : N \to I$. N is considered to be ordered with the standard ordering of the natural numbers.

- $N_c = \{n : N | P(n)_1 = c\}$ — is the restriction of the index set to instructions on c, for all $c : C$.

We are interested in the partial strict-ordering $\prec$ on N where $n_1 \prec n_2$ means 'the AVOp at n_1 *necessarily* completes before the AVOp at n_2'. The reflexive closure of this is frequently known as 'happens-before relation' (Flanagan and Godefroid 2005) or 'Mazurkiewicz's trace' (Mazurkiewicz 1987; Godefroid 1996).

For a fixed core, $c : C$, the instructions in the program assigned to c are completed in order, so we demand that, for each core $c : C$, the restriction of $<$ to N_c is included in $\prec$:

$$\forall c : C . \forall n_1, n_2 : N_c . n_1 < n_2 \Rightarrow n_1 \prec n_2 \tag{4.2}$$

There is no a priori order between AVOps assigned to distinct cores, and this may lead to unsound programs like the example in Section 4.3.2. Ordering is determined by transition dependencies, which in our case means blocking communication.

It would be sufficient to calculate $\prec$ completely, but only a small portion of this partial order is required at a time. Since AVOps on a single core are totally ordered, it is easy to calculate the entire relation $\prec$ if, for each AVOp index n and distant core, the latest AVOp known to complete before n and the earliest AVOp known to complete after n are stored.

4.3.5 Φ Slices

We will see that it is sufficient to store only the latest instruction known to complete before the current instruction, and convenient to repeat information for all cores, in the form of a Φ slice indexed by $n \in N$:

$$F = C \times C \to N$$
$$\Phi_n : F \tag{4.3}$$
$$\Phi_n(c_1, c_2) = \max_{\leq} \{m : N_{c_1} \mid \forall n' : N_{c_2} . n \leq n' \Rightarrow m \prec n'\}$$

index	core 1	core 2	core 3
1	SendSignal $s_1 \to c_2$		
2		WaitSignal s_1	
3		$\Phi_3(c_1, c_2) = 1$	RunComputation

FIGURE 4.2

Φ is defined for other cores. For convenience, $\Phi_n(c_1, c_2)$ is defined for all n, even if $n \notin N_{c_2}$, to be the index of the last instruction on c_1 known to complete before an instruction on c_2 up to and including index n.

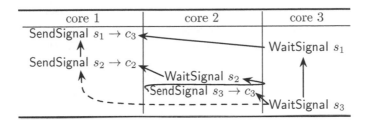

FIGURE 4.3

Visualization of Φ dependency

Note, in particular, that if $n \notin N_{c_2}$, this definition does not say that it is the latest instruction on c_1 such that all future instructions on c_2 follow m. Figure 4.2 shows that the instruction at index $\Phi_n(c_1, c_2)$ is always a send instruction on core c_1.

The distinction between the Φ map and a slice Φ_n is that a slice corresponds to a particular AVOp index $n \in N$ whereas the full Φ map has entries for all AVOp indices which are valid for a given program. We make this distinction to minimize the amount of memory used as slices are stored in other datastructures.

4.3.6 Discussion of the Follows Map (Φ)

Since Φ is at the core of our verification algorithm, the details of the definition given above are reiterated here. Φ keeps track of dependency information between each core. Specifically, it keeps track of how far execution is guaranteed to have progressed on a given core given an instruction on another core.

These dependencies are visualized for a very simple program in Figure 4.3. We see here an execution trace of a program running on three cores. The execution is broken down with a single AVOp executing at any given time. An AVOp a points to another AVOp b if b must complete execution before a can complete execution.

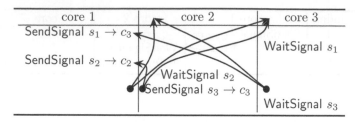

FIGURE 4.4

Φ map at instruction $SendSignal\ s_3 \rightarrow c_3$

Direct dependencies are visualized by arrows. In the diagram we see two different cases. First, there is a dependency between $WaitSignal\ s_1$ and its corresponding $SendSignal$, as can be seen by the topmost arrow. Second, there is an obvious dependency, made explicit, between AVOps executing on the same core, as can be seen by the curved arrow on core 2. The second type of dependency is not actually kept track of in the Φ map as it only tracks dependencies between different cores.

Indirect dependencies are visualized by the dotted arrow. These dependencies are computed by merging Φ maps stored at different points in the program's execution. In this example we can see that $WaitSignal\ s_3$ cannot complete execution before the corresponding $SendSignal$ on core 2, or the vertical arrows on cores 1 and 3. However, $SendSignal\ s_3 \rightarrow c_3$ cannot complete execution before $SendSignal\ s_2 \rightarrow c_2$ by following the dependency arrows. By deduction we can draw the inference that $WaitSignal\ s_3$ cannot complete before $SendSignal\ s_2 \rightarrow c_2$ completes execution. Indirect dependency analysis and Φ merging is performed by the $\widetilde{\bigvee}$ function defined in the next section.

The Φ map at any point in the execution will contain all dependencies, both direct and indirect, between any two cores. Equivalent information can be obtained by starting at any point and tracing all arrows between the two cores of interest.

Figures 4.4 and 4.5 provide a look at two Φ maps at different locations in the execution of the program. Each core is marked with a dot and has dependency arrows going to all the other cores. Despite an AVOp executing on only one core, Φ is defined for all pairs of cores at that point. Additionally, the Φ map is initialized to -1 at the beginning of the program for all pairs of cores, which signifies that there have been no dependency inducing AVOps yet.

Each $SendSignal$ saves the Φ map at that point in the σ map and we make use of it when processing the corresponding $WaitSignal$. The Φ map from Figure 4.4 is used here to infer the dependency on core 1 from core 3. This exact computation occurs in the $\widetilde{\bigvee}$ function when it merges the Φ map stored by the SendSignal and the previous instruction's Φ map (in this simple case they happen to be the same Φ maps).

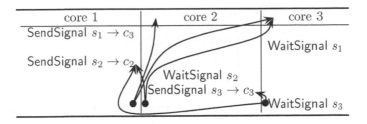

FIGURE 4.5
Φ map at instruction *WaitSignal* s_3

While a SendSignal stores the Φ map at that point in memory, a WaitSignal will look up this information and use it to strengthen dependency information when updating the Φ map.

4.3.7 Merging Φ Slices to Strengthen Our Partial Order

For the verification method, we need to define a function

$$\widetilde{\bigvee} : (C \times N \times F) \times (C \times F) \to F$$

where F models a slice of Φ. The function $\widetilde{\bigvee}$ combines the slices of the partial order at a corresponding pair of SendSignal / WaitSignal AVOps that induce an order on the completion of WaitSignal AVOps on *other cores*. This potentially strengthens what we know about $\Phi_{n_2}(c, c_{\text{wait}})$ for all c, specifically, about the last AVOps known to complete on all other cores before the WaitSignal at index n_2 and core c_{wait} can complete.

To define $\widetilde{\bigvee}$ we make use of the map modification notation introduced in Section 4.2.1. One thing to notice is how the outermost modification (c_1, c_2) will override the inner modification (c', c_2). The left argument (subscript 1) is the information we had at the time of a SendSignal and the right argument (subscript 2) is the information at the time of the WaitSignal that is being currently processed. See the definition of WaitSignal in Section 4.6.3.

$$\widetilde{\bigvee}((c_1, n_1, \Phi_{n_1}), (c_2, \Phi_{n_2})) =$$
$$\begin{aligned}[(c_1, c_2) &\to \max\{\Phi_{n_2}(c_1, c_2), n_1\}](\\ [(c', c_2) &\to \max\{\Phi_{n_2}(c', c_2), \Phi_{n_1}(c', c_1)\}](\Phi_{n_2}))\end{aligned} \quad (4.4)$$

As a more complete example of the notation we use, the above definition is equivalent to the one below using case notation.

$$\widetilde{\bigvee}((c_1, n_1, \Phi_{n_1}), (c_2, \Phi_{n_2}))(c', c'') =$$

$$\begin{cases} \max\{\Phi_{n_2}(c', c''), n_1\}, & c' = c_1, c'' = c_2 \\ \max\{\Phi_{n_2}(c', c''), \Phi_{n_1}(c', c_1)\}, & c'' = c_2 \\ \Phi_{n_2}(c', c''), & \text{otherwise.} \end{cases}$$

Consider the situation where we know that some core c is executing the program at index n_1 (on core c_1) which is further advanced than the current instruction on core c_2. $\widetilde{\bigvee}$ will then incorporate this strengthened dependency information from Φ_{n_1}, incorporating it into the current slice Φ_{n_2} and producing the slice at Φ_{n_2+1}.

4.3.8   State

In addition to Φ_n, our method of verification constructs the following function inductively in the instruction index, n.

$$G = C \times S \to \begin{pmatrix} C \times N \times F \\ \oplus \\ N \end{pmatrix} \tag{4.5}$$

$$\sigma_n : G \tag{4.6}$$

where $\oplus$ means disjoint union (reviewed in Section 4.2.3).

To calculate X_{n+1}, where $X_n = (\sigma_n, \Phi_n)$, it is only necessary to have access to X_n, because we make copies of slices of Φ which are needed in future steps.

The disjoint unions can be understood in the following way for a given core c and signal s:

1. Receive pending (signal sent) — $\sigma(c, s)$ records the (core c', index n' and follows map slice Φ') on a SendSignal. Thus, signal s is expected to arrive at core c from core c'.

2. Signal cleared (signal received) — $\sigma(c, s)$ records the (index n) at which the signal was received by a WaitSignal. Thus, signal s was received on core c at instruction n.

We initialize all entries in σ_0 to -1 (all signals have last been received at index -1) at the start of program verification. Similarly, all entries in Φ_0 are initialized to -1, indicating that nothing is known about ordering.

4.3.9   Verification Step

The maps σ_{n+1} and Φ_{n+1} are constructed from the maps and AVOp executing at index n according to the universal rule $\Phi_{n+1}(c, c) = (n)$, and specific rules depending on the executing AVOp:

SendSignal c c' s' We need to verify that there is no current use of this signal in any program order. Divide into two cases according to the form of $\sigma_n(c', s')$; the first case excludes programs in which the signal is in use in the presentation order. This is not sufficient because the presentation order is not the only possible execution order, so we need to use the follows relation to check this.

If $\sigma_n(c', s') = (c'', n', \Phi')$, this is an error, because this signal s' may be sent by both cores (from c at the present instruction n and from c'' at n') at the same time.

If $\Phi_n(c', c) \geq \sigma_n(c', s')$, then we know that the last instruction to consume the signal s' on c' at $\sigma_n(c', s')$ completes before $\Phi_n(c', c)$ on c' which (by definition) occurs before the current instruction on c. (This includes the case that this signal has never been used, in which case $\sigma_n(c', s') = -1$.) Let $\Phi_{n+1} := \Phi_n$, $\Delta_{n+1} := \Delta_n$, and $\sigma_{n+1}(c', s') := (c, n, \Phi_n)$, $\sigma_{n+1}(c'', s'') := \sigma_n(c'', s'')$ for other values of (c'', s'').

If the inequality is not satisfied, the program is not sound, because we do not have $n' \prec n$, meaning that this signal could arrive before the previous WaitSignal s' on c', which is the case of the bad example program.

WaitSignal c s' If $\sigma_n(c, s') - (c'', n', \Phi')$, then we know that signal s' will come from core c'', and we can use the fact that the wait completes after the send to update the follows map. Let $\Phi_{n+1} := \widetilde{V}((c'', n', \Phi'), (c, \Phi_n))$, $\Delta_{n+1} := \Delta_n$, $\sigma_{n+1}(c, s') := (n)$ otherwise $\sigma_{n+1}(c''', s''') := \sigma_n(c''', s''')$.

If $\sigma_n(c, s') = (n')$, this is an error, or the program is not locally sequential, as required for this analysis.

Additional AVOps are covered in Section 4.7.1.

Theorem 4.1 *A locally sequential program is order independent* iff *the inductive verification described in this section terminates without error.*

The full proof of this theorem is presented in Anand and Kahl (2008).

4.4 The Loop AVOp

We introduce a loop AVOp to specify repetition in our DSL. This has a number of benefits: first, to reduce the communication overhead when transmitting AVOps to their corresponding cores. As the number of cores in the system increases, the program, which lists AVOps for all cores, becomes more and more unwieldy. Thus, loops provide a compression mechanism for the program (see Section 4.8.3). Second, the DSL becomes more expressive and it becomes easier to specify complex patterns of communication across many cores.

4.4.1   Loop AVOp Indexing

Within a program AVOps are assigned unique indices. The ordering between indices corresponds to the ordering of execution of the AVOps. Loops complicate the requirements for indices. We need to be able to refer to a specific AVOp within the body of the loop as well as a specific execution of that AVOp in some iteration of the loop. Nested loops also need to be supported. To achieve this, the index type, N, needs to be extended with a case for loops.

$$N : \begin{pmatrix} Index \\ \oplus \\ Index \times Iteration \times N \end{pmatrix} \tag{4.7}$$

$$Index : \mathbb{Z} \tag{4.8}$$

$$Iteration : \mathbb{Z} \tag{4.9}$$

AVOps that are executed outside of any loop are numbered by their *Index* as before. Once a loop is encountered, it is assigned an *Index* just like any other AVOp and it uniquely identifies the loop. The *Iteration* specifies which iteration of the loop the AVOp is executing. This recursive definition will nest once for each level of loop nesting. The 3rd AVOp in the 10th iteration of a loop which itself is the 1st AVOp of the program would have index $(0, 9, 2)$ (remember 0-based numbering). An example (an unrolling of Figure 4.6) is presented in Figure 4.7. We define a lexicographic ordering of the indices that corresponds to the ordering in the presentation order.

$$i_1 < i_2 = \text{usual ordering of } i_1 \text{ and } i_2 \text{ in } \mathbb{Z} \tag{4.10}$$

$$i_1 < (i_2, j_2, n_2) = \begin{cases} \text{true} & \text{if } i_1 \le i_2 \\ \text{false} & \text{otherwise} \end{cases} \tag{4.11}$$

$$(i_1, j_1, n_1) < i_2 = \begin{cases} \text{true} & \text{if } i_1 < i_2 \\ \text{false} & \text{otherwise} \end{cases} \tag{4.12}$$

$$(i_1, j_1, n_1) < (i_2, j_2, n_2) = \begin{cases} \text{true} & \text{if } i_1 < i_2 \\ \text{true} & \text{if } i_1 = i_2, j_1 < j_2 \\ \text{true} & \text{if } i_1 = i_2, j_1 = j_2, n_1 < n_2 \\ \text{false} & \text{otherwise} \end{cases} \tag{4.13}$$

We define a helper function ninr which retrieves the innermost loop index from an index. For example, $\text{ninr}(0, 1, (2, 3, 4)) = (2, 3, 4)$ and not 4, which would eliminate the context of the innermost loop. In general this is useful for removing the context that a loop is in and focusing only on its component in the index. It is defined as follows:

$$\text{ninr}(i) = i \tag{4.14}$$

$$\text{ninr}(i, j, i') = (i, j, i') \tag{4.15}$$

$$\text{ninr}(i, j, n) = \text{ninr}(n) \tag{4.16}$$

index	AVOp
0	SendSignal c_1 $s \rightarrow c_2$
1	Loop 2 [
$(1,*,0)$	$\quad$ WaitSignal c_2 s,
$(1,*,1)$	$\quad$ SendSignal c_1 $s \rightarrow c_2$
	]
2	...

FIGURE 4.6
Example using the Loop AVOp. The *Iteration* component of indices is specified by $*$ as it cannot be represented in the text of the program.

index	AVOp
0	SendSignal c_1 $s \rightarrow c_2$
(1,0,0)	WaitSignal c_2 s
(1,0,1)	SendSignal c_1 $s \rightarrow c_2$
(1,1,0)	WaitSignal c_2 s
(1,1,1)	SendSignal c_1 $s \rightarrow c_2$
2	...

FIGURE 4.7
Example of an unrolled loop from Figure 4.6

We define a helper function to modify the *Iteration* component of the innermost loop index by x.

$$\text{nchg}_x(n) = \text{nchg}_x{}'(\text{ninr}(n)) \tag{4.17}$$

$$\text{nchg}_x{}'(i, j, n) = \text{ninr}(i, j + x, n) \tag{4.18}$$

Finally, we also define a function (denoted by $+1$) for getting the next successive AVOp in an AVOp stream.

$$n + 1 = \min_{\leq}\{n' : N | n' > n\} \tag{4.19}$$

4.4.2   Loop Definition

Loops are introduced as a new AVOp: Loop k β; where k is the number of iterations the loop should perform and β is the body of the loop, consisting of a sequence of instructions. Below is an example of a loop where $k = 2$ and $\beta = [(c_2, \text{WaitSignal } s), (c_1, \text{SendSignal}c_2 \ s)]$. We refer to this loop as starting at index $l = 1$. We present the same loop, but with the AVOps unrolled as if they had been executed. Notice that this lets us fill in the *Iteration* component of the index.

Definition 4.1 IterIxs — *We let* IterIxs(β, i) *denote the set of indices of all AVOps of β at the ith iteration of the loop.*

In Figure 4.6, $\text{IterIxs}(\beta, 0) = \{(1,0,0),(1,0,1)\}$ and $\text{IterIxs}(\beta, 1) = \{(1,1,0),(1,1,1)\}$. We also assign an index to the loop itself. This is important when dealing with nested loops as $\text{IterIxs}(\beta, *)$ will give the outermost indices (it does not penetrate into an inner loop). (See Figure 4.8.)

index	AVOp
0	SendSignal $c_1\ s \to c_2$
1	Loop 10 $\beta_1 = [$
$(1,*,0)$	WaitSignal $c_2\ s$,
$(1,*,1)$	Loop 20 $\beta_2 = [$
$(1,*,(1,*,0))$	SendSignal $c_2\ s \to c_3$,
$(1,*,(1,*,1))$	WaitSignal $c_3\ s$
	],
$(1,*,2)$	SendSignal $c_1\ s \to c_2$
	]
2	...

FIGURE 4.8
Example with nested loops. $\text{IterIxs}(\beta_1, *)$ treats the inner loop as a single AVOp so that $\text{IterIxs}(\beta_1, 5) = \{(1,5,0),(1,5,1),(1,5,2)\}$. The inner loop is assigned a single index of $(1,5,1)$ as if it were a single AVOp.

4.5 Efficient Verification of Looping Programs

If we were to naively extend the existing verification algorithm to support loops we would have to completely unroll the loop to verify the program. This would mean that verification would take time proportionate to the execution of the program, which could be extremely large for a program running on a supercomputer or cluster. We are able to do much better by taking advantage of the features of our AVOp stream language.

4.5.1 Locally Sequential Loops

As a reminder, a stream of AVOps is *locally sequential* if every $(c_1, \text{SendSignal } c_2\ s)$ is followed by a corresponding $(\text{WaitSignal } c_2 s)$.

Definition 4.2 *An AVOp stream with loops is* locally sequential *iff the unrolled version of the stream is locally sequential.*

Lemma 4.1 *If a loop is unrolled twice and remains locally sequential, then it will be locally sequential when fully unrolled.*

Proof 4.1 *We begin with a loop at index l.*

If a loop is unrolled twice and is not locally sequential, it is clear that the fully unrolled version will not be locally sequential either.

If a twice-unrolled loop is locally sequential, $\forall c \in C, \forall s \in S, \sigma(c, s)$ can be in two states at the end of the second iteration.

Case I: $\sigma(c, s) = n$

We have most recently encountered a $(c,$ WaitSignal $s)$ AVOp.

If $n < l$ then we did not encounter a WaitSignal in the body of the loop. Since the body was executed twice, there could also not have been a SendSignal c s or we would have two SendSignal s without a WaitSignal between them. This in turn would mean the twice-unrolled loop would not be locally sequential. Thus, the unrolled version of the loop will not change the state of $\sigma(c, s)$.

If $n \geq l$ then we encounter a WaitSignal in the body of the loop. Since the body was executed twice, there must be a $(*,$ SendSignal c $s)$ in the body of the loop. Otherwise, the first two executions of the loop would not have been locally sequential because of back to back WaitSignal s. Subsequent iterations execute the same body and will also keep the sends and waits in balance.

Since a WaitSignal was last encountered, the unrolled loop will terminate with σ set to a single integer as will the twice-unrolled version. Thus, the unrolled version will also be locally sequential.

Case II: $\sigma(c, s) = (c', n', \Phi')$

We have most recently encountered a $(c',$ SendSignal c $s)$ AVOp. A similar argument to the one above applies.

4.5.2   Bumping State to the Next Iteration via Λ

We define a map Λ for specifying the transformation of one iteration's verification state into the next iteration's. It is defined in two equations: (4.21) and (4.22), chosen based on the format of the argument. There is one equation for each item in the product of the type. Notice that these types correspond to the result types of Φ and σ.

$$\Lambda : \begin{pmatrix} N \to N \\ \times \\ C \times N \times F \to C \times N \times F \end{pmatrix} \tag{4.20}$$

In defining Λ_l we have:

- l is the index of the beginning of the loop.

- n is the index of the AVOp currently being verified.

An index is transformed to the next iteration's corresponding index by incrementing the iteration component of the index. Notice that if an index is before the beginning of the loop we do not modify it.

Suppose we have Φ_n and want to transform it to the corresponding $\Phi_{\mathrm{nchg}_1(n)}$ after 1 iteration. This transformation is given by $\Lambda_l \circ \Phi_n = \Phi_{\mathrm{nchg}_1(n)}$ given that Φ_n is stabilized in the sense that any entry that is going to be updated during the execution of the loop has been updated.

$$\Lambda_l(n) = \begin{cases} n & \text{if } n < l \\ (i', j'+1, n') & \text{otherwise, where } (i', j', n') = \mathrm{ninr}(n) \end{cases} \quad (4.21)$$

Stored in σ are Φ slices which have to be updated. Once again we can decide whether a transformation is required based on whether the index was before the beginning of the loop. If the Φ slice was stored before the beginning of the loop (by a SendSignal) then we do not update it. The Φ slice is modified by composing it with Λ, which effectively moves it one iteration over.

$$\Lambda_l(c, n, \Phi) = \begin{cases} (c, n, \Phi) & \text{if } n < l \\ (c, \Lambda_l(n), \Lambda_l \circ \Phi) & \text{otherwise} \end{cases} \quad (4.22)$$

4.5.3   Loop Short Circuit Theorem

We can verify loops by running the verifier on an unrolled version. However, this would take time proportional to the number of iterations and we can do better. If we notice that one iteration of the loop can be related to the previous iteration via Λ, then we can extrapolate the effects on Φ and σ to the last iteration of the loop.

Theorem 4.2 *Given* Loop k β *at index* l, *if* $\exists i : \mathbb{N}_0, i > 0, \forall n \in \mathrm{IterIxs}(\beta, i)$,

$$\Phi_n = \Lambda_l \circ \Phi_{\mathrm{nchg}_{-1}(n)}$$

$$\sigma_n = \Lambda_l \circ \sigma_{\mathrm{nchg}_{-1}(n)}$$

Then $\forall n' \in \mathrm{IterIxs}(\beta, k-1)$ *where* $(*, *, i) = \mathrm{ninr}(n) = \mathrm{ninr}(n')$,

$$\Phi_{n'} = \Lambda_l^{(k-1)-i} \circ \Phi_n$$

$$\sigma_{n'} = \Lambda_l^{(k-1)-i} \circ \sigma_n$$

Before we proceed to prove this theorem, we will need to introduce two additional Lemmas: first, in Lemma 4.2, we establish a property for relating the Φ map across loop iterations. Second, in Lemma 4.3, we sketch a proof for the corresponding property relating the σ map. After establishing these lemmas we proceed to prove Theorem 4.2 in Proof 4.4.

Lemma 4.2 *Given* **Loop** k β *at index* l, *if* $\exists i : \mathbb{N}_0, i > 0, \forall n \in \text{IterIxs}(\beta, i)$,

$$\Phi_n = \Lambda_l \circ \Phi_{\text{nchg}_{-1}(n)}$$

$$\sigma_n = \Lambda_l \circ \sigma_{\text{nchg}_{-1}(n)}$$

Then, $\forall n \in \text{IterIxs}(\beta, i), \Phi_{\text{nchg}_1(n)} = \Lambda_l \circ \Phi_n$.

Proof 4.2 *Take preconditions for Lemma 4.2 as given.*

Suppose $\exists n \in \text{IterIxs}(\beta, i), \exists c, c' \in C : \Phi_{\text{nchg}_1(n)}(c, c') \neq \Lambda_l(\Phi_n(c, c'))$.

We assume that no hazards are detected while verifying AVOps up to and including n. If hazards are detected, we report the error at that point and stop the verification procedure. There are two cases to consider for Φ with no hazards detected:

Case I: $\Phi_n(c, c') < l$

> *There are no $(c, $ **SendSignal** $c's)$ or $(c', $ **WaitSignal** $s)$ instructions in β because Φ_n would have been updated since $i > 0$. If there are no such instructions, then $\Phi_{\text{nchg}_1(n)}$ will not be updated in any future iteration. Thus $\Phi_n(c, c') = \Phi_{\text{nchg}_1(n)}(c, c')$.*

Case II: $\Phi_n(c, c') \geq l$

> *There must be instructions of both forms $(c, $ **SendSignal** c' $s)$ and $(c', $ **WaitSignal** $s)$ in β, otherwise Φ_n would not have been updated. Since $\Phi_{\text{nchg}(n)} \geq \Phi_n$ we have that $\Phi_{\text{nchg}_1(n)} \geq l$ as well.*

> *Notice that both Φ_n and $\Phi_{\text{nchg}_1(n)}$ refer to the same instruction in the text of the loop since $(*, *, i) = \text{ninr}(n) = \text{ninr}(\text{nchg}_1(n))$.*

> *However, $\Phi_n \neq \Lambda_l \circ \Phi_{\text{nchg}_1(n)}$ means that the preceding SendSignal is at a different instruction in the text of the loop across the two iterations. Since Φ_n and $\Phi_{\text{nchg}_1(n)}$ are defined over the same loop body β, the presentation order is locally sequential and no hazards came up, there is exactly one SendSignal in β which corresponds to this wait signal.*

> *This is a contradiction.*

Therefore, no such n can exist, and $\forall n \in \text{IterIxs}(\beta, i), \Phi_{\text{nchg}_1(n)} = \Lambda_l \circ \Phi_n$.

Lemma 4.3 *Given* **Loop** k β *at index* l, *if* $\exists i : \mathbb{N}_0, i > 0, \forall n \in \text{IterIxs}(\beta, i)$,

$$\Phi_n = \Lambda_l \circ \Phi_{\text{nchg}_{-1}(n)}$$

$$\sigma_n = \Lambda_l \circ \sigma_{\text{nchg}_{-1}(n)}$$

Then $\forall n \in \text{IterIxs}(\beta, i), \sigma_{\text{nchg}_1(n)} = \Lambda_l(\sigma_n)$.

Proof 4.3 *Take preconditions for Lemma 4.3 as given.*

Suppose $\exists n \in \mathrm{IterIxs}(\beta, i), \exists c \in C : \sigma_{\mathrm{nchg}_1(n)}(c) \neq \Lambda_l(\sigma_n(c))$.

There are three cases to consider:

Case I: $\sigma_n(c) = n$

> *This case follows a similar argument to that for Φ since it does not have to worry about the Φ slices stored in σ.*

Case II: $\sigma_n(c) = (c', n', \Phi')$ and $n' < l$

> *This case follows a similar argument to case 1 for Φ as none of the arguments is modified by Λ and n' is before the loop index l.*

Case III: $\sigma_n(c) = (c_1, n_1, \Phi_1)$ and $n_1 \geq l$

> *Let $(c_2, n_2, \Phi_2) = \sigma_{\mathrm{nchg}_1(n)}(c)$, then we have $c_1 = c_2$. If $c_1 \neq c_2$ then there would have to be a different* **SendSignal** *AVOp in the loop text to bind a different core. This is impossible since the loop is locally sequential, both AVOps correspond to the same AVOp in the loop text β, and no hazard was found.*

> *Similarly, we have that $n_1 = \Lambda_l(n_2)$.*

> *Finally, we have that $\Phi_2 = \Lambda \circ \Phi_1$, as proved in the first proof section on Φ.*

Therefore, no such n can exist, and $\forall n \in \mathrm{IterIxs}(\beta, i), \sigma_{\mathrm{nchg}_1(n)} = \Lambda_l(\sigma_n)$.

Proof 4.4 *Given* **Loop** *k β at index l, if $\exists i : \mathbb{N}_0, i > 0, \forall n \in \mathrm{IterIxs}(\beta, i)$,*

$$\Phi_n = \Lambda_l \circ \Phi_{\mathrm{nchg}_{-1}(n)}$$
$$\sigma_n = \Lambda_l \circ \sigma_{\mathrm{nchg}_{-1}(n)}$$

By Lemmas 4.2 and 4.3 we establish the following properties:

$$\forall n \in \mathrm{IterIxs}(\beta, i), \Phi_{\mathrm{nchg}_1(n)} = \Lambda_l \circ \Phi_n$$
$$\forall n \in \mathrm{IterIxs}(\beta, i), \sigma_{\mathrm{nchg}_1(n)} = \Lambda_l(\sigma_n)$$

We take the given properties as the base case and the properties established by the lemmas as the recursive case. By induction, we see that we can use Λ to compute the state of Φ at the final iteration of the loop, from Φ at an iteration where the preconditions hold. Likewise for relating the state of σ.

4.5.4   Verifying Nested Loops

When verifying a loop, any nested loops are treated as a single AVOp in the context of the outer loop. To treat an inner loop as a single AVOp we must run through the initial iterations of the inner loop, checking for hazards, and then apply the short-circuit theorem to compute Φ and σ after the inner loop completes. We then proceed to check the following AVOps in the outer loop with the updated Φ and σ maps.

Confusion may arise in our use of the ninr function in the definition of Λ. Remember that an inner loop's index contains information about all the loops that it is nested in. The potential problem is that we will get an index that is more deeply nested than we want when working on a loop which has nested loops in its body. Because we treat nested loops as single AVOps in the context of the outer loop, this problem does not actually materialize.

4.5.5 Verifier Run Time

Since we always try to apply the Short Circuit Theorem (Theorem 4.2) when verifying loops, we are interested in how long it takes until the preconditions of the theorem are met.

Definition 4.3 *Stabilized — A loop is stabilized at iteration j if the preconditions for Theorem 4.2 are met at iteration j.*

Although it may initially appear that a constant number of iterations is required until the loop is stabilized, there are some hidden dependency chains lurking in σ. Because old Φ-slices may be stored in σ and then used to update the information of future Φ entries, information does not necessarily propagate immediately. Instead, dependency information can travel from one core to another until it finally reaches its final destination.

Looking at Figure 4.9 we see an example with such a dependency chain set up. The example switches between strengthening Φ ($\widetilde{\bigvee}$ merges a Φ-slice stored in σ with the current Φ) and then storing the strengthened Φ-slice in σ. This process iterates until we hit the last core. It is clear that there is no *direct* dependency on c_0 from c_3. However, there is an indirect dependency, and it takes multiple iterations to propagate this information. It is important to notice that this delay is a one-time event at the beginning of the loop. As the initial propagation is happening, with every iteration a new one is started. Thus, if the diagram showed one more iteration, we would see $\Phi(0,3) = (0,1,1)$ at the end of iteration 5.

Theorem 4.3 *The maximum number of loop iterations executed before the preconditions for Theorem 4.2 are met is $|C| + 1$.*

Proof 4.5 *Given Loop k β at index l.*

Suppose $\exists c', c : C$ such that $\Phi(c', c) < l$ after iteration $|C| - 1$ (iterations are zero based). Then there is no pair of AVOps SendSignal c' c s and WaitSignal c s in β. If there were, then by the second iteration $\Phi(c', c) > l$, which is a contradiction. If $\Phi(c', c) < l$ for all iterations, then it does not impact the preconditions.

index	core 0	core 1	core 2	core 3	Φ
(0,0,0)		W s			
(0,0,1)	S $s \to c_1$				$\Phi(*,*) = -1$
(0,0,2)			W s		
(0,0,3)		S $s \to c_2$			
(0,0,4)				W s	
(0,0,5)			S $s \to c_3$		
(0,0,6)	W s				
(0,0,7)				S $s \to c_0$	
(0,1,0)		W s			$\Phi(0,1) = (0,0,1)$
(0,1,1)	S $s \to c_1$				
(0,1,2)			W s		
(0,1,3)		S $s \to c_2$			$\Phi(0,1) = (0,0,1)$ stored in σ
(0,1,4)				W s	
(0,1,5)			S $s \to c_3$		
(0,1,6)	W s				
(0,1,7)				S $s \to c_0$	
(0,2,0)		W s			
(0,2,1)	S $s \to c_1$				
(0,2,2)			W s		$\Phi(0,2) = (0,0,1)$ by merge
(0,2,3)		S $s \to c_2$			
(0,2,4)				W s	
(0,2,5)			S $s \to c_3$		$\Phi(0,2) = (0,0,1)$ stored in σ
(0,2,6)	W s				
(0,2,7)				S $s \to c_0$	
(0,3,0)		W s			
(0,3,1)	S $s \to c_1$				
(0,3,2)			W s		
(0,3,3)		S $s \to c_2$			
(0,3,4)				W s	$\Phi(0,3) = (0,0,1)$ by merge
(0,3,5)			S $s \to c_3$		
(0,3,6)	W s				
(0,3,7)				S $s \to c_0$	$\Phi(0,3) = (0,0,1)$ stored in σ

FIGURE 4.9

Non-rewritable loop example where Φ keeps changing across many iterations, requiring a variable number of iterations before stabilizing. Pairs of send/wait instructions (connected by arrows) induce ordering on the AVOps, which are recorded in Φ. The particular chain of dependencies which leads to $\Phi(0,3)$ being set for the first time in the fourth iteration is highlighted as dashed arrows.

Suppose $\Phi(c',c) > l$ after some iterations. Since there must be an indirect dependency, there must be some $c'' : C$ and $s : S$ such that there is a SendSignal $c''\ c\ s$ that will eventually store a Φ-slice such that $\Phi(c',c'') > l$ (and the corresponding WaitSignal $c\ s$). Notice that $c'' \neq c'$, otherwise it would be a direct dependency. This must happen after iteration $|C| - 2$ otherwise $\Phi(c',c) > l$ at iteration $|C| - 1$. That is, $\Phi(c',c'') < l$ at iteration $|C| - 2$.

Therefore, there must be some $c''' : C$ and $s' : S$ such that there is a

SendSignal $c'''c''s'$ that will store $\Phi(c',c''') > l$ but not by iteration $|C| - 3$. Similarly, $c''' \neq c'' \neq c'$ or we would have $\Phi(c',c) > l$ at iteration $|C| - 2$.

In the worst case, we go through all the cores one by one until we reach c'''' at iteration 1 (zero based second iteration) that has the relevant AVOps but $\Phi(c',c'''') < l$. This is a direct contradiction since this is a direct dependency and by the second iteration all WaitSignal will be hit by their SendSignal or otherwise the program will not verify.

Now, since all entries of Φ that will be greater than l (initialized) are greater than l at iteration $|C| - 1$. All σ entries will contain initialized Φ-slices by iteration $|C|$ that is the $(|C| + 1)^{th}$ iteration. All other components of σ are updated directly from the previous iteration.

We also state an additional lemma to make explicit the fact that it is safe to skip iterations via the Short Circuit Theorem (Theorem 4.2) after $|C| + 1$ itcrations.

Lemma 4.4 *In order for the preconditions of Theorem 4.2 to be met, the program must verify correctly. Thus, if an error occurs it must occur in the first $|C| + 1$ iterations.*

Proof 4.6 *This follows directly from Theorems 4.2 and 4.3.*

4.6 Rewritable Loops

Loops as presented so far are only slightly useful. The problem is that they repeat exactly the same action, which restricts the communication patterns that are implementable. This is analogous to variables staying constant across all iterations of a loop in a conventional language. To make loops more expressive we introduce a rewrite which modifies which cores and signals are being referred to after each loop iteration.

This kind of AVOp modification scheme was initially inspired by the rotating register windows of the SPARC instruction set (Weaver and Germond 1994). However, the register rotation mechanism of the IA-64 architecture (Intel Corporation 2010a) is much closer in spirit to what we are trying to accomplish. Because we are not limited by hardware constraints, we allow a much more general expression of these ideas than implemented in either the SPARC or the IA-64. The main reason for such a scheme is to enable the amortization of latencies by working on independent data via software pipelining (Lam 1988).

We implement this by extending the looping AVOp to Loop k ρ β where

index	AVOp	comment
0	SendSignal c_0 $s \to c_2$	*prologue*
1	SendSignal c_1 $s \to c_0$	*prologue*
2	Loop 4 ($\rho_C = (0\ 1\ 2)$) [	
$(2, *, 0)$	WaitSignal c_2 s	
$(2, *, 1)$	SendSignal c_2 $s \to c_1$,	
	]	
3	WaitSignal c_2 s	*epilogue*
4	WaitSignal c_0 s	*epilogue*

FIGURE 4.10
Motivating example for loop rewriting. This example illustrates the synchronization issues but contains no computation.

ρ is a rewrite. We implement rewrites as permutations since we need a bijective map from one set (for example, cores) onto itself. This restriction makes sure that we never reach a state where we collapse multiple cores onto one, losing the ability to address some of the cores. It also ensures that we know deterministically which core is being referred to.

The map $\rho_C : C \to C$ maps cores to cores and is specified as a permutation. The rewrite system consists of two rewrite maps of exactly the same type. The first, $\hat{\rho}$, is the global rewrite that is applied to each AVOp whether or not it is within a loop and starts as the identity map at the start of the program. The second, ρ, is a loop rewrite that modifies the global rewrite after the end of each iteration. A global rewrite is required to allow consistent reasoning after the end of a loop.

4.6.1 Motivating Example

We present an example of a loop with rewrites. The loop rewrite is (given first in cycle notation, then as an explicit map):

$$\rho_C = (0\ 1\ 2) = \begin{pmatrix} 0 \mapsto 1 \\ 1 \mapsto 2 \\ 2 \mapsto 0 \end{pmatrix}$$

This type of loop pattern could be used for buffered computation in which each SendSignal signifies that a condition is met and each WaitSignal waits for that condition. In the meantime, as much work as possible is scheduled to ensure that by the time the WaitSignal is hit the condition has already been met. This introductory example presents the signaling pattern only (Figure 4.10).

Notice that we need to execute the AVOps in the epilogue with a $\hat{\rho}$ which is the composition of all the rewrites encountered during the execution of the loop. For this reason we need to have a global $\hat{\rho}$ that will apply outside

index	core 0	core 1	core 2
0	SendSignal $s \to (c_2 = 2)$		
1		SendSignal $s \to (c_0 = 0)$	
$(2,0,1)$			WaitSignal s
$(2,0,0)$			SendSignal $s \to (c_1 = 1)$
$(2,1,1)$	WaitSignal s		
$(2,1,0)$	SendSignal $s \to (c_1 = 2)$		
$(2,2,1)$		WaitSignal s	
$(2,2,0)$		SendSignal $s \to (c_1 = 0)$	
$(2,3,1)$			WaitSignal s
$(2,3,0)$			SendSignal $s \to (c_1 = 1)$
3	WaitSignal s		
4		WaitSignal s	

FIGURE 4.11
Motivating example unrolled

(after) the scope of a loop. The two WaitSignal AVOps immediately after the loop will always wait for the correct outstanding sent signals no matter how many iterations of the loop are executed. Unrolling the loop while applying the rewrite at each iteration would result in the AVOp sequence shown in Figure 4.11.

In the above execution trace we see that c_1 stands for different cores based on the state of $\hat{\rho}$. This is because each AVOp implicitly applies $\hat{\rho}$ at each invocation.

4.6.2   The Rewrite Map: ρ

The rewrite map, $\hat{\rho}$, performs rewrites for each AVOp of the executing program. After each iteration of a loop is complete, $\hat{\rho}$ is modified by composing it with the loop's rewrite, ρ. Suppose we are given Loop $k\ \rho\ \beta$ and that $\hat{\rho}_j$ denotes $\hat{\rho}$ at the beginning of iteration j; the modification performed is $\hat{\rho}_{j+1} := \rho \circ \hat{\rho}_j$. Both types of maps have the following type:

$$\rho : R \tag{4.23}$$

$$\hat{\rho} : R \tag{4.24}$$

$$R = C \times S \to C \times S \tag{4.25}$$

Although we present definitions in terms of ρ, equivalent ones are used for $\hat{\rho}$. In later definitions we use two projections directly: first, the projection relevant to rewriting signals; second, the projection relevant to rewriting cores.

$$\rho_S : C \times S \to S \tag{4.26}$$

$$\rho_C : C \to C \tag{4.27}$$

To define these two projected versions precisely, we first introduce the projections themselves. Each one selects the relevant item of the tuple.

$$\pi_C(c, s) = c \tag{4.28}$$

$$\pi_S(c, s) = s \tag{4.29}$$

Rewriting signals is defined as a direct projection.

$$\rho_S = \pi_S \circ \rho \tag{4.30}$$

Rewriting cores is more complicated. In defining ρ_C we make use of a commutative diagram that must be satisfied between ρ_C, ρ, and π_C.

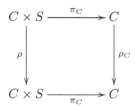

Notice that this places a restriction on ρ in that all entries for any given core must map to the same core.

$$\forall c \in C \ \ \forall s, s' \in S \ \ \pi_C(\rho(c, s)) = \pi_C(\rho(c, s')) \tag{4.31}$$

Then ρ_C is just ρ but with the domain and range restricted to C.

4.6.3   Formulation of AVOps as Functions on the State

We redefine the effect of AVOps when updating Φ and σ here. This is an extension of the definitions laid out in Section 4.3.9 to take into account the global rewrite map $\hat{\rho}$. The AVOps are specified in such a way that one application of their interpretation as a function on (Φ_n, σ_n) corresponds to executing a single AVOp of that type.

$$(\hat{\rho}_{n+1}, (\Phi_{n+1}, \sigma_{n+1})) = (\mathsf{SendSignal}\ c\ c'\ s')(\hat{\rho}_n, (\Phi_n, \sigma_n))$$

We specify the type of each AVOp as a function on the Φ and σ maps:

$$\mathsf{SendSignal} : C \times C \times S \to (R, (F, G)) \to (R, (F, G))$$
$$\mathsf{WaitSignal} : C \times S \times N \to (R, (F, G)) \to (R, (F, G))$$
$$\mathsf{Loop} : \mathbb{N}_0 \times R \times [AVOp] \to (R, (F, G)) \to (R, (F, G))$$

If an error is detected, the verification algorithm stops immediately and reports the error. An error would be detected after $\hat{\rho}$ is applied, which corresponds to the $\mathsf{SendSignal}'$ part of the definition. Let us define the $\mathsf{SendSignal}$ in the case of no errors. The definition consists of two parts. First, we apply ρ

to rewrite the cores and signals appropriately. Second, we define the semantics of the AVOp as an equation.

$$(\text{SendSignal } c \ c' \ s')(\hat{\rho}, (\Phi, \sigma)) = (\hat{\rho}, (\text{SendSignal}' \ \hat{\rho}_C(c) \ \hat{\rho}_C(c') \ \hat{\rho}_S(c', s'))(\Phi, \sigma))$$
$$(4.32)$$

$$(\text{SendSignal}' \ c \ c' \ s')(\Phi, \sigma) = (\Phi, [(c', s') \to (c, n, \Phi)](\sigma))$$

We present a similar definition for WaitSignal. The index of the WaitSignal is n and is interpreted including the iteration of all loops it is nested in (even though this is an execution time property).

$$(\text{WaitSignal } c \ s \ n)(\hat{\rho}, (\Phi, \sigma)) = (\hat{\rho}, (\text{WaitSignal}' \ \hat{\rho}_C(c) \ \hat{\rho}_S(c, s) \ n)(\Phi, \sigma))$$
$$(4.33)$$

$$(\text{WaitSignal}' \ c \ s \ n)(\Phi, \sigma) = \left(\widetilde{\bigvee}(\sigma(c, s), (c, \Phi)) \ , \ [(c, s) \to n](\sigma) \right)$$

We define Loop inductively on the number of iterations performed by the loop with the base case at $k = 0$. Otherwise, we compose all the AVOps of the body of the loop and perform our rewrite at the end of an iteration.

$$(\text{Loop } k \ \rho \ \beta)(\hat{\rho}, (\Phi, \sigma)) = \begin{cases} (\hat{\rho}, (\Phi, \sigma)) & k \le 0 \\ ((\text{Loop } (k-1) \ \rho \ \beta) \circ \text{rwr}_\rho \circ \beta)(\hat{\rho}, (\Phi, \sigma)) & k > 0 \end{cases}$$
$$(4.34)$$

where we make use of the body of the loop, β, acting on the state. Here, β_i denotes the ith AVOp of the loop body; thus

$$\beta(\hat{\rho}, (\Phi, \sigma)) = (\beta_{|\beta-1|} \circ \cdots \circ \beta_1 \circ \beta_0)(\hat{\rho}, (\Phi, \sigma))$$

We also made use of the function rwr, which applies the loop rewrite after the body of the loop is executed.

$$\text{rwr}_\rho(\hat{\rho}, (\Phi, \sigma)) = (\rho \circ \hat{\rho}, (\Phi, \sigma))$$

4.6.4 Induced Rewrites and Rewriting AVOps

In the definitions of Section 4.6.3 it was clear that only the Loop AVOp had any effect on the global rewrite $\hat{\rho}$. We refer to this as inducing a rewrite and introduce the concept here formally.

Definition 4.4 *Induced Rewrite $\tilde{\rho}$ — The rewrite induced by an AVOp a, given*
$a(\hat{\rho}, (\Phi, \sigma)) = (\hat{\rho}', (\Phi', \sigma'))$, *is a rewrite $\tilde{\rho}$ such that $\hat{\rho}' = \tilde{\rho} \circ \hat{\rho}$.*

index	AVOp
0	Loop k ρ $\beta = [$
$(0, *, 0)$	Loop k' ρ' β',
$(0, *, 1)$	$\tilde{\rho}(\mathsf{SendSignal}\ c_0\ s \to c_1)$,
$(0, *, 2)$	$\tilde{\rho}(\mathsf{WaitSignal}\ c_1\ s)$
	$]$

FIGURE 4.12
The effects of an inner loop can be taken into account by rewriting following AVOps by the inner loop's induced rewrite $\tilde{\rho}$. The outer loop's induced rewrite can also be calculated by Lemma 4.5.

We introduce the explicit wording of an induced rewrite because the full effect of a loop on the global rewrite is not equivalent to just the loop rewrite. Given a Loop k ρ β that has no nested loops, we have that $\tilde{\rho} = \rho^k$. That is, we perform the rewrite as many times as there are loop iterations and that is the full effect of the loop on $\hat{\rho}$. The rewrite induced by a loop with nested loops is more complicated.

Lemma 4.5 *Given* Loop k ρ β, *where* β *has* m *nested loops and the rewrite induced by the* i^{th} *nested loop is* $\tilde{\rho}_i$ *then the rewrite induced by* Loop *is*

$$(\rho \circ (\tilde{\rho}_{m-1} \circ \tilde{\rho}_{m-2} \circ \cdots \circ \tilde{\rho}_0))^k$$

Although the effect of nested loops on $\hat{\rho}$ can easily be computed by Lemma 4.5, the nested loops also affect all AVOps following them in the body of the loop. To take this into account we introduce the concept of rewriting AVOps themselves.

Definition 4.5 *Rewriting an AVOp* a *by* ρ — *we define the effect of applying a rewrite on an AVOp. This changes the effect of the AVOp on the state maps* Φ *and* σ *but does not modify its effect on* $\hat{\rho}$.

$$\rho(a)(\hat{\rho}, (\Phi, \sigma)) = \big(\mathrm{fst}(a(\hat{\rho}, (\Phi, \sigma)))\ ,\ \mathrm{snd}(a(\rho \circ \hat{\rho}, (\Phi, \sigma)))\big)$$

where we make use of the functions fst *and* snd *which return the corresponding element of a tuple, that is,* $\mathrm{fst}(\hat{\rho}, (\Phi, \sigma)) = \hat{\rho}$.

By taking both Lemma 4.5 and Definition 4.5 into account we can completely take into account the effects of a nested loop, both on following AVOps and on the rewrite induced by the outer loop, as shown in Figure 4.12.

4.6.5   Short Circuit Theorem for Rewritable Loops

Verifying rewritable loops is quite complicated unless we first simplify to the case without rewrites. This can be done by unrolling the loop $|\rho|$ times where $|\rho|$ is the order of ρ (that is, $\forall x. \rho^{|\rho|}(x) = x$).

index	AVOp
0	SendSignal c_2 $s \to c_0$
1	Loop k $\rho_C = (0\ 1\ 2)$ $\beta = [$
$(1, *, 0)$	WaitSignal c_0 s
$(1, *, 1)$	SendSignal c_0 $s \to c_1$
	$]$

FIGURE 4.13
Loop with rewriting which can be verified without fully unrolling

Suppose we are given the program in Figure 4.13 with a rewritable loop and want to verify its correctness. Presented in Figure 4.14 are the effects of unrolling this loop by $|\rho| = 3$ times. The main thing to notice is that there is a correspondence between indices $|\rho|$ apart instead of 1 iteration apart as was the case for loops without rewrites. In terms of the bumping function Λ we have $\Lambda_l^{|\rho|}$ instead of Λ_l.

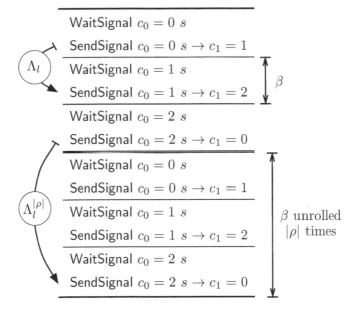

FIGURE 4.14
Rewritable loop unrolled into a loop without a rewrite

We proceed to state the Short Circuit Theorem for Rewritable Loops, taking into account this difference in unrolling. Since we do not know that k is a multiple of $|\rho|$ we may also need to perform extra iterations of the loop at the end to compute the full effects of the loop on the state $(\hat{\rho}, (\Phi, \sigma))$.

Theorem 4.4

Given **Loop** k ρ β *at index* l, *if* $\exists i : \mathbb{N}_0, i > 0, \forall n \in \bigcup\limits_{j=0}^{|\rho|-1} \text{IterIxs}(\beta, i|\rho| + j),$

$$\Phi_n = \Lambda_l^{|\rho|} \circ \Phi_{\text{nchg}_{-|\rho|}(n)}$$

$$\sigma_n = \Lambda_l^{|\rho|} \circ \sigma_{\text{nchg}_{-|\rho|}(n)}$$

Then $\forall n' \in \text{IterIxs}(\beta, k-1)$, $\forall n'' \in \text{IterIxs}(\beta, i|\rho|)$ *where* $(*, *, i) = \text{ninr}(n'') = \text{ninr}(n')$, $X = \left\lfloor \frac{(k-1)-i|\rho|}{|\rho|} \right\rfloor$, *and* $x \equiv (k-1) \mod |\rho|$ *where* $0 \le x < |\rho|$,

$$\Phi_{n'} = \rho^x(\beta) \circ \rho^{x-1}(\beta) \circ \cdots \circ \rho^0(\beta) \circ \Lambda_l^{X|\rho|} \circ \Phi_{n''}$$

$$\sigma_{n'} = \rho^x(\beta) \circ \rho^{x-1}(\beta) \circ \cdots \circ \rho^0(\beta) \circ \Lambda_l^{X|\rho|} \circ \sigma_{n''}$$

Proof 4.7 *The proof is by reducing the rewritable loop to a non-rewritable loop and checking any remaining iterations.*

For each AVOp a in β we apply the global rewrite $\hat{\rho}$ followed by $\tilde{\rho}_i$ for each of the n nested loops preceding a in β. That is, we convert a to a' by the following equation:

$$a' = (\tilde{\rho}_{n-1} \circ \tilde{\rho}_{n-2} \circ \cdots \circ \tilde{\rho}_0 \circ \hat{\rho})(a)$$

Once all the relevant rewrites have been applied we proceed to unroll $|\rho|$ times. Let m be the total number of nested loops in β. To calculate the AVOp a_j, corresponding to a' at iteration j, we apply the following rewrite:

$$a_j = (\tilde{\rho}_{n-1} \circ \tilde{\rho}_1 \circ \cdots \circ \tilde{\rho}_0 \circ \rho \circ \tilde{\rho}_{m-1} \circ \cdots \circ \tilde{\rho}_{n+1} \circ \tilde{\rho}_n)^j(a')$$

Notice that we have to apply the induced rewrites of the nested loops following a before applying the loop rewrite ρ and then applying the induced rewrites preceding a. This is because composition of permutations is not a commutative operation.

At this point we may apply Theorem 4.2 to the unrolled loop body. We then proceed to bump as many iterations as possible, but we need to take into account the unrolling. As shown in Figure 4.14, we must bump by $\Lambda_l^{|\rho|}$ when bumping from one iteration of an unrolled loop to the next.

After this, all that remains is to apply the remaining iterations of β since k is not always a multiple of $|\rho|$.

4.6.6   Verifier Run Time on Rewritable Loops

We examine the run time of the verification algorithm on rewritable loops. Some results from Section 4.5.5 are required.

Theorem 4.5 *The verification algorithm needs to examine at most $|\rho| \cdot (|C| + 2)$ iterations of a rewritable loop in order to verify correctness and compute the state after the loop.*

Proof 4.8 *In order to verify a rewritable loop, it is unrolled $|\rho|$ times. After unrolling, it is equivalent to a non-rewritable loop, with a $|\rho|$-times bigger body, which requires $|C| + 1$ iterations before establishing the Short Circuit Theorem. After the short circuit is applied, there remain at most another $|\rho|$ iterations that must be checked without unrolling since the iteration count k is not always divisible by the unrolling factor $|\rho|$.*

4.6.7 Memory Requirements

In our treatment of Φ-slices one may think that the memory requirement for a Φ-slice is $\mathcal{O}(|\text{cores}|^2)$. However, in implementation, it is possible to store the slice specific to a given core. That is, we fix the core $c : C$ that performed the SendSignal and for all other $c' : C$ we store $\Phi(c', c)$. This results in more efficient memory usage of $\mathcal{O}(|\text{cores}|)$.

In Anand and Kahl (2008) the memory requirements for the verification algorithm were incorrectly stated as $\mathcal{O}(|\text{cores}|^2)$. This happened, most likely, because only the memory requirements of Φ were taken into account. However, it gets more complicated as Φ-slices may themselves be stored in σ. In the worst case, σ will map all cores and signals to a Φ-slice, which results in the memory usage of $\mathcal{O}(|\text{cores}|^2 \cdot |\text{signals}|)$. Once we take all AVOps into account (see Section 4.7) we need to take δ into account, which may also store Φ-slices or DMA Tags. Thus, the final memory usage is $\mathcal{O}(C^2 \cdot \max(S, B) + C \cdot B \cdot D|)$ where:

- C — is the number of cores.

- S — is the number of signals per core.

- B — is the number of data buffers per core.

- D — is the number of DMA tags.

4.7 Extension to Full AVOp Set

The AVOp language that we have covered so far is only a subset of the full language which deals directly with synchronization across cores. To be useful, we also need constructs for addressing main memory, sending data between cores, and running kernel computations.

4.7.1   Extension of AVOp State and Hazard Checking

To keep track of data transfers we need to extend the state from $(\hat{\rho}, (\Phi, \sigma))$ to $(\hat{\rho}, (\Phi, \sigma, \Delta))$, where Δ keeps track of the status of data transfers and makes use of the following types of entries:

- L — is the name of a data buffer, an area of memory local to a given core.

- D — is the set of DMA tags. These reference a specific DMA transfer operation so that we may be notified about transfer completion.

The type of Δ is as follows (where 2^D is the power set of D, and $\oplus$ denotes disjoint union):

$$\Delta_n : C \times L \to \begin{pmatrix} C \times N \times F \\ \oplus \\ N \\ \oplus \\ 2^D \\ \oplus \\ D \end{pmatrix} \tag{4.35}$$

Each part of the disjoint union of Δ may be interpreted as follows (we list the AVOps that modify entries Δ into this state):

1. An incoming transfer from core in C started at instruction in N with known partial order in F is expected (SendData – destination).

2. The data was last used at an instruction in N (WaitData, WaitDma, RunComputation).

3. The data buffer is the source of one or more DMA transfers whose list of (completion) tags which have *not yet* been consumed by a WaitDMA are in 2^D, and the data tag is in T (SendData – source, StoreMemory, WaitDma).

4. An incoming DMA with tag in D was started. Such a transfer can only come from main memory (LoadMemory).

For the initial Δ map, $\forall c \in C, \forall l \in L : \Delta_0(c, l) = -1$ (case 2).

We make use of definitions from Anand and Kahl (2008) in this section; however, the data tag type T has been abstracted away in this presentation of the language. In practice, it is an extra boolean tacked onto the memory area of a buffer and is the last thing to be modified by a DMA. By repeatedly checking whether or not this flag is set, it is possible to determine whether or not the DMA has completed on the receiving core.

The non-rewrite part of the state $(\Phi_{n+1}, \sigma_{n+1}, \Delta_{n+1})$ at index $n+1$ is updated from the previous state at index n as follows by each of the following AVOps (including checking for hazards):

SendData c l' λ c'' l'' d **At the Source:** We need to ensure that the source
data is not about to be rewritten and that the use of d does not
conflict with another use. If d appears in $\Delta_n(c, \tilde{l})$ for some $\tilde{l}$, then the
DMA tag is already in use, and the completion of this DMA would
be indistinguishable from the completion of another DMA already
originated from this core; this is an error. If d occurs in $\Delta_n(c, \tilde{l})$ for
some $\tilde{l}$, signal an error.

If $\Delta_n(c, l') = (n'')$ then no DMAs (Synergistic Processing Unit (SPU)
to SPU transfers or transfers from main memory) are pending for this
buffer, so record the send by letting $\Delta_{n+1}(c, l') := (\{d\})$ (with Δ_{n+1}
equal to Δ_n at all other values).

If $\Delta_n(c, l') = (d')$, then on the sending core c an incoming DMA is in
progress from main memory with DMA tag d'. Therefore it is not safe
to start an outgoing DMA as this data could be overwritten while it
is being sent; this is an error.

If $\Delta_n(c, l') = (D')$ for some subset $D' \subset D$, then there are already
outgoing DMAs in progress, which is OK if we do not try to reuse
DMA tags.

Otherwise, record its use by $\Delta_{n+1}(c, l') := (D' \cup \{d\})$, and copy the
other maps from n to $n + 1$.

Otherwise, this buffer is the target of a DMA from another core,
which is an error.

At the Destination: Between cores, execution order does not have to
follow presentation order. We therefore need to check that all use of
the previous contents of the target buffer must have completed before
the current instruction.

If $\Delta_n(c'', l'') = (n'')$ and $n'' \le \Phi_n(c'', c)$, then the destination buffer
has no pending DMAs after instruction n'' and the current instruction
is known to complete ($\Phi_n(c'', c)$) after the last instruction (n'') to
have used the target buffer, so it is safe thus far. Because we assume
an implementation of SendData and WaitData with a hardware based
signal mechanism for each data tag, we cannot perform a SendData
unless there is no incoming SendData transfer on the destination SPU
with the same data tag (because signal usage cannot overlap). So if
we have that for any l''', $\Delta_n(c'', l''') := (c''', n''', \Phi_n)$ where $c''' \in C$
and $n''' \in N$, then we return an error because this data tag signal
may not be consumed by the time this signal arrives.

Then we record that $\Delta_n(c'', l'') := (c, n, \Phi_n)$. Otherwise, report
the error that multiple incompatible DMAs with source and target
(c'', l'') were detected, or the data tag is illegally reused.

WaitData c l' If $\Delta_n(c, l') = (c'', n', \Phi')$, then we are waiting for this data and
have stored the follows map at the send instruction, which we can combine

with the map at the current instruction: $\Phi_{n+1} := \tilde{V}((c'', n', \Phi'), (c, \Phi_n))$. Let $\sigma_{n+1} := \sigma_n$, and $\Delta_{n+1} := \Delta_n$ except $\Delta_{n+1}(c, l') := (n)$, which indicates that this buffer is safe to use after the current instruction.

Otherwise, no incoming DMA from another core preceded this instruction, so the program is not locally sequential, or this DMA overlaps DMAs initiated locally, which is unsafe.

LoadMemory $c \; l' \; \lambda \; g' \; d$ Check that d does not appear in the image of $\Delta_n(\{c\} \times L)$, otherwise the uses of this tag overlap, which is an error.

If $\Delta_n(c, l') = (n')$, then the buffer has no pending I/O, and we can initiate an incoming DMA by setting $\Delta_{n+1} := \Delta_n$ except $\Delta_{n+1}(c, l') := (d)$, $\sigma_{n+1} := \sigma_n$ and $\Phi_{n+1} := \Phi_n$.

Otherwise, the target buffer is the source or target for transfer which may not have completed, and this will produce an indeterminate result.

StoreMemory $c \; l' \; \lambda \; g' \; d$ Check that d does not appear in the image of $\Delta_n(\{c\} \times L)$, otherwise the uses of this tag overlap, which is an error.

If $\Delta_n(c, l') = (n')$, then no other I/O is pending, and the store is safe. Set $\Delta_{n+1} := \Delta_n$ except $\Delta_{n+1}(c, l') := (\{d\})$ $\sigma_{n+1} := \sigma_n$ and $\Phi_{n+1} := \Phi_n$.

If $\Delta_n(c, l') = (D')$, other outgoing DMAs are pending, which is allowed. So set $\Delta_{n+1} := \Delta_n$ except $\Delta_{n+1}(c, l') := (D' \cup \{d\})$ (to record the new DMA tag we are waiting for), $\sigma_{n+1} := \sigma_n$ and $\Phi_{n+1} := \Phi_n$.

Otherwise, the source buffer is the target for a pending transfer, and this will produce an indeterminate result.

WaitDMA $c \; d$ The DMA tag could be associated with incoming or outgoing I/O. In either case $\exists l$ such that $\Delta_n(c, l) = (d)$ (incoming), or $\Delta_n(c, l) = (D')$ and $d \in D'$ (outgoing), and l is unique. Otherwise, there is an error, because no DMAs are pending using this tag. This indicates either an unsound program or a non-locally sequential presentation.

If $\Delta_n(c, l') = (\{d\})$ then $\Delta_{n+1}(c, l') := (n)$.

If $\Delta_n(c, l') = (D')$ then $\Delta_{n+1}(c, l') := (D' \setminus \{d\})$.

If $\Delta_n(c, l') = (d)$ then $\Delta_{n+1}(c, l') := (n)$.

RunComputation $c \; x \; (l', l'', \ldots) \; (\tilde{l}', \tilde{l}'', \ldots) \; (p', p'', \ldots)$ **Inputs:** If for every $l \in \{l', l'', \ldots\}$, $\Delta_n(c, l) = (n')$ or (D') then $\Delta_{n+1}(c, l) := (n)$ or (D') (respectively), otherwise, the buffer has pending incoming I/O, which makes this computation unsound.

Outputs: If for every $\tilde{l} \in \{\tilde{l}', \tilde{l}'', \ldots\}$, $\Delta_n(c, \tilde{l}) = (n')$ then $\Delta_{n+1}(c, \tilde{l}) := (n)$, otherwise, there is pending I/O and this modification makes the program unsound.

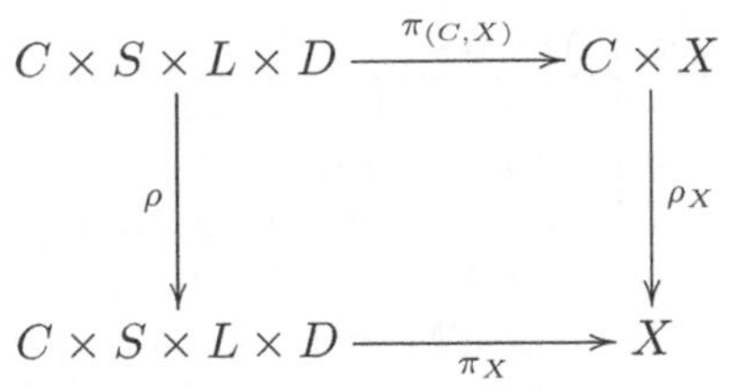

$$C \times S \times L \times D \xrightarrow{\ \pi_{(C,X)}\ } C \times X$$

$$\rho \downarrow \qquad\qquad\qquad \downarrow \rho_X$$

$$C \times S \times L \times D \xrightarrow[\ \pi_X\ ]{} X$$

FIGURE 4.15
Defining diagram for projected rewrite ρ_X where X may be S, L or D. We made use of a helper function $\pi_{(C,X)}(y) = (\pi_C(y), \pi_X(y))$.

4.7.2 Extension of the Rewrite Map

The rewrite maps need to be extended to take into account some of these new datatypes. The type of a rewrite map thus becomes:

$$\rho : R$$
$$\hat{\rho} : R$$
$$R = C \times S \times L \times D \to C \times S \times L \times D$$

We define two additional projected rewrite maps, on top of the extension of ρ_C and ρ_S from Section 4.6.2.

$$\rho_C : C \to C$$
$$\rho_S : C \times S \to S$$
$$\rho_L : C \times L \to L$$
$$\rho_D : C \times D \to D$$

To define these two projected versions precisely, we first introduce the projections themselves. Each one selects the relevant item of the tuple.

$$\pi_C(c, s, l, d) = c$$
$$\pi_S(c, s, l, d) = s$$
$$\pi_L(c, s, l, d) = l$$
$$\pi_D(c, s, l, d) = d$$

Because of the extension of the domain and codomain, the definition of all projected rewrite maps takes on the form shown in the commutative diagram of Figure 4.15, which enforces the following equation for all $X = S$:

$$\forall c \in C \ \ \forall l, l' \in L \ \ \forall d, d' \in D \ \ \forall s \in S \ \ \pi_S(\rho(c, s, l, d)) = \pi_S(\rho(c, s, l', d'))$$

A similar equation holds for each of the other cases.

index	AVOp
-	$\gamma_r(g_a) = [0, 1, 2, 3, 4]$
-	$\gamma_r(g_b) = [5, 6, 7, 8, 9]$
-	$\gamma_w(g_c) = [10, 11, 12, 13, 14]$
0	Loop 5 $\rho_C = (0\ 1\ 2\ 3\ 4)$ $\beta = [$
(0,*,0)	LoadMemory $c_0\ l\ \lambda\ g_a\ d$
(0,*,1)	LoadMemory $c_0\ l'\ \lambda\ g_b\ d'$
(0,*,2)	WaitDma $c_0\ d$
(0,*,3)	WaitDma $c_0\ d'$
(0,*,4)	RunComputation$_+$ $c_0\ (l, l')\ (l''')$
(0,*,5)	StoreMemory $c_0\ l'''\ \lambda\ g_c\ d'''$
(0,*,6)	WaitDma $c_0\ d'''$
-	]

FIGURE 4.16

Accessing global memory to perform a vector addition $A + B = C$ that is parallelized across 5 cores. The read-only global memory buffer list γ_r and write-once buffer list γ_w are specified at the beginning, $\lambda = 1$ denotes buffer size, l denotes a location in local memory, d denotes a DMA tag.

4.7.3   Extension of Loops for Accessing Global Memory

In addition to the changes required to the rewrite map of a loop, there is also a larger problem with how to handle global memory access. If the same locations in memory were to be accessed with every iteration of the loop, the set of useful programs that could be expressed would be limited. To this end we introduce a list of global memory buffers γ, where from each list we assign a memory location to every instance of a symbol g, in unrolled presentation order. To be precise, for a single global buffer g in γ we need $|\gamma(g)| \geq \kappa$ where κ is the number of times g is referenced in the unrolled program. Notice that κ can be much larger than the number of iteration counts of a single loop, especially with nested loops.

The global memory access list is further broken down into γ_r for memory locations that can only be read from and γ_w for memory locations that can be written to only once. This is a key property which allows arbitrary reads and no data hazards such as read-on-write to occur.

In an implementation, it would not be practical to have a list with one entry per execution of the loop body. More efficient approaches such as functional lazy lists or generators are envisioned instead.

In Figure 4.16 we present an example of computing the sum of two vectors and storing the result in a third vector.

It should be noted that all verification burden is on the user. She needs to prove beforehand that all of the global addresses listed in any γ_w are used only once and are not listed in any γ_r. Among the issues that need to be addressed is the limitation of the buffer size parameter λ so that overlap checks can be performed (if an arbitrary and varying λ were allowed the user could not check that any given global location does not extend and overlap a later global location).

4.7.4 Extension of Verification Algorithm

There are no further extensions necessary to be able to apply the Short Circuit Theorem in the presence of this extended AVOp set. Just as SendSignal and WaitSignal cause a dependency between cores, so do SendData and WaitData. In fact, in many ways they are interchangeable.

4.8 Future Work

We cover extensions of results presented in this chapter to stronger postulated results. Also covered are highly relevant, but different, research directions required to make our approach practical on large scale supercomputers.

4.8.1 Restricted Global Memory Access for Verification

We can extend γ_w to memory locations which can be read or written to from only a single core over the whole program. We make this single core restriction because the AVOps (thus the reads and writes) of a single core are totally ordered, hence none of the data hazards — read-after-write, write-after-read, write-after-write — can occur. However, we would have to verify that each memory location is accessed from only that core. Allowing read and write access from arbitrary cores would introduce indirect communication via global memory, which is inefficient due to synchronization overhead, and difficult to verify.

Another possible approach would be to introduce barriers that seperate the program into phases with writing to a given global memory location being allowed only during a single phase. This write would then be necessarily ordered to finish before any further reads in the next phase.

Finally, the global memory access patterns themselves could be abstracted over in our AVOp language into an algebra of strided memory access. Access patterns could then be combined to generate more intricate patterns and so on while the verifier could inspect the access pattern to ensure correctness.

4.8.2 Stronger Short Circuit Theorem for Rewritable Loops

We postulate that the effect of executing the loop body on the state $(\hat{\rho}, (\Phi, \sigma))$ is equivalent to applying the rewrite and bumping to the next iteration.

$$\rho \circ \Lambda = \beta \tag{4.36}$$

In fact, it was during the analysis of a weaker (unrolled) property that we came up with the current formulation of Theorem 4.4. We can take advantage of the commutativity of Λ and ρ to unroll the loop body by the order of ρ.

$$(\rho \circ \Lambda)^{|\rho|} = \rho^{|\rho|-1}(\beta) \circ \cdots \circ \rho(\beta) \circ \beta \qquad \text{Unrolling (4.36) } |\rho| \text{ times}$$

$$\rho^{|\rho|} \circ \Lambda^{|\rho|} = \rho^{|\rho|-1}(\beta) \circ \cdots \circ \rho(\beta) \circ \beta \quad \text{By the commutativity of } \Lambda \text{ and } \rho$$

$$\Lambda^{|\rho|} = \rho^{|\rho|-1}(\beta) \circ \cdots \circ \rho(\beta) \circ \beta \qquad \text{Since } \rho^{|\rho|} \text{ is the identity}$$

Thus, we can see that by working with a loop body unrolled $|\rho|$ times the rewrite term is eliminated.

Using (4.36) would save work when verifying the later loop iterations. However, since we already need to check at least $2|\rho|$ iterations at the beginning of the loop to stabilize the loop, an additional $|\rho|$ iterations at the end, in the worst case, does not impact the asymptotic run time. For this reason we did not pursue this idea further but anyone working on practical implementations may be interested in this development to further optimize the actual run time and simplify the implementation.

4.8.3 Breaking up AVOp Streams to Increase AVOp Feed Rate

Some supercomputer architectures provide a tree structure to broadcast and potentially collect information back up the tree; for example, the IBM Blue Gene supercomputer provides this type of communication pattern (Adiga et al. 2002) for both broadcast and collection of information. One well known example of just the collection is the reduce part in the map/reduce method that powers the Google search engine (Dean and Ghemawat 2008). We are especially interested in the other direction of tree communication – broadcast.

This type of tree structured communication pattern is important in feeding potentially many thousands of cores with AVOps. Since the entire program, split across all cores, is presented as a single listing, this listing must then be forwarded to the relevant cores for computation. Having a single core feed all the other cores, the approach envisioned for the IBM Cell Processor, does not scale up to larger sizes. Many cores would be waiting for the main core, starved of AVOps to execute. What needs to be done is for the program to be progressively split into only the AVOps that are relevant to that subset of cores.

Loops are important for scaling because they allow computation across many cores to be highly compressed into a single loop body. Future work will have to examine efficient methods of progressively breaking up the loops as the program is transferred down the tree. With each step down the tree there are fewer relevant cores, which restricts the original program to the sub-program relevant to only those cores.

4.8.4 Verifying the Verifier

The verification system we have presented has the virtues of low complexity and specific error messages in case a hazard is detected. It is, however, sufficiently complicated to implement, that for safety- or mission-critical applications it should itself be verified.

One approach is to use a fundamentally different verification method and exhaustively test all the programs of a small size. The approach we have implemented is symbolic execution. This has the virtue that a computation graph representing the result is recovered and can be checked against a specification, but it is much more expensive than our verification algorithm. In practice, only programs of tens of AVOps can be checked. We are working to prove that if the symbolic execution is correct, and it agrees with the verifier on a certain set of programs, then the verifier must also be correctly implemented.

Another approach which we have mooted is the formalization of the verification state and verification algorithm and the generation of a formal proof which can be model checked. We have not attempted this approach, but would like to see it completed.

4.8.5 Modifications for Cached Memory Multicore Processors

The scheme we have outlined is adapted to tightly coupled distributed memory systems. Programming such systems in conventional languages is difficult because it requires explicit memory transfer operations, which were not conceived of in the formulation of such languages. Not coincidentally, global memory-space multicore systems with hardware-controlled caching are more common. In such a system, explicit data transfer instructions and hardware signals are absent. To adapt our model to a cached architecture, we can greatly simplify the presentation of the verification state and algorithm. This simplification is deceiving, however, as we now need to check for conflicts between all possibly concurrent memory accesses on all cores, since any core can hazardously read or write to any address. The complexity goes from $\mathcal{O}(|\text{cores}|^2 \cdot \max(|\text{signals}|, |\text{buffers}|) + |\text{cores}| \cdot |\text{buffers}| \cdot |\text{dma tags}|)$ to $\mathcal{O}(|\text{cores} \cdot \text{buffers}|^2)$, which will be significantly larger in practical cases, due to the squaring of the buffer term.

Signals could be synthesized using ordinary memory accesses to reduce the number of possibly concurrent memory accesses, but this may significantly degrade performance. Synthesizing signals is inherently more expensive on cached memory systems than on systems with hardware signals because the cache coherency hardware does not know that a cache line is only written by one core and read by another, and every time the signal is checked or set coherency traffic will be generated.

We are implementing a simplified run-time system for cached multicore systems to investigate these issues.

4.8.6   Scheduling

We have not addressed the issue of scheduling AVOps with loops for optimal performance. This is a key part of the out-of-order execution on CPUs that our approach mimics. The idea is to rearrange a program's AVOps to hide communication latency while preserving order independence (that given the same memory contents at input locations, memory will be the same at output locations). All software pipelining methods for instruction loops can be applied to the AVOp scheduling problem.

Scheduling AVOp streams without loops is covered in Anand and Kahl (2008).

4.8.7   Fault Tolerance

Hardware failure is unavoidable on supercomputers or large clusters due to the sheer amount of hardware involved. No fault tolerance results have been presented so far (by us or anyone else) for the AVOp language. This is an important area for future research in order to make the AVOp language applicable in real settings.

4.9   Conclusion

We have demonstrated how to extend AVOp streams with rewritable loops. Such a construct enables us to express complex communication patterns and hide communication latencies in an approach similar to software pipelining of loops. Rewritable loops also allow the AVOp streams to be highly compressed. This has especially important implications for the future of distributing AVOp stream computations across a large collection of cores.

Furthermore, the verification procedure for AVOp streams was extended to the case with rewritable loops. The new verification procedure runs independently of the run time of the program being verified (for programs with loops repeating many times). This is a direct result of compressing the program text. Without loops, the program text is proportional to program execution. We need to check $\mathcal{O}(|\text{cores}| \cdot |\text{signals}| \cdot |\text{buffers}| \cdot |\text{tags}|)$ iterations of a rewritable loop body in order to verify it. In practice, this may be much better depending on the program text.

We have also refined the memory requirements needed for the verification algorithm to take into account two things. First, the fact that some state entities store other state entities and this needs to be taken into account for an accurate memory bound. Second, the fact that we can use a more restrictive version of some maps to minimize the memory used in the actual implementation. The memory bound for our verification procedure is $\mathcal{O}(|\text{cores}|^2 \cdot \max(|\text{signals}|, |\text{buffers}|) + |\text{cores}| \cdot |\text{buffers}| \cdot |\text{dma tags}|)$.

AVOp streams themselves only use non-blocking communication, which allows our approach to scale to large systems and this approach is maintained with our extension to the language.

5

Accelerating Critical Section Execution with Asymmetric Multicore Architectures

M. Aater Suleman
Department of Electrical and Computer Engineering, The University of Texas at Austin, Austin, TX, USA

Onur Mutlu
Department of Electrical and Computer Engineering, Carnegie Mellon University, Pittsburgh, PA, USA

CONTENTS

5.1   Introduction

Improving the performance of a single program using multiple processor cores requires that the application be partitioned into *threads* that execute concurrently on multiple cores. The principle of *mutual exclusion* dictates that threads cannot be allowed to update shared data concurrently; thus, accesses to shared data are encapsulated inside *critical sections*. Only one thread executes a critical section at a given time; other threads wanting to execute the same critical section must wait. Critical sections can serialize threads, thereby reducing performance and scalability (i.e., the number of threads at which performance saturates). This performance loss due to critical sections can be reduced by shortening the execution time inside critical sections.

This chapter describes the *Accelerated Critical Sections (ACS)* mechanism to reduce performance degradation due to critical sections. ACS is based on the Asymmetric Chip Multiprocessor (ACMP), which consists of at least one large, high-performance core and many small, power-efficient cores (Morad et al. 2006; Suleman et al. 2007; Hill and Marty 2008; Annavaram, Grochowski, and Shen 2005). The ACMP was originally proposed to run Amdahl's serial bottleneck (where only a single thread exists) more quickly on the large core and the parallel program regions on the multiple small cores. In addition to Amdahl's bottleneck, ACS also runs selected critical sections on the large core, which runs them faster than the other smaller cores. By accelerating critical section execution, ACS reduces serialization: it lowers the likelihood of threads waiting for a critical section to finish.

ACS dedicates the large core exclusively to run critical sections (and Amdahl's bottleneck). In conventional systems, when a core encounters a critical section, it acquires the lock for the critical section, executes the critical section, and then releases the lock. In ACS, when a small core encounters a critical section, the small core sends a request to the large core for execution of that critical section and stalls. The large core executes the critical section in turn and notifies the small core once it has completed the critical section. The small core then resumes execution.

Our evaluation on a set of 12 critical-section-intensive workloads shows that ACS reduces the average execution time by 34% compared to an equal-area 32-core symmetric CMP and by 23% compared to an equal-area ACMP. Moreover, for 7 of the 12 workloads, ACS also increases scalability (i.e., the number of threads at which performance saturates).

5.2 The Problem

A multithreaded application consists of two parts: the serial part and the parallel part. The serial part is the classical Amdahl's bottleneck (Amdahl 1967) where only one thread exists. The parallel part is where multiple threads execute concurrently. Threads operate on different portions of the same problem and communicate via shared memory. The principle of *mutual exclusion* dictates that multiple threads shall not be allowed to update shared data concurrently. Thus, accesses to shared data are encapsulated in regions of code guarded by synchronization primitives (e.g. locks) called *critical sections*. Only one thread can execute a particular critical section at any given time. Critical sections are different from Amdahl's serial bottleneck: during the execution of a critical section, other threads that do not need to execute the same critical section can make progress. In contrast, no other thread exists in Amdahl's serial bottleneck. We use a simple example to show the performance impact of critical sections.

Figure 5.1(a) shows the code for a multithreaded kernel where each thread dequeues a work quantum from the priority queue (PQ) and attempts to process it. If the thread cannot solve the problem, it divides the problem into sub-problems and inserts them into the priority queue. This is a common technique for parallelizing many branch-and-bound algorithms (Lawler and Wood 1966). In our benchmarks, this kernel is used to solve the popular 15-puzzle problem (Wikipedia). The kernel consists of three parts. The initial part A and the final part E are the serial parts of the program. They comprise Amdahl's serial bottleneck since only one thread exists in those sections. Part B is the parallel part, executed by multiple threads. It consists of code that is both inside the critical section (C1 and C2, both protected by lock X) and outside the critical section (D1 and D2). Only one thread can execute the critical section at a given time, which can cause serialization of the parallel part and reduce overall performance.

Figure 5.1(b) shows the execution time of the kernel shown in Figure 5.1(a) on a 4-core CMP. After the serial part A, four threads (T1, T2, T3, and T4) are spawned, one on each core. Once part B is complete, the serial part E is executed on a single core. We analyze the serialization caused by the critical section in the steady state of part B. Between times t_0 and t_1, all threads execute in parallel. At time t_1, T2 starts executing the critical section while T1, T3, and T4 continue to execute the code independent of the critical section. At time t_2, T2 finishes the critical section, which enables three threads (T1, T3, and T4) to contend for the critical section – T3 wins and enters the critical section. Between times t_2 and t_3, T3 executes the critical section while T1 and T4 remain idle, waiting for T3 to finish. Between times t_3 and t_4, T4 executes the critical section while T1 continues to wait. T1 finally gets to execute the critical section between times t_4 and t_5.

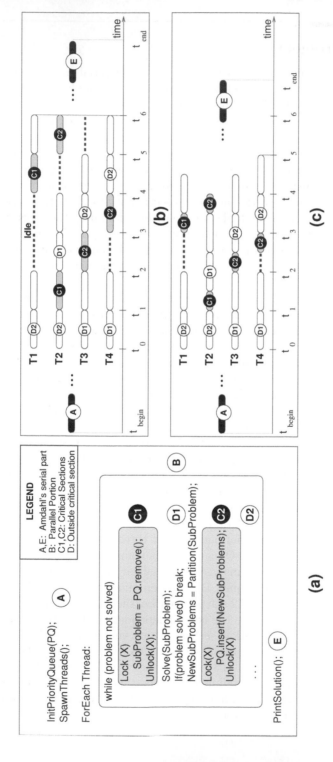

FIGURE 5.1

Amdahl's serial part, parallel part, and the critical section in a multithreaded 15-puzzle kernel: (a) code example, and execution timelines on (b) the baseline CMP and (c) accelerated critical sections

This example shows that the time taken to execute a critical section significantly affects not only the thread that executes it but also the threads that are waiting to enter the critical section. For example, between t_2 and t_3 there are two threads (T1 and T4) waiting for T3 to exit the critical section, without performing any useful work. Therefore, accelerating the execution of the critical section not only improves the performance of T3 but also reduces the wasteful waiting time of T1 and T4. Figure 5.1(c) shows the execution of the same kernel assuming that critical sections take half as long to execute. Halving the time taken to execute critical sections reduces thread serialization, which significantly reduces the time spent in the parallel portion. Thus, accelerating critical sections can provide significant performance improvement.

On average, the critical section shown in Figure 5.1(a) executes 1.5 K instructions. During an insert, the critical section accesses multiple nodes of the priority queue (implemented as a heap) to find a suitable place for insertion. Due to its lengthy execution, this critical section incurs high contention. When the workload is executed with 8 threads, on average 4 threads wait for this critical section at a given time. The average number of waiting threads increases to 16 when the workload is executed with 32 threads. In contrast, when this critical section is accelerated using ACS, the average number of waiting threads reduces to 2 and 3 for 8- and 32-threaded execution, respectively.

The task of shortening critical sections has been traditionally left to programmers. However, tuning code to minimize critical section execution requires substantial programmer time and effort, which are often scarce resources. A mechanism that can shorten the execution time of critical sections transparently in hardware, without requiring programmer support, can significantly impact both the software and the hardware industry.

5.3 Accelerated Critical Sections (ACS)

This chapter proposes *Accelerated Critical Sections (ACS)*, a new approach to handle critical sections by accelerating their execution. In traditional CMPs, when a core encounters a critical section, it acquires the lock associated with the critical section, executes the critical section, and releases the lock. In ACS, when a core encounters a critical section, it requests a high-performance core to execute that critical section. The high-performance core acquires the lock, executes the critical section, and notifies the requesting core when the critical section is complete.

5.3.1 Design

The ACS mechanism is implemented on a homogeneous-ISA, heterogeneous-core CMP that provides hardware support for cache coherence. ACS is based

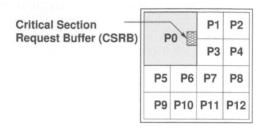

FIGURE 5.2
Accelerated Critical Sections (ACS)

on the ACMP architecture (Morad et al. 2006; R. Kumar et al. 2005; Ipek et al. 2007; Suleman et al. 2007), which was proposed to handle Amdahl's serial bottleneck. ACS consists of one high-performance core and several small cores. The serial part of the program and the critical sections execute on the high-performance core, whereas the remaining parallel parts execute on the small cores.

Figure 5.2 shows an example ACS architecture implemented on an ACMP consisting of one large core (P0) and 12 small cores (P1–P12). ACS executes the parallel threads on the small cores and dedicates P0 for the execution of critical sections (as well as serial program portions). P0 is augmented with a *critical section request buffer (CSRB)* that buffers the critical section execution requests from the small cores. ACS introduces two new instructions, CSCALL and CSRET, which are inserted at the beginning and end of a critical section, respectively.

Figure 5.3 contrasts ACS with a conventional system. In conventional locking (Figure 5.3a), when a small core encounters a critical section, it acquires the lock that protects the critical section, executes the critical section, and releases the lock. In ACS (Figure 5.3b), when a small core is about to encounter a critical section, it executes a CSCALL instruction. The small core sends a *critical section execution request (CSCALL)* to P0 and stalls until it receives a response. When P0 receives a CSCALL request, it buffers it in the CSRB. P0 starts executing the requested critical section at the first opportunity and continues normal processing until it encounters a CSRET instruction, which signifies the end of the critical section. When P0 executes the CSRET instruction, it sends a *critical section done (CSDONE)* signal to the requesting small core. Upon receiving the CSDONE signal, the small core resumes normal execution.

Our 2009 paper (Suleman et al.) describes the ACS architecture in detail, focusing on hardware/ISA/compiler/library support, interconnect extensions, OS support, handling nested critical sections, handling interrupts/exceptions, and accommodating multiple large cores and multiple parallel applications.

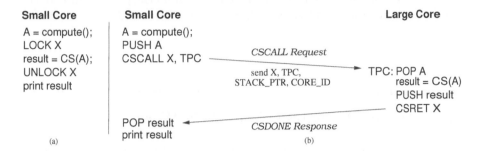

FIGURE 5.3
Source code and its execution: (a) baseline (b) ACS

5.3.2   False Serialization

ACS, as we have described it thus far, executes all critical sections on a single dedicated core. This can have the following negative effect: ACS can serialize the execution of independent critical sections which could have executed concurrently in a conventional system. We call this effect *false serialization*. To reduce false serialization, ACS can dedicate **multiple execution contexts** for critical section acceleration. This can be achieved by making the large core a simultaneous multithreading (SMT) engine or by providing multiple large cores on the chip. With multiple contexts, different contexts operate on independent critical sections concurrently, thereby reducing false serialization.

Some workloads continue to experience false serialization even when more than one context is available for critical section execution. For such workloads, we propose *Selective Acceleration of Critical Sections (SEL)*. SEL tracks false serialization experienced by each critical section and disables the accelerated execution of critical sections that experience high false serialization.

False serialization is estimated by augmenting the CSRB with a table of saturating counters (one per critical section) for tracking false serialization. We quantify false serialization by counting the number of critical sections present in the CSRB for which the LOCK_ADDR is different from the LOCK_ADDR of the incoming request. If this count is greater than 1 (i.e., if there are at least two independent critical sections in the CSRB), the estimation logic adds the count to the saturating counter corresponding to the LOCK_ADDR of the incoming request. If the count is 1 (i.e., if there is exactly one critical section in the CSRB), the corresponding saturating counter is decremented. If the counter reaches its maximum value, all small cores are notified that ACS has been disabled for the particular critical section. Future CSCALLs to that critical section are executed locally at the small core.[1] Our 2009 paper

---

[1] Our implementation of SEL hashes lock addresses into 16 sets and uses 6-bit counters. The total storage overhead of SEL is 36 bytes: 16 counters of 6 bits each and 16 ACS_DISABLE bits for each of the 12 small cores.

(Suleman et al.) shows that SEL completely recovers the performance loss due to false serialization, and ACS with SEL outperforms ACS without SEL by 15%.

5.4   Trade-off Analysis: Why ACS Works

ACS makes three performance trade-offs compared to conventional systems:

1. **Faster critical sections versus fewer threads:** ACS has fewer threads than conventional systems because ACS dedicates a large core, which could otherwise be used for more threads, for accelerating critical sections. ACS improves performance when the benefit of accelerating critical sections is greater than the loss due to the unavailability of more threads. ACS's performance improvement becomes more likely when the number of cores on the chip increases for two reasons. First, the marginal loss in parallel throughput due to the large core is smaller (for example, if the large core replaces 4 small cores, then it eliminates 50% of the smaller cores in a 8-core system but only 12.5% of cores in a 32-core system). Second, more cores (threads) increase critical section contention, thereby increasing the benefit of faster critical section execution.

2. **CSCALL/CSDONE signals versus lock acquire/release:** ACS requires the communication of CSCALL and CSDONE transactions between a small core and a large core, an overhead not present in conventional systems. However, ACS can compensate for the overhead of CSCALL and CSDONE by keeping the lock at the large core, thereby reducing the cache-to-cache transfers incurred by conventional systems during lock acquire operations (Rajwar and Goodman 2002). ACS actually has an advantage in that the latency of CSCALL and CSDONE can be overlapped with the execution of another instance of the same critical section. On the other hand, in conventional locking, a lock can only be acquired after the critical section has been completed, which *always* adds a delay before critical section execution.

3. **Cache misses due to private versus shared data:** In ACS, private data that is referenced in the critical section is transferred from the cache of the small core to the cache of the large core. Conventional locking does not incur this overhead. However, conventional systems incur overhead in transferring shared data: the shared data 'ping-pongs' between caches as different threads execute the critical section. ACS eliminates the transfers

of shared data by keeping it at the large core,[2] which can offset the misses it causes to transfer private data into the large core. In fact, ACS decreases cache misses if the critical section accesses more shared data than private data. Note that ACS can improve performance even if there are equal or greater numbers of accesses to private data than shared data because the large core can still (1) improve performance of other instructions and (2) hide the latency of some cache misses using latency tolerance techniques like out-of-order execution.

In summary, the performance benefits of ACS (faster critical section execution, improved lock locality, and improved shared data locality) outweigh its overheads (reduced parallel throughput, CSCALL and CSDONE overhead, and reduced private data locality). This is supported by our extensive experimental results on a wide variety of systems and quantitative analysis of the performance trade-offs ACS makes. These results are presented in Sections 6 and 7 of our 2009 paper (Suleman et al.).

5.5 Results

We evaluate three CMP architectures: a symmetric CMP (SCMP) consisting of all small cores, an asymmetric CMP (ACMP) with one large core with 2-way SMT and remaining small cores, and an ACMP augmented with support for the ACS mechanism (ACS). All comparisons are at an equal-area budget. We use 12 critical-section-intensive parallel applications, including sorting, MySQL databases, data mining, IP packet routing, web caching, and SPECjbb. Details of our methodology can be found in our paper (Suleman et al. 2009).

5.5.1 Performance at Optimal Number of Threads

We evaluate ACS when the number of threads is set equal to the optimal number of threads for each application-configuration pair. Figure 5.4 shows the execution time of SCMP and ACS normalized to ACMP at area budgets of 8, 16, and 32. ACS improves performance compared to ACMP at all area budgets. At an area budget of 8, SCMP outperforms ACS because ACS is unable to compensate for the loss in throughput due to fewer parallel threads. However, this loss diminishes as the area budget increases. At area budget 32,

---

[2]By keeping shared data in the large core's cache, ACS reduces the cache space available to shared data compared to conventional locking (where shared data can reside in any on-chip cache). This can increase cache misses. However, we find that such cache misses are rare and do not degrade performance because the private cache of the large core is usually large enough.

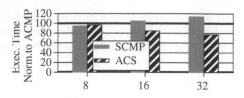

FIGURE 5.4

Execution time (normalized to ACMP) when number of threads is equal to optimal number of threads for each application

ACS improves performance by 34% compared to SCMP, and 23% compared to ACMP.

5.5.2 Performance When the Number of Threads Equals the Number of Contexts

When an estimate of the optimal number of threads is unavailable, most systems spawn as many threads as there are available thread contexts. Having more threads increases critical section contention, thereby further increasing ACS's performance benefit. Figure 5.5 summarizes the execution time of SCMP and ACS normalized to ACMP when the number of threads is equal to the number of contexts for area budgets of 8, 16, and 32. ACS improves performance compared to ACMP for all area budgets. ACS improves performance over SCMP for all area budgets except 8, where accelerating critical sections does not make up for the loss in throughput. For an area budget of 32, ACS outperforms both SCMP and ACMP by 46% and 36%, respectively.

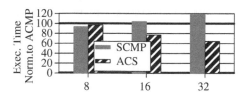

FIGURE 5.5

Execution time (normalized to ACMP) when number of threads is equal to number of thread contexts

5.5.3 Application Scalability

We evaluate the impact of ACS on application scalability (the number of threads at which performance saturates). Figure 5.6 shows the speedup curves of ACMP, SCMP, and ACS over one small core as the area budget is varied from 1 to 32. The curves for ACS and ACMP start at 4 because they require

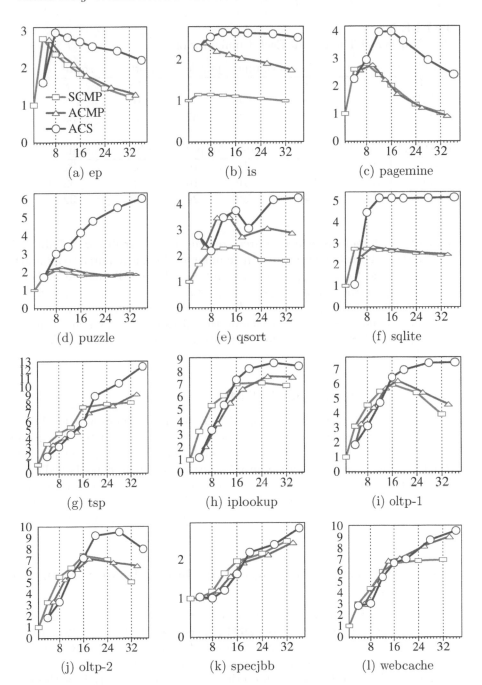

FIGURE 5.6

Speedup over a single small core. The horizontal axis shows area (small core); the vertical axis shows the speedup versus a small core.

TABLE 5.1
Best Number of Threads for Each Configuration

Workload	ep	is	pagemine	puzzle	qsort	sqlite
SCMP	4	8	8	8	16	8
ACMP	4	8	8	8	16	8
ACS	4	12	12	32	32	32

Workload	tsp	iplookup	oltp-1	oltp-2	specjbb	webcache
SCMP	32	24	16	16	32	32
ACMP	32	24	16	16	32	32
ACS	32	24	32	24	32	32

at least one large core which is area-equivalent to 4 small cores. Table 5.1 summarizes Figure 5.6 by showing the number of threads required to minimize execution time for each application using ACMP, SCMP, and ACS. For 7 of the 12 applications (is, pagemine, puzzle, qsort, sqlite, oltp-1, and oltp-2) ACS improves scalability. For the remaining applications, the scalability is not affected. We conclude that if thread contexts are available on the chip, ACS uses them more effectively compared to both ACMP and SCMP.

5.5.4   ACS on Symmetric CMP

Part of the performance benefit of ACS is due to the improved locality of shared data and locks. This benefit can be realized even in the absence of a large core. A variant of ACS can be implemented on a symmetric CMP, which we call *symmACS*. In symmACS, one of the small cores is dedicated to executing critical sections. On average, symmACS reduces execution time by only 4%, which is much lower than the 34% performance benefit of ACS. We conclude that most of the performance improvement of ACS comes from using the large core.

5.5.5   ACS versus Techniques to Hide Critical Section Latency

Several proposals try to hide the latency of a critical section, e.g., transactional memory (TM) (Herlihy and Moss 1993), speculative lock elision (SLE) (Rajwar and Goodman 2001), transactional lock removal (TLR) (Rajwar and Goodman 2002), and speculative synchronization (SS) (Martínez and Torrellas 2002). These proposals execute critical sections speculatively with other instances of the same critical section *as long as they do not have data conflicts with each other*. In contrast, ACS accelerates *all* critical sections regardless of their length and the data they are accessing. Furthermore, unlike TM, ACS

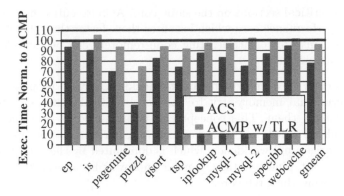

FIGURE 5.7
ACS versus TLR performance

(1) does not require code modification, (2) incurs a significantly lower hardware overhead, and (3) reduces the cache misses for shared data, a feature unavailable in all previous schemes.

We compare the performance of ACS and TLR. Figure 5.7 shows the execution time of an ACMP augmented with TLR[3] and the execution time of ACS normalized to ACMP (area budget is 32 and number of threads set to the optimal number for each system). ACS outperforms TLR on all benchmarks. TLR reduces average execution time by 6% while ACS reduces it by 23%. We conclude that ACS is a compelling, higher-performance alternative to state-of-the-art mechanisms that transparently aim to reduce the overhead of critical sections.

5.6 Contributions and Impact

This chapter proposes the idea of accelerating critical sections. The proposed mechanism promises higher performance and power-efficient execution of multithreaded workloads that include critical sections in their parallel regions. In addition to introducing the ACS architecture, this paper introduces the following concepts that can impact future research and development in parallel programming and CMP architectures.

A new approach to handling critical sections: Current systems treat critical sections as a part of the normal thread code and execute them in place, i.e., on the same core where they are encountered. Instead of exe-

---

[3]TLR was implemented as described in Rajwar and Goodman (2002). We added a 128-entry buffer to each small core to handle speculative memory updates.

cuting critical sections on the same core, ACS executes them on a remote high-performance core which executes them faster than the other smaller cores. This idea of accelerating code remotely can be extended to other program portions.

The concept of *accelerating* bottlenecks: Current proposals, such as transactional memory, try to *parallelize* critical sections. Instead, ACS *accelerates* them. This concept can be extended to other critical paths in multithreaded programs. For example, performance of pipeline parallel (i.e., streaming) workloads, which is dominated by the execution speed of the slowest pipeline stage, can be improved by accelerating the slowest stage using the large core.

The trade-off between shared and private data: Many believe that executing a part of the code at a remote core will increase the number of cache misses. ACS shows that this assumption is not always true. Remotely executing code can actually reduce the number of cache misses if more shared data is accessed than private data (as in the case of most critical sections). This trade-off can be used by compiler/run-time schedulers to decide whether to run a code segment locally or remotely.

Impact on programmer productivity: Correctly inserting critical sections, and shortening them, is among the most challenging tasks in parallel programming. By executing critical sections faster, ACS reduces the programmers' burden: programmers can write code with longer critical sections (which is easy to get correct) and rely on hardware for critical section acceleration.

Impact on parallel algorithms: ACS could enable the use of more efficient parallel algorithms that are traditionally not used in favor of algorithms with shorter critical sections. For example, ACS could make practical the use of memorization tables (data structures to cache computed values for avoiding computation redundancy) that are avoided in parallel programs because their use introduces long critical sections with poor shared data locality.

Impact on CMP architectures: Our previous work proposed ACMP with a single large core and that the large core should run the serial Amdahl's bottleneck (Suleman et al. 2007). Thus, ACMP, without ACS, is inapplicable for parallel programs with non-existent (or small) serial portions (e.g., servers). ACS provides a mechanism to leverage one or more large core(s) in the parallel program portions, which makes the ACMP applicable to programs with or without serial portions. This makes ACMP practical for multiple market segments.

Alternate ACS implementations: We proposed a combined hardware/ software implementation of ACS. Future research can develop other implementations of ACS for systems with different trade-offs/requirements.

For example, ACS can be implemented solely in software as a part of the runtime/library. The software-only ACS requires no hardware support and can be used in today's systems, but it has a higher overhead for sending CSCALLs to the remote core(s). Other implementations of ACS can be developed for SMP and multisocket systems.

In summary, ACS provides a high-performance and simple-to-implement substrate that enables the more productive and easier development of parallel programs, a key problem facing multicore systems and computer architecture today. As such, our proposal opens up a new research area while enabling the software and hardware industries to make an easier transition to many-core engines. We believe that ACS will not only impact future CMP designs but also make parallel programming more accessible to the average programmer.

5.7   Related Previous Work

The most closely related work to ACS is the numerous proposals to optimize the implementation of lock acquire/release operations and the locality of shared data in critical sections using operating system and compiler techniques. We are not aware of any work that speeds up the execution of critical sections using more aggressive execution engines.

5.7.1   Improving Locality of Shared Data and Locks

Sridharan et al. (2006) propose a thread scheduling algorithm to increase shared data locality. When a thread encounters a critical section, the operating system migrates the thread to the processor that has the shared data. While ACS also improves shared data locality, it does not require the OS intervention and the thread migration overhead incurred by their scheme. Moreover, ACS accelerates critical section execution, a benefit unavailable in Sridharan et al. (2006). Trancoso and Torrellas (1996) and Ranganathan et al. (1997) improve locality in critical sections using software prefetching. These techniques can be combined with ACS for improved performance. Primitives (e.g., Test&Test&Set, Compare&Swap) implement lock acquire and release operations (Culler, Singh, and Gupta 1999) efficiently but they do not increase the speed of critical section processing or the locality of shared data.

5.7.2   Hiding the Latency of Critical Sections

Several schemes (Rajwar and Goodman 2001, 2002; Martínez and Torrellas 2002) hide critical section latency. We compared ACS with TLR, a mechanism that hides critical section latency by overlapping multiple instances of the same critical section as long as they access disjoint data. ACS largely outperforms TLR because, unlike TLR, ACS accelerates critical sections whether or not they have data conflicts.

5.7.3   Asymmetric CMPs and CoreFusion

Several researchers (Morad et al. 2006; Suleman et al. 2007; Hill and Marty 2008; Annavaram, Grochowski, and Shen 2005) show that there is potential in improving the performance of the serial part of an application using an Asymmetric Chip Multiprocessor (ACMP). We use the ACMP to accelerate critical sections as well as the serial part in multithreaded workloads. Ipek et al. (2007) show that multiple small cores can be combined, i.e., fused, to form a powerful core to speed up the serial program portions. Our technique can be combined with their scheme where the powerful core accelerates critical sections. R. Kumar et al. (2005) use heterogeneous cores to reduce power and increase throughput for multiprogrammed workloads, not multithreaded programs.

5.7.4   Remote Procedure Calls

The idea of executing critical sections remotely on a different processor resembles the *Remote Procedure Call (RPC)* mechanism (Birrell and Nelson 1984) used in network programming. RPC is used to execute client subroutines on remote server computers. ACS is different from RPC: ACS runs critical sections remotely on the *same chip* within the *same address space*, not on a remote computer.

5.8   Conclusion

We propose *Accelerated Critical Sections (ACS)* to speed up the execution of critical sections in multithreaded applications. We believe that ACS will not only impact future CMP designs but also make parallel programming more accessible to the average programmer.

Part III

Memory Systems

6

TMbox: A Flexible and Reconfigurable Hybrid Transactional Memory System

Nehir Sonmez, Oriol Arcas, and Osman S. Unsal

Barcelona Supercomputing Center, Universitat Politècnica de Catalunya, Barcelona, Spain

Adrián Cristal

Barcelona Supercomputing Center/CSIC — Spanish National Research Council, Spain

Satnam Singh

Google Inc., Mountain View, CA, USA

CONTENTS

In this chapter, we present the design and implementation of TMbox: A multiprocessor system-on-chip (MPSoC) built to explore trade-offs in multicore design space and to evaluate recent parallel programming proposals such as Transactional Memory (TM). For this work, we evaluate a 16-core Hybrid

Transactional Memory implementation based on the TinySTM-ASF (Software Transactional Memory – Advanced Synchronization Facility) proposal on a Virtex-5 FPGA and we accelerate three benchmarks written to investigate TM trade-offs. Our flexible system, composed of MIPS R3000 compatible cores, is easily modifiable to study different architecture, library, or operating system extensions.

6.1   Introduction

Since the first 'core mitosis' in the processor market in 2005, the multicore era has implied a drastic change in computer architecture, demanding better use of thread-level parallelism (TLP) and new ways of providing concurrency in a chip multiprocessor (CMP).

With the always-increasing frequencies of typical uniprocessors, the investigation of architectural schemes has been realized by software-based microarchitectural simulators, such as Simplescalar, Simics, M5, and Ptlsim (Austin, Larson, and Ernst 2002; Magnusson et al. 2002; Binkert et al. 2006; Yourst 2007). Although these sequential simulators are more expressive and it's relatively easy and fast to make changes to the system in a high-level environment, little effort has been made to parallelize or accelerate these programs, which turn out to be slow, to simultaneously simulate the multiple cores of a typical multiprocessor of the current era of chip multicores. This has caused the computer architecture community to consider performing emulations on reconfigurable fabrics instead of using software-based simulations.

6.1.1   FPGAs for Architectural Investigation

A recent alternative for exploring new generations of multicores is based on building a multiprocessor system-on-chip (MPSoC). This approach enables the emulation of large parallel architectures on top of a reconfigurable FPGA platform whose speed and process technology (currently 28 nm) are evolving faster than application-specific integrated circuits (ASICs). Today's FPGA systems can integrate multiple hard/soft processor cores, multiported static RAM (SRAM) blocks, high-speed digital signal processor (DSP) units, and programmable I/O interfaces along with the configurable fabric of logic cells.

Indeed, FPGAs can be good alternatives to implement complex computer circuitry. On-chip block RAM (BRAM) resources on an FPGA which are optionally pre-initialized or with built-in error correction (ECC) can be used in many configurations, such as RAM or SRAM (for implementing on-chip instruction/data cache, direct mapped, or set associative; cache tags, cache coherence bits, snoop tags, register file, multiple contexts, branch target caches, return address caches, branch history tables, etc.), content addressable memory (CAM) (for reservation stations, out-of-order instruction issues, fully associative translation look-aside buffers (TLBs), etc.), ROM (bootloader, lookup tables, etc.) or asynchronous FIFO (to buffer data between processors, peripherals, or coprocessors)(Jan Gray 1998). BRAM capacity, which does not occupy general-purpose Look-Up Table (LUT) space or flip-flops could be used to implement debug support tables for breakpoint address/value registers, count registers, or memory access history. Special on-chip DSP blocks can be cascaded to form large multipliers/dividers or floating-point units. Complete architectural inspection of the memory and processor subsystems can be performed using statistics counters embedded in the FPGAs without any overhead.

Many vendors provide large FPGA programming boards and high-end FPGA prototyping boxes with preferential pricing for academia. FPGAs have already been proposed to teach computer architecture courses for simple designs as well as for more advanced topics (Jan Gray 2000; Manzke and Brennan 2004; Thacker 2010a).

Nowadays it is possible to prototype large architectures in a full-system environment, which allows for faster and more productive hardware research than software simulation. Over the past decade, the RAMP project has already established a well-accepted community vision and various novel FPGA architecture designs (Chiou et al. 2006; Chung et al. 2008; Dave, Pellauer, and Emer 2006; Krasnov et al. 2007; Njoroge et al. 2007; Tan et al. 2010). Another advantage of FPGA emulation over software simulation is out-of-the-way profiling and the possibility for a variety of debugging options.

Another advantage of using FPGAs are the already-tested and readily available models of open IP cores. There exists a plethora of open-source synthesizable Register-Transfer Level (RTL) models of various x86, MIPS, PowerPC, SPARC architectures that can already include detailed specifications for multilevel cache hierarchy, out-of-order issues, speculative execution, and Floating Point (FP) cores, and branch prediction can run at up to 100 MHz. These open processor soft cores are excellent resources to start building a credible multicore system for any kind of architectural or systems research.

One direction is to choose a well-known architecture like MIPS and enjoy the commonly available toolchains and library support. Although supporting a minimal OS might be acceptable, a deeper software stack could have many advantages by providing memory protection, performing scheduling, aiding debugging, file system support, etc. Full OS support can be accomplished by highly detailed design implementations, as well as with hybrid approaches where a nearby host computer serves system calls and exceptions, instead of implementing them in the FPGA model (Chung et al. 2008).

6.1.2 Transactional Memory

A proposal that has drawn considerable attention for programming shared-memory CMPs has been the use of Transactional Memory (TM), an attractive paradigm for deadlock-free execution of parallel code without using locks. Locks are prone to deadlock or priority inversion while TM provides optimistic concurrency by executing atomic transactions in an all-or-none manner. The programmer encapsulates critical sections inside the `atomic{}` construct and the underlying TM mechanism automatically detects data inconsistencies and aborts and restarts one or more transactions. If there are no inconsistencies, all the side effects of a transaction are committed as a whole.

Transactional Memory can be implemented in hardware (HTM) (Chafi et al. 2007; Moore et al. 2006), which is fast but resource-bounded while it might require changes to the caches and the instruction set architecture (ISA); or software (STM) (Felber, Fetzer, and Riegel 2008) which can be flexible, run on off-the-shelf hardware, albeit at the expense of lower performance. To have the best of both worlds, there are intermediate Hybrid TM (HyTM) proposals where transactions first attempt to run on hardware, but are backed off to software when hardware resources are exceeded, and hardware-assisted STM (HaSTM) which by architectural means aims to accelerate a TM implementation that is controlled by software. By leaving the policy to software, different experimentations on contention management, deadlock and livelock avoidance, data granularity, and nesting can be accomplished. Hardware-accelerated STMs may provide performance some way between HTM and STM. HaSTM does not implement any TM semantics in hardware, but provides mechanisms that accelerate an STM, which may have uses beyond TM. Architectural support can also be included to accelerate all transactions, including those that exceed the cache size, those that span OS scheduling quanta, nested transactions, and those that were interrupted by the garbage collector.

6.1.3   Contributions

This flexible experimental systems platform on an FPGA offers a multiprocessor System-on-Chip (SoC) implementation that (i) can be configurable for integrating various Instruction Set Architecture (ISA) options and hardware organizations, (ii) fit and scale well for large designs, (iii) offer high enough performance, (iv) run at least some minimal OS, and (v) must provide credible results. The applications that run on real hardware should provide the researcher with acceptably fast, wide-ranging exploration of hardware/software options and head-to-head comparisons to determine the trade-offs between different implementations. Despite the fact that FPGA emulators of many complex architectures of various ISAs have been proposed, only a few of these are on research on TM, and only up to a small number of cores. Furthermore, the majority of these proposals are based on proprietary or hard processor cores, which imply rigid pipelines that can prevent an architect from modifying the ISA and the microarchitecture of the system.

In this chapter, we present TMbox, a shared-memory CMP prototype with Hybrid TM support. More specifically, our contributions are as follows:

- A description of the first 16-core implementation of a Hybrid TM that is completely modifiable from top to bottom. This implies convenience to study hardware/software tradeoffs in emerging topics like TM.

- We describe how we construct a multicore with MIPS R3000 compatible cores, interconnect the components in a bi-directional ring with backwards invalidations, and adapt the TinySTM-ASF hybrid TM on our infrastructure.

- We present experimental results for three TM benchmarks designed to investigate trade-offs in TM.

The next section presents the TMbox architecture, Section 6.3 explains the Hybrid TM implementation, and Section 6.4 the results of running three benchmarks on TMbox. Related work can be found in Section 6.5, and Section 6.6 concludes this work.

6.2   The TMbox Architecture

The basic processing element of TMbox is the Honeycomb CPU core, a heavily modified and extended version of the Plasma soft core (Rhoads 2001). The synthesizable MIPS R2000-compatible soft processor core Plasma was designed for embedded systems and written in VHDL. It features a configurable 2/3 stage pipeline, a 4 KB direct-mapped write-through L1 cache, and can address up to 64 MB of RAM. It was designed to run at a clock speed of 25 MHz, and it includes UART and Ethernet IP cores. It was chosen to be used in this work because it is based on the popular MIPS architecture, it is complete, and it has a relatively small area footprint on the FPGA. Such reduced instruction set computer (RISC) architectures with simpler pipelines are more easily customizable and require fewer FPGA resources compared to a deeply pipelined superscalar processor, so they are more appropriate to be integrated into a larger multiprocessor SoC.

To effectively upgrade the MIPS R2000-compatible Plasma to our MIPS R3000-compatible Honeycomb, we made several changes to the Plasma soft core:

- Design and implementation of two coprocessors: CP0, which provides support for virtual memory using a Translation Lookaside Buffer (TLB), and CP1 encapsulating an FPU,

- Optimization of the cores to make better use of the resources on our Virtex-5 FPGAs, where it can run at twice the frequency (50 MHz),

- Memory architecture modifications to enable virtual memory addressing for 4 GB and caches of 8 KB,

- Implementation of extra instructions to better support exceptions and thread synchronization (load-linked and store-conditional),

- Development of system libraries for memory allocation, I/O, and string functions, as in Sonmez et al. (2011).

The Honeycomb core (without a floating-point unit (FPU) and the double data rate (DDR) controller) occupies 5,827 LUTs (Table 6.1) on a Virtex-5 FPGA including the ALU, MULT/DIV and Shifter units, the coherent L1 cache and the UART controller: a size comparable to the Xilinx Microblaze soft core. A DDR2 controller that occupies a small portion of the FPGA (around 2%) performs calibration and serves requests (Thacker 2009). Using one controller provides sequential consistency for our multicore since there is only one address bus, and the loads are blocking and stall the processor pipeline.

TABLE 6.1
LUT Occupation of Components of the Honeycomb Core

Component	6-LUTs	Component	6-LUTs
PC_next	138	Mem_ctrl	156
Control	139	Reg_File	147
Bus_mux	155	ALU	157
Shifter	201	MULT	497
Pipeline	112	Cache	1985
TLB	202	TM_unit	1242
Bus_node	619	DDR_ctrl	1119
UART	77		
		TOTAL	**6946**

6.2.1   Interconnection

To interconnect the cores, we designed and implemented a bi-directional ring, as shown in Figure 6.1. Arranging the components on a ring rather than a shared bus requires shorter wires, which eases placement on the chip, relaxing constraints, and is a simple and efficient design choice to diminish the complexities that arise in implementing a large crossbar on FPGA fabric. Apart from increased place and route time, longer wires would lead to more capacitance, longer delay, and higher dynamic power dissipation. Using a ring will also enable easily adding and removing shared components such as an FPU or any application-specific module; however, this is out of the scope of this work.

CPU requests move counterclockwise: they go from the cores to the bus controller, e.g., $CPU_i - CPU_{i-1} - \ldots - CPU_0 - BusCtrl$. Requests may be in the form of a read or a write, carrying a type field, a 32-bit address, a

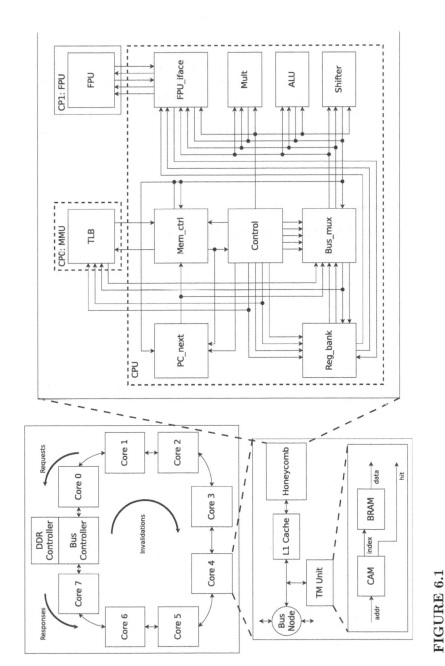

FIGURE 6.1

An 8-core TMbox infrastructure showing the ring bus, the TM Unit, and the processor core

CPU ID, and a 128-bit data field, which is the data word size in our system. Memory responses also move in the same direction, from the bus controller to the cores, e.g., $BusCtrl - CPU_n - CPU_{n-1} - \ldots - CPU_{i+1} - CPU_i$. They use the same channel as requests, carrying responses to the read requests that are served by the DDR Ctrl.

On the other hand, moving clockwise are backwards invalidations caused by the writes to memory, which move from the Bus Ctrl toward the cores in the opposite direction, e.g., $BusCtrl - CPU_0 - \ldots - CPU_{i-1} - CPU_i$. These carry only a 32-bit address and a CPU ID field. When a write request meets an invalidation to the same address on any node, it gets cancelled. Moreover, the caches on each core snoop and discard the lines corresponding to the invalidation address, providing systemwide cache coherency. We detail how we extend this protocol for supporting Hybrid Transactional Memory in the next section.

6.3   Hybrid TM Support for TMbox

TinySTM (Felber, Fetzer, and Riegel 2008) is a lightweight and efficient word-based STM library implementation in C and C++. It differentiates from other STMs such as TL2 and Intel STM mainly by its time-based algorithm and lock-based design. By default, it compiles and runs on 32- or 64-bit x86 architectures, using the atomic_ops library to implement atomic operations, which we modified to include Compare and Swap (CAS) and Fetch and Add (FAA) primitives for the MIPS architecture, using load-linked and store conditional (LL/SC) instructions. TinySTM-ASF is a hybrid port that enables TinySTM to be used with AMD's HTM proposal, ASF (Christie et al. 2010), which we modified to work with TMbox. Our hardware design closely follows the ASF proposal with the exception of nesting support.

A new processor model (-march=honeycomb) was added by modifying GCC and GAS (the GNU Assembler). This new ISA includes all the R3000 instructions plus RFE (Return from Exception), LL, SC, and the transactional instructions in Figure 6.2. All GNU tools (GAS, ld, objdump) were modified to work with these new instructions.

This version starts the transactions in hardware mode and jumps to software if (i) the hardware capacity of the TM unit is exceeded, (ii) there is too much contention, causing many aborts, or (iii) the application explicitly requires it, e.g., in the case of a system call or I/O inside of a transaction.

To enable hardware transactions, we extended our design with a per-core TM unit that contains a transactional cache that only admits transactional loads and stores. By default it has a capacity of 16 data lines (256 bytes). If the TM cache capacity is exceeded, the transaction aborts and sets the TM

TABLE 6.2

HTM Instructions for TMbox

Instruction	Description
XBEGIN (addr)	Starts a transaction and saves the abort address (addr) in TM register $TM0. Also saves the contents of the $sp (stack pointer) to TM register $TM1.
XCOMMIT	Commits a transaction. If it succeeds, it continues execution. If it fails, it rolls back the transaction, sets TM register $TM2 to ABORT_CONFLICT, restores the $sp register, and jumps to the abort address.
XABORT (20-bit)	Used by software to explicitly abort the transaction. Sets TM register $TM2 to ABORT_SOFTWARE, restores the $sp register, and jumps to the abort address. The 20-bit code is stored in the TM register $TM3.
XLW, XSW	Transactional load/store of words (4 bytes).
XLH, XSH	Transactional load/store of halfwords (2 bytes).
XLB, XSB	Transactional load/store of bytes.
MFTM (reg, TM_reg)	Move From TM: Reads from a TM register and writes to a general purpose register.

register $TM2 to ABORT_FULL (explained in the next section), after which the transaction reverts to software mode and restarts.

The transactions are first started in hardware mode with a *XBEGIN* instruction. A transactional load/store causes a cache line to be written to the special TM cache. *XCOMMIT* ends the atomic block and starts committing it to memory. An invalidation of any of the lines in the TM cache causes the current transaction to be aborted. Modifications made to the transactional lines are not sent to memory until the whole transaction successfully commits. The TM unit provides single-cycle operations on the transactional read/write set stored inside. A Content Addressable Memory (CAM) is built using LUTs both to enable asynchronous reads and since BRAM-based CAMs grow superlinearly in resources. Two BRAMs store the data that is accessed by an index provided by the CAM. Additionally, the TM unit can serve LD/ST requests on an L1 miss if the line is found on the TM cache.

In software mode, the STM library is in charge of that transaction, keeping track of read and write sets and managing commits and aborts. This approach enables using the fast TM hardware whenever it is possible, but meanwhile to have an alternative way of processing transactions that are more complex or too large to make use of the TM hardware.

Setting the TM cache size to 64 lines per core causes an extra usage of 1800

LUTs per core on average. Although bigger caches are desirable, we opted for fitting more cores on the FPGA and kept the TM cache size at 16 for our experiments.

The hardware TM implementation supports lazy commits: Modifications made to the transactional lines are not sent to memory until the whole transaction is allowed to successfully commit. However, TinySTM supports switching between eager and lazy committing and locking schemes with software transactions.

6.3.1 Instruction Set Architecture Extensions

To support HTM, we augmented the MIPS R3000 ISA with the new transactional instructions listed in Table 6.2. We have also extended the register file with four new transactional registers, which can only be read with the MFTM (move from TM) instruction. The $TM0 register contains the abort address, $TM1 has a copy of the stack pointer for restoring when a transaction is restarted, $TM2 contains the bit field for the abort (overflow, contention, or explicit), and $TM3 stores a 20-bit abort code that is provided by TinySTM, e.g., abort due to malloc/syscall/interrupt inside a transaction, or maximum number of retries reached, etc.

Aborts in TMbox are processed like an interrupt, but they do not cause any traps, instead they jump to the abort address and restore the $sp (stack pointer) in order to restart the transactions. Regular loads and stores should not be used with addresses previously accessed in transactional mode; therefore it is left to the software to provide isolation of transactional data if desired. Load Linked and Store Conditional instructions can be used simultaneously with TM instructions provided that they do not access the same address.

Figure 6.2 shows an atomic increment in TMbox MIPS assembly. In this simple example, the abort code is responsible for checking if the transaction has been retried a maximum number of times, and if there is a hardware overflow (the TM cache is full), and in this case jumps to an error handling code (not shown).

6.3.2 Bus Extensions

To support HTM, two new bus messages were introduced. We added a new type of request, namely, *COMMIT_REQ*, and a new response type, *LOCK_BUS*. When a commit request arrives at the DDR, it causes a backwards *LOCK_BUS* message on the ring, which destroys any incoming write requests from the opposite direction, and locks the bus to grant exclusive access to perform a serialized commit action. All writes are then committed through the 'channel' created, after which the bus is unlocked with another *LOCK_BUS* message, resuming normal operation. More efficient schemes can be supported in the future to enable parallel commits (Chafi et al. 2007).

```
LI     $11, 5            //set max. retries = 5
LI     $13, HW_OFLOW     //reg 13 has err. code
J      $TX

$ABORT:
  MFTM  $12, $TM2         //check error code
  BEQ   $12, $13, $ERR    //jump if HW overflow
  ADDIU $10, $10, 1       //retries++
  SLTU  $12, $10, $11     //max. retries?
  BEQZ  $12, $ERR2        //jump if max. retries

$TX:
  XBEGIN($ABORT)          //provide abort address
  XLW   $8, 0($a0)        //transactional LD word
  ADDi  $8, $8, 1         //a++
  XSW   $8, 0($a0)        //transactional ST word
  XCOMMIT                 //if abort go to $ABORT
```

FIGURE 6.2
TMbox MIPS assembly for `atomic{a++}` (No Operations (NOPs) and branch delay slots are not included for simplicity)

6.3.3   Cache Extensions

The cache state machine reuses the same hardware for transactional and non-transactional loads and stores; however, a transactional bit dictates if the line should go to the TM cache or not. Apart from regular cached load/store, uncached accesses are also supported, as shown in Figure 6.3. Cache misses first make a memory read request to bring the line from the DDR and to wait in *WaitMemRD* state. In the case of a store, the *WRback* and *WaitMemWR* states manage the memory write operations. While in these two states, if an invalidation arrives to the same address, the write will be cancelled. In case of a store-conditional instruction, the write will not be re-issued, and the LL/SC will have failed. Otherwise, the cache FSM will re-issue the write after such a write-cancellation-on-invalidation.

While processing a transactional store inside of an atomic block, an incoming invalidation to the same address causes an abort and possibly the restart of the transaction. Currently our HTM system supports lazy version management: The memory is updated at commit-time at the end of transactions, as opposed to having in-place updates and keeping an undo log for aborting. On the other hand, data inconsistencies in TMbox are detected only during transaction execution (between *XBEGIN* and *XCOMMIT/XABORT*). When the speculative data is being committed to memory, each transactional write committed causes an invalidation signal, which traverses the ring, aborting

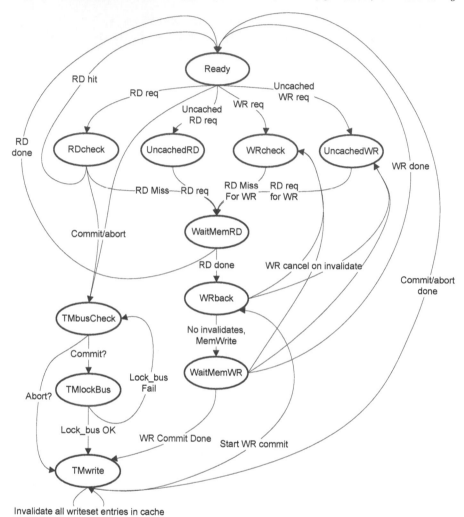

FIGURE 6.3
Cache state diagram. Some transitions (LL/SC) are not shown for visibility.

the transactions that already have those lines in the TM cache. So, once a transaction gets to commit phase, it is sure to commit its speculative changes to memory. To support HTM, the cache state machine is extended with three new states, *TMbusCheck*, *TMlockBus*, and *TMwrite*. One added functionality is to dictate the locking of the bus prior to committing and granting exclusive access of the bus. Another duty is performing burst writes during a success-ful commit which runs through the *TMwrite-WRback-WaitMemWR-TMwrite* loop. The *TMwrite* state is also responsible for the gang clearing of all entries in the TM cache and those TM cache entries that are also found in the L1 cache after a commit/abort. To enable this, address entries that are read from the TM unit are sent to the L1 cache as invalidation requests, after which the TM cache is cleared in preparation for a new transaction.

6.4   Experimental Evaluation

TMbox can fit 16 cores in a Virtex-5 FPGA, occupying 86,797 LUTs (95% of total slices) and 105 BRAMs (49%). In this section, we first examine the trade-offs of our implementation and then discuss the results of executing three TM benchmarks. We used Xilinx ISIM and ModelSim for offline functional simulation and Chipscope Pro for real-time debugging, for which we apply various triggers and hardware debug registers. All results were obtained using a 64-bit Xilinx ISE 12.2 running on RHEL5.

6.4.1   Architectural Benefits and Drawbacks

On the TM side, the performance of our best-effort Hybrid TM is bounded by the size of the transactional cache of the TM unit. Although for this work we chose to use a small, 16-entry TM cache, larger caches can certainly be supported on the TMbox on larger FPGAs (keeping in mind the extra area overhead introduced).

In pure HTM mode, all 16 lines of the TM cache can be used for running the transaction in hardware; however, the benchmark cannot run to completion if there are larger transactions that do not fit in the TM cache, since there is no hardware or software mechanism to decide what to do in this case. The largest overhead related to STMs is due to keeping track of each transactional load/store in software. The situation can worsen when the transactions are large and there are many aborts in the system.

In Hybrid TM mode, it is desired to commit as many transactions as possible on dedicated hardware; however, when this is not possible, it is also important to provide an alternative path using software mechanisms. All transactions that overflow the TM cache will be restarted in software, implying all work done in hardware TM mode is wasted in the end. Furthermore, to enable hybrid execution, TinySTM-ASF additionally keeps the lock variables inside the TM cache. This results in allowing a maximum of 8 variables in the read/write sets of each transaction as opposed to 16 for pure HTM. Of course this is true provided that neither the transactional variables nor the lock variables share the cache lines, in which case, in some executions we observed some transactions having a read/write set of 9 or 10 entries successfully committing in hardware TM mode.

On the network side, the ring is an FPGA-friendly option: We have reduced the place and route time of an 8-core design to less than an hour using the

TABLE 6.3
TM Benchmarks Used

TM Benchmark	Description
Eigenbench[a]	Highly tunable microbenchmark for TM with orthogonal characteristics. We have used this benchmark (2000 loops) with (i) r1=8, w1=2 to overflow the TM cache and vary contention (by changing the parameters a1 and a2) from 0–28%, and (ii) r1=4 and w1=4 to fit in the TM cache and vary the contention between 0–35%.
Intruder[b]	Network intrusion detection. A high abort rate benchmark, contains many transactions dequeuing elements from a single queue. We have used this benchmark with 128 attacks.
SSCA2[c]	An efficient and scalable graph kernel construction algorithm. We have used problem scale = 12.

[a]Hong et al. (2010)
[b]Minh et al. (2008)
[c]Minh et al. (2008)

ring network, whereas it took more than two hours using a shared crossbar for interconnection and we could not fit more than 8 cores (Sonmez et al. 2011). However, each memory request has to travel as many cycles as the total number of nodes on the ring plus the DDR2 latency, during which the CPU is stalled. This is clearly a system bottleneck: Using write-back caches or relaxed memory consistency models might be key in reducing the number of messages that travel on the ring to improve system performance.

On the processor side, the shallow pipeline negatively affects the operating frequency of the CPU. Furthermore, larger L1 caches that cannot fit on our FPGA could be supported on larger, newer generation FPGAs, which would help the system to better exploit data locality. Having separate caches for instructions and data would also be a profitable enhancement.

6.4.2   Experimental Results

Table 6.3 summarizes the three benchmarks that were used to evaluate TM-box. Eigenbench is a synthetic benchmark that can be tuned to discover TM bottlenecks. As Figure 6.4 shows, the transactions in Eigenbench with 2 read − 8 write variables overflow (since TinySTM-ASF keeps the lock variables in the transactional cache) and get restarted in software, exhibiting worse performance than STM. However, the 4 read − 4 write variable version fits in the cache and shows a clear improvement over STM.

In the SSCA2 results presented in Figure 6.5, we get an 1% − 8% improvement over STM because this benchmark contains small transactions that fit in the transactional cache. Although Intruder (Figure 6.6) is a benchmark that is

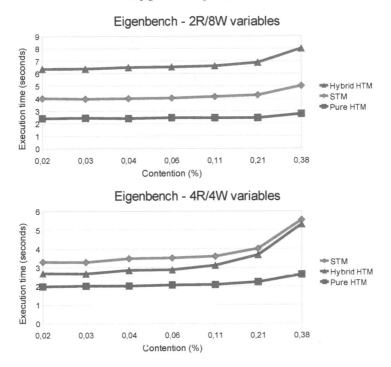

FIGURE 6.4
Eigenbench results on 1–16 cores

frequently used for TM, it is not a particularly TM-friendly benchmark, causing a high rate of aborts and non-scalable performance. However, especially with 16 cores, our scheme (i) discovers conflicts early and (ii) commits 48.7% of the total transactions in hardware, which results in almost 5× superior performance compared to direct-update STM, which has to undo all changes on each abort. This benchmark cannot be run on pure HTM because it contains memory operations like malloc/free inside transactions that are complex to run under HTM and are not supported on TMbox.

These three benchmarks can benefit from our hybrid scheme because they do not run very large transactions, so most of the fallbacks to software caused are due to repeated aborts or mallocs inside transactions. For SSCA2, we see good scalability for up to 8 cores, and for Intruder for up to 4 cores. The performance degradations in STM for Intruder are caused by the fact that the STM directly updates the memory and as the abort rates increase, its performance drastically decreases. Furthermore, the system performance is benchmark-dependent: compared to the sequential versions, the TM versions can perform in the range of 0.2× (Intruder) to 2.4× (SSCA2). We will be looking more into overcoming the limitations of the ring bus and improving on the TM implementation (serialized commits) and the coherency mechanism.

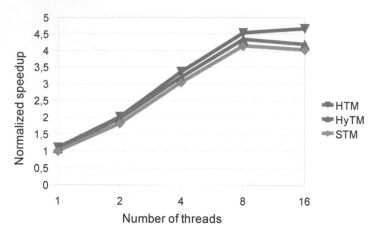

FIGURE 6.5
SSCA2 benchmark results on 1–16 cores

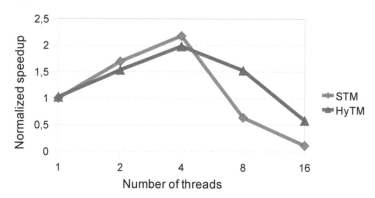

FIGURE 6.6
Intruder benchmark results on 1–16 cores

6.5   Related Work

Few, mostly initial, works have been published in the context of studying Transactional Memory on FPGA prototypes. ATLAS is the first full-system prototype of an 8-way CMP system with PowerPC hard processor cores, buffers for read/write sets, and per-CPU caches augmented with transactional read-write bits and TCC-type HTM support, with a ninth core for running Linux and serving OS requests from other cores (Njoroge et al. 2007).

Pusceddu et al. (2010) describe a TM implementation for embedded systems which can work without caches, using a central transactional controller on four Microblaze cores (Kachris and Kulkarni 2007). TM is used as a simple synchronization mechanism that can be used with higher level CAD tools like EDK for non-cache coherent embedded MPSoC. The proposal occupies a small area on chip, but it is a centralized solution that would not scale as we move up to tens of cores. Similarly, the compact TM proposal, composed of off-the-shelf cores with a software API managing transactions, can be useful for early validation of programs to TM.

Recent work that also utilizes MIPS soft cores focuses on the design of the conflict detection mechanism that uses Bloom filters for an FPGA-based HTM (Labrecque, Jeffrey, and Steffan 2010). Application-specific signatures are compared to detect conflicts in a single pipeline stage. The design takes little area, reducing false conflicts. The underlying bit-level parallelism used for signatures makes this approach a good match for FPGAs. This proposal was the first soft core prototype with HTM albeit only with 2 cores; it is not clear what is done in case of overflow or how the design would scale. Another approach that uses Bloom filters on FPGAs, but to accelerate STMs on commodity machines, was presented by Casper et al. (2011).

Ferri et al. (2010) proposed an energy-efficient HTM on a cycle-accurate software simulator, where transactions can overflow to a nearby victim cache. It is a realistic system with cache coherence and non-centralized TM support, running a wide range of benchmarks on various configurations; however, bus-based snoopy protocol would not scale with more cores, the simulator is not scalable and it would suffer from modeling larger numbers of processors, and no ISA changes are possible to the ARM hard CPU core.

Recently, an HTM was proposed by Thacker (2010b) for the Beehive system. In case of overflow the entire transaction is run under a commit lock without using the transactional hardware. We believe that software transactions might have more to offer. The Beehive design also uses a uni-directional ring where messages are added to the head of a train with the locomotive at the end (Thacker 2010b). Ring networks are suggested as a better architecture for shared memory multiprocessors by Barroso and Dubois (1991) and a cache coherent bi-directional ring was presented by Oi and Ranganathan (1999), but as far as we know, using backwards-propagating write-destructive invalidations is a novel approach. Unlike some of the proposals above, our system features a large number of processors and is completely modifiable, which enables investigating different interconnects, ISA extensions, or coherency mechanisms.

6.6 Conclusions

We have presented a 16-core Hybrid TM design on an FPGA, providing hardware support and accelerating a modern TM implementation and running benchmarks that are widely used in TM research.

The results agree with our insights and findings from other works (Minh et al. 2008): hybrid TM works well when hardware resources are sufficient, providing better performance than software TM. However, when hardware resources are exceeded, performance can fall below the pure software scheme in certain benchmarks. The good news is that Hybrid TM is flexible; a smart implementation should be able to decide what is best by dynamic profiling. We believe that this is a good direction for further research.

We have also shown that a ring network fits well on an FPGA fabric and using smaller cores can help in building larger prototypes. Newer generations of FPGAs will continue to present multicore researchers with interesting possibilities, having become sufficiently mature as to permit investigating credible large-scale systems architecture. We are looking forward to extending the TMbox with a memory directory to utilize all four FPGAs on the BEE3 board.

TMbox can enable the study of many other topics such as shared memory versus distibuted memory, message passing, homogeneous versus heterogeneous processing on different memory models, utilizing various interconnect architectures or ISA extensions. The TMbox is available at `http://www.velox-project.eu/releases`. Furthermore, a low overhead profiling and visualization environment for TMbox was described in Arcas et al. (2012).

Acknowledgments

This work is supported by the cooperation agreement between the Barcelona Supercomputing Center and Microsoft Research, by the Ministry of Science and Technology of Spain and the European Union (FEDER funds) under contract TIN2007-60625 and TIN2008-02055-E, by the European Network of Excellence on High-Performance Embedded Architecture and Compilation (HiPEAC), and by the European Commission FP7 project VELOX (216852).

7

EM²: A Scalable Shared Memory Architecture for Large-Scale Multicores

Omer Khan

Department of Electrical and Computer Engineering, University of Connecticut, Storrs, CT, USA

Mieszko Lis, Keun Sup Shim, Myong Hyon Cho, and Srinivas Devadas

Massachusetts Institute of Technology, Cambridge, MA, USA

CONTENTS

We introduce the concept of deadlock-free migration-based coherent shared memory to the Non-Uniform Cache Access (NUCA) family of architectures. Migration-based architectures move threads among cores to guarantee sequential semantics in large multicores. Using the Execution Migration Machine (EM^2), we achieve performance comparable to directory-based architectures without using directories: avoiding automatic data replication significantly reduces cache miss rates, while a fast hardware, network-level thread migration scheme takes advantage of shared data locality to reduce remote cache accesses that limit traditional NUCA performance.

EM^2 area and energy consumption are very competitive, and, on the average, it outperforms a directory-based baseline by 1.3× and a traditional Shared-NUCA design by 1.2×. We argue that with EM^2, scaling performance has much lower cost and design complexity than in directory-based coherence and traditional NUCA architectures: by merely scaling network bandwidth from 256- to 512-bit flits, the performance of our architecture improves by an additional 13%, while the baselines show negligible improvement.

7.1   Background

Current trends in processor design clearly indicate an era of multicores for the 2010s. As transistor density continues to grow geometrically, processor manufacturers are already able to place a hundred cores on a chip (e.g., Tilera Tile-Gx), with massive multicore chips on the horizon; many industry observers are predicting 1000 or more cores by the middle of this decade (Borkar 2007). Will the current architectures and their memory subsystems scale to hundreds of cores, and will these systems be easy to program?

The main barrier to scaling current memory architectures is the *off-chip memory bandwidth wall* (Borkar 2007; Hardavellas et al. 2009): off-chip bandwidth grows with package pin density, which scales much more slowly than on-die transistor density (*The International Technology Roadmap for Semiconductors: Assembly and Packaging* 2007). Today's multicores integrate very large shared last-level caches on chip to reduce the number of off-chip memory accesses (Rusu et al. 2010); interconnects used with such shared caches, however, do not scale beyond relatively few cores, and the power requirements of large caches (which grow quadratically with size) exclude their use in chips on a 1000-core scale — for example, the Tilera Tile-Gx does not have a large shared cache.

For massive-scale multicores, then, we are left with relatively small per-core caches. Since a programming model that relies exclusively on software-level message passing among cores is inconvenient and so has limited applicability, programming complexity considerations demand that the per-core caches must present a unified addressing space with coherence among caches managed automatically at the hardware level.

On scales where bus-based mechanisms fail, the traditional solution to this dilemma is directory-based cache coherence: a logically central directory coordinates sharing among the per-core caches, and each core cache must negotiate shared (read-only) or exclusive (read/write) access to each line via a complex coherence protocol. In addition to protocol complexity and the associated design and verification costs, directory-based coherence suffers from three other problems: (a) directory sizes must equal a significant portion of the *combined* size of the per-core caches, as otherwise directory evictions will limit performance (Gupta, Weber, and Mowry 1992); (b) automatic replication of shared data significantly decreases the effective total on-chip cache size because, as the core counts grow, a lot of cache space is taken by replicas and fewer lines in total can be cached, which in turn leads to costly protocol-related latencies and sharply increased off-chip access rates; and (c) frequent writes to shared data can result in repeated cache invalidations and the attendant long delays due to the coherence protocol.

Two of these shortcomings have been addressed by S-NUCA (Kim, Burger, and Keckler 2002) and its variants (Fensch and Cintra 2008). These architectures unify the per-core caches into one large shared cache, in their pure form keeping only one copy of a given cache line on chip and thus steeply reducing off-chip access rates compared to directory-based coherence. In addition, because only one copy is ever present on chip, cache coherence is trivially ensured and a coherence protocol is not needed. This comes at a price, however, as accessing data cached on a remote core requires a potentially expensive two-message round-trip: where a coherence protocol would take advantage of spatial and temporal locality by making a copy of the block containing the data in the local cache, S-NUCA must repeat the round-trip *for every access* to ensure sequential memory semantics. Various NUCA and hybrid proposals have therefore leveraged data migration and replication techniques previously explored in the Non-Uniform Memory Access (NUMA) context (e.g., Verghese et al. (1996)) to move private data to its owner core and replicate read-only shared data among the sharers at the operating system level (Cho and Jin 2006; Hardavellas et al. 2009; Awasthi et al. 2009) or aided by hardware (Zhang and Asanović 2005; Chaudhuri 2009; Sudan et al. 2010), but while these schemes improve performance on some kinds of data, they still do not take full advantage of spatio-temporal locality and require either coherence protocols or repeated remote accesses to access read/write shared data.

To address this limitation and take advantage of available data locality in a memory organization where there is only one copy of data, we propose to allow computation threads to migrate from one core to another at a fine-grain instruction level. When several consecutive accesses are made to data assigned to a given core, migrating the execution context allows the thread to make a sequence of local accesses on the destination core rather than pay the performance penalty of the corresponding remote accesses. While computation migration, originally considered in the context of distributed multiprocessor architectures (Garcia-Molina, Lipton, and Valdes 1984), has recently re-emerged at the single-chip multicore level, (e.g., Michaud 2004; Kandemir et al. 2008; Chakraborty, Wells, and Sohi 2006), for power management and fault-tolerance, we are unique in using migrations to provide cache coherence. We also propose a hybrid EM^2 architecture that includes support for NUCA-style remote access.

Specifically, in this chapter we:

1. introduce the idea of using instruction-level execution migration (EM^2) to ensure memory coherence and sequential consistency in directoryless multicore systems with per-core caches;

2. combine execution migration (EM^2) with NUCA-style remote memory accesses (RA) to create a directoryless shared-memory multicore architecture which takes advantage of data locality;

3. utilize a provably deadlock-free hardware-level migration algorithm (Cho et al. 2011) to move threads among the available cores with unprecedented efficiency and generalize it to be applicable to the hybrid EM^2 architecture.

7.2   Migration-Based Memory Coherence

The essence of traditional distributed cache management in multicores is bringing data to the locus of the computation that is to be performed on it: when a memory instruction refers to an address that is not locally cached, the instruction stalls while either the cache coherence protocol brings the data to the local cache and ensures that the address can be safely shared or exclusively owned (in directory protocols), or a remote access is sent and a reply received (in S-NUCA).

Migration-based coherence brings the *computation* to the data: when a memory instruction requests an address not cached by the current core, the execution context (architecture state and translation look-aside buffer (TLB) entries) moves to the core that is *home* for that data. As in traditional NUCA architectures, each address in the system is assigned to a unique core where it may be cached: the physical address space in the system is partitioned among the cores, and each core is responsible for caching its region.

Because each address can be accessed in at most one location, many operations that are complex in a system based on a cache coherence protocol become very simple: sequential consistency and memory coherence, for example, are ensured by default. (For sequential consistency to be violated, multiple threads must observe multiple writes in different order, which is only possible if they disagree about the value of some variable, for example, when their caches are out of sync. If data is never replicated, this situation never arises.) Atomic locks work trivially, with multiple accesses sequentialized on the core where the lock address is located, and no longer ping-pong among core local caches as in cache coherence.

In what follows, we first discuss architectures based purely on remote accesses and purely on execution migrations, and then combine them to leverage the strengths of both.

7.2.1   Remote-Access-Only (RA) Architecture

In the remote-access (RA) architecture, equivalent to traditional S-NUCA, all non-local memory accesses cause a request to be transmitted over the interconnect network, the access to be performed in the remote core, and the data (for loads) or acknowledgment (for writes) to be sent back to the requesting core: when a core C executes a memory access for address A, it must:

1. compute the *home* core H for A (e.g., by masking the appropriate bits);

2. if $H = C$ (a *core hit*),

    (a) forward the request for A to the cache hierarchy (possibly resulting in a dynamic random access memory (DRAM) access);

3. if $H \neq C$ (a *core miss*),

    (a) send a remote access request for address A to core H,

    (b) when the request arrives at H, forward it to H's cache hierarchy (possibly resulting in a DRAM access),

    (c) when the cache access completes, send a response back to C,

    (d) once the response arrives at C, continue execution.

To avoid interconnect deadlock,[1] the system must ensure that all remote requests must always eventually be served. This is accomplished by using an independent virtual network for cache to home core traffic and another for cache to memory controller traffic. Next, within each such subnetwork, the

---

[1]In the deadlock discussion, we assume that events not involving the interconnect network, such as cache and memory controller internals, always eventually complete, and that the interconnect network routing algorithm itself is deadlock-free or can always eventually recover from deadlock.

reply must have higher priority than the request. Finally, network messages between any two nodes within each subnetwork must be delivered in the order in which they were sent.

7.2.2   The Execution Migration Machine (EM^2)

In the execution-migration-only variant (EM^2), all non-local memory accesses cause the executing thread to be migrated to the core where the relevant memory address resides and to be executed there.

What happens if the target core is already running another thread? One option is to allow each single-issue core to round-robin execute several threads, which requires duplicate architectural state (register file, TLB); another is to evict the executing thread and migrate it elsewhere before allowing the new thread to enter the core. Our design features two execution contexts at each core: one for the core's *native* thread (i.e., the thread originally assigned there and holding its private data), and one for a *guest* thread. When an incoming guest migration encounters a thread running in the guest slot, this thread is evicted to its native core.

Thus, when a core C running thread T executes a memory access for address A, it must:

1. compute the *home* core H for A (e.g., by masking the appropriate bits);

2. if $H = C$ (a *core hit*),

   (a) forward the request for A to the cache hierarchy (possibly resulting in a DRAM access);

3. if $H \neq C$ (a *core miss*),

   (a) interrupt the execution of the thread on C (as for a precise exception),

   (b) migrate the microarchitectural state to H via the on-chip interconnect:

      i. if H is the native core for T, place it in the native context slot;
      ii. otherwise:
         a. if the guest slot on H contains another thread T', evict T' and migrate it to its native core N';
         b. move T into the guest slot for H;

   (c) resume execution of T on H, requesting A from its cache hierarchy (and potentially accessing DRAM).

Deadlock avoidance requires that the following sequence always eventually completes:

1. migration of T from $C \rightarrow H$,

2. possible eviction of T' from $H \rightarrow N'$,

3. possible cache $\rightarrow$ DRAM request $H \rightarrow M$, and

4. possible DRAM $\rightarrow$ cache response $M \rightarrow H$.

As with the remote-access-only variant from Section 7.2.1, cache $\leftrightarrow$ memory controller traffic (steps 3 and 4) travels on one virtual network with replies prioritized over requests, and migration messages travel on another. Because DRAM $\rightarrow$ cache responses arrive at the requesting core, a thread with an outstanding DRAM request cannot be evicted until the DRAM response arrives; because this will always eventually happen, however, the eviction will eventually be able to proceed. Eviction migrations will always complete if (a) each thread T' has a unique native core N' which will always accept an eviction migration,[2] and (b) eviction migration traffic is prioritized over migrations caused by core misses. Since core-miss migrations can only be blocked by evictions, they will also always eventually complete, and the migration protocol is free of deadlock. Finally, to avoid migration livelock, it suffices to require each thread to complete at least one CPU instruction before being evicted from a core.

Because combining two execution contexts in one single-issue core may result in round-robin execution of the two threads, when two threads are active on the core they both experience a serialization effect: each thread is executing only 50% of the time. Although this seems like a relatively high overhead, observe that most of the time threads access private data and are executing on their native cores, so in reality the serialization penalty is not a first-order effect.

7.2.3 Hybrid EM² Architecture

In the hybrid EM² architecture, each core-miss memory access may either perform the access via a remote access as in Section 7.2.1 or migrate the current execution thread as in Section 7.2.2. The hybrid architecture is illustrated in Figure 7.1.

For each access to memory cached on a remote core, a decision algorithm determines whether the access should migrate to the target core or execute a remote access. Because this decision must be taken on every access, it must be implementable as efficient hardware. In this chapter, we consider and evaluate a simple heuristic scheme:[3] the DISTANCE scheme. If the migration destination is the native core, the distance scheme always migrates; otherwise, it evaluates the hop distance to the home core. It migrates execution if the distance exceeds some threshold d; otherwise it makes a round-trip remote cache access.

In order to avoid deadlock in the interconnect, migrations must not be blocked by remote accesses and vice versa; therefore, a total of three virtual

---

[2]In an alternate solution, where T' can be migrated to a non-native core such as T's previous location, a domino effect of evictions can result in more and more back-and-forth messages across the network and, eventually, deadlock.

[3]Our future work will evaluate sophisticated hardware-implementable decision schemes.

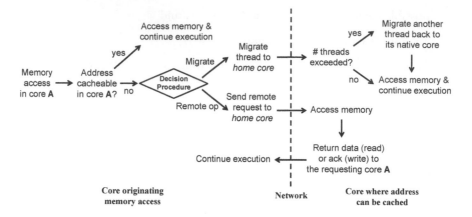

FIGURE 7.1
In the hybrid migration/remote-access architecture, memory accesses to addresses not assigned to the local core cause the execution context to be migrated to the core, or may result in a remote data access.

subnetworks (one for remote accesses, one for migrations, and one for memory traffic) are required. At the protocol level, evictions must now also wait for any outstanding remote accesses to complete in addition to waiting for DRAM → cache responses.

7.2.4   Hardware-Level Migration Framework

The novel architectural component we introduce here is fast, hardware-level migration of execution contexts between two cores via the on-chip interconnect network. Since the core-miss cost is dominated by the remote access cost and the migration cost, it is critical that the migrations be as efficient as possible. Therefore, unlike other thread-migration approaches (such as Thread Motion (Rangan, Wei, and Brooks 2009), which uses special cache entries to store thread contexts and leverages the existing cache coherence protocol to migrate threads), our architecture migrates threads directly over the on-chip interconnection network to achieve the shortest possible migration latencies.

Per-migration bandwidth requirements, although larger than those required by cache-coherent and remote-access-only designs, are not prohibitive by on-chip standards: in a 32-bit x86 processor, the relevant architectural state amounts to about 1.5 Kbits including TLB (Rangan, Wei, and Brooks 2009). In some cases, one may want to migrate additional state, such as branch prediction state and floating point registers, and therefore, we consider both a 1.5 Kbit and 4 Kbit context in Section 7.5.4.

Figure 7.2 shows the differences needed to support efficient execution mi-

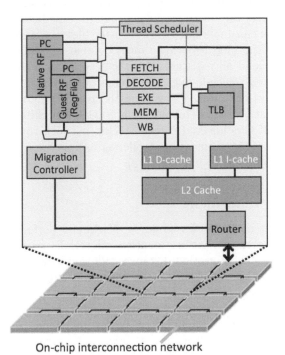

(a) Microarchitecture of a tile with support for a 2-way single-issue multithreaded core for EM²

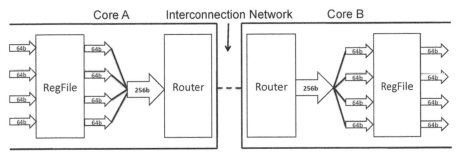

(b) Microarchitecture for a single context transfer in EM²

FIGURE 7.2

Efficient execution migration in a five-stage CPU core. (a) A single-issue five-stage pipeline with efficient context migration; differences from a single-threaded pipeline are shaded. (b) For a context transfer, the register file of the originating core is unloaded onto the router, transmitted across the network, and finally loaded onto the home core's register file via the router.

gration in a single-threaded five-stage CPU core. When both context slots (native and guest) are filled, execution round-robins between them to ensure that all threads can make progress. Register files now require wide read and write ports, as the migration logic must be able to unload all registers onto the network or load all registers from the network in relatively few cycles; to enable this, extra multiplexing logic connects the register files directly with the on-chip network router. The greater the available network bandwidth, the faster the migration. As with traditional S-NUCA architectures, the memory subsystem itself is connected to the on-chip network router to allow for accesses to the off-chip memory controller as well as for reads and writes to a remote cache (not shown in the figure).

7.2.5   Data Placement

The assignment of addresses to cores affects the performance of EM^2 and RA in three ways: (a) because context migrations pause thread execution and therefore longer migration distances will slow down performance; (b) because remote accesses also pause execution and longer round trips will also limit performance; and (c) indirectly by influencing cache performance. On the one hand, spreading frequently used addresses evenly among the cores ensures that more addresses are cached in total, reducing cache miss rates and, consequently, off-chip memory access frequency; on the other hand, keeping addresses accessed by the same thread in the same core cache reduces migration rate and network traffic.

As in standard S-NUCA architectures, the operating system controls memory-to-core mapping via the existing virtual memory mechanism: when a virtual address is first mapped to a physical page, the OS chooses where the relevant page should be cached by mapping the virtual page to a physical address range assigned to a specific core. Since the OS knows which thread causes a page fault, more sophisticated heuristics are possible: for example, in a first-touch-style scheme, the OS can map the page to the thread's *native* core, taking advantage of data access locality to reduce the migration rate while keeping the threads spread among cores.

In migration and remote-access architectures, data placement is key, as it determines the frequency and distance of remote accesses and migrations. Although placement has been studied extensively in the context of NUMA architectures (e.g., Verghese et al. 1996) as well as more recently in the NUCA context (e.g., Hardavellas et al. 2009), we wish to concentrate here on the potential of the EM^2 and remote-access architectures and implement none of them directly. Instead, we combine a first-touch data placement policy (Marchetti et al. 1995), which maps each page to the first core to access it, with judicious profiling-based source-level modifications to our benchmark suite (see Section 7.4.3 and Shim et al. (2011)) to provide placement and replication on a par or better than that of available automatic methods.

7.3   Analytical Models: Directory Coherence versus EM²

Intuitively, replacing off-chip memory traffic due to (hopefully infrequent) cache misses with possibly much more frequent on-chip thread migration traffic or remote accesses may not seem like an improvement. To gain some intuition for where migration-based architectures can win on performance, we consider the average memory latency (AML), a metric that dominates program execution times with today's fast cores and relatively slow memories.

This section presents an analytical model (Figure 7.3) comparing the Modified Shared Invalid (MSI) directory-based protocol (*DirCC* below) with our execution migration-only NUCA variant (*EM²* below). For clarity, we analyze the relatively simple MSI protocol and show that, for the OCEAN_CONTIGUOUS benchmark (from the SPLASH-2 suite (Woo et al. 1995)), it results in memory accesses 1.6× slower than the migration-only architecture; naturally, a suitably more complicated form of the same analysis applies to more complex protocols such as MOSI,[4] but still shows them to be slower than even a basic, migration-only design. The system parameters used here are summarized in Tables 7.1 and 7.2.

7.3.1   Interconnect Traversal Costs

Both protocols incur the cost of on-chip interconnect transmissions to retrieve data from memory, migrate thread contexts (in EM²), and communicate among the caches (in DirCC). In the interconnect network model we assume a 16×16 mesh with two-cycle-per-hop 256-bit flit pipelined routers, an average distance of 12 hops with network congestion overheads, consistent with what we observed for OCEAN_CONTIGUOUS, making the network transit cost

$$cost_{\rightarrow, DirCC} = 24 + 12 = 36, \text{ and } cost_{\rightarrow, EM2} = 24 + 12 = 36. \quad (7.17)$$

Delivering a message adds a load/unload latency dependent on the packet size: for example, transmitting the 1.5 Kbit EM² context requires

$$cost_{context\ xfer} = 36 + \frac{1536 \text{ bits}}{256 \text{ bits}} + 3 = 45. \quad (7.18)$$

By the same token, in both DirCC and EM², transmitting a single-flit request takes 37 cycles ($cost_{core \rightarrow dir}$ and $cost_{core \rightarrow mem}$) and transferring a 64-byte cache line needs 38 cycles ($cost_{dir \rightarrow core}$ and $cost_{mem \rightarrow core}$).

7.3.2   Off-Chip Memory Access Costs

In addition to the DRAM latency itself, off-chip accesses may experience a queueing delay due to contention for the DRAM itself; moreover, retrieving

---

[4]Modified Owned Shared Invalid, a variant of the MSI cache coherence protocol.

$$AML_{DirCC} = cost_{\$\ access,\ DirCC} + rate_{\$\ miss,\ DirCC} \times cost_{\$\ miss,\ DirCC} \quad (7.1)$$

$$cost_{\$\ access} = cost_{L1} + rate_{L1\ miss} \times cost_{L2} \quad (7.2)$$

$$cost_{rdI,wrI,rdS} = cost_{core \to dir} + cost_{dir\ lookup} + cost_{DRAM}$$
$$+ cost_{dir \to core} + cost_{\$\ insert} \quad (7.3)$$

$$cost_{wrS} = cost_{core \to dir} + cost_{dir\ lookup} + cost_{dir \to core} + cost_{\$\ invalidate}$$
$$+ cost_{core \to dir} + cost_{DRAM} + cost_{dir \to core} + cost_{\$\ insert} \quad (7.4)$$

$$cost_{rdM} = cost_{core \to dir} + cost_{dir\ lookup} + cost_{dir \to core} + cost_{\$\ flush}$$
$$+ cost_{core \to dir} + cost_{DRAM} + cost_{dir \to core} + cost_{\$\ insert} \quad (7.5)$$

$$cost_{wrM} = cost_{core \to dir} + cost_{dir\ lookup} + cost_{dir \to core} + cost_{\$\ flush}$$
$$+ cost_{core \to dir} + cost_{dir \to core} + cost_{\$\ insert} \quad (7.6)$$

$$cost_{\$\ miss,\ DirCC} = rate_{rdI,wrI,rdS} \times cost_{rdI,wrI,rdS} + rate_{wrS} \times cost_{wrS}$$
$$+ rate_{rdM} \times cost_{rdM} + rate_{wrM} \times cost_{wrM} \quad (7.7)$$

$$cost_{DRAM,\ DirCC} = cost_{DRAM\ latency}$$
$$+ cost_{DRAM\ serialization}$$
$$+ cost_{DRAM\ contention} \quad (7.8)$$

$$cost_{message\ xfer} = cost_{\to,\ DirCC} + \left\lceil \frac{pkt\ size}{flit\ size} \right\rceil \quad (7.9)$$

$$cost_{\to,\ DirCC} = \#\ hops \times cost_{per\text{-}hop} + cost_{congestion,\ DirCC} \quad (7.10)$$

(a) MSI cache coherence protocol

$$AML_{EM2} = cost_{\$\ access,\ EM2} + rate_{\$\ miss,\ EM2} \times cost_{\$\ miss,\ EM2}$$
$$+ rate_{core\ miss,\ EM2} \times cost_{context\ xfer} \quad (7.11)$$

$$cost_{\$\ access\ EM2} = cost_{L1} + rate_{L1\ miss,\ EM2} \times cost_{L2} \quad (7.12)$$

$$cost_{\$\ miss,\ EM2} = cost_{core \to mem} + cost_{DRAM} + cost_{mem \to core} \quad (7.13)$$

$$cost_{DRAM,\ EM2} = cost_{DRAM\ latency}$$
$$+ cost_{DRAM\ serialization}$$
$$+ cost_{DRAM\ contention} \quad (7.14)$$

$$cost_{message\ xfer} = cost_{\to,\ EM2} + \left\lceil \frac{pkt\ size}{flit\ size} \right\rceil + cost_{Pipeline\ insertion} \quad (7.15)$$

$$cost_{\to,\ EM2} = \#\ hops \times cost_{per\text{-}hop} + cost_{congestion,\ EM2} \quad (7.16)$$

(b) EM^2

FIGURE 7.3

Average memory latency (AML) costs for our MSI cache coherence protocol
and for EM^2. The significantly less complicated description for EM^2 suggests
that EM^2 is easier to reason about and implement. The description for a
protocol such as Modified Owned Shared Invalid (MOSI) or Modified Owned
Exclusive Shared Invalid (MOESI) would be significantly bigger than for MSI.

TABLE 7.1
Various Parameter Settings for the Analytical Cost Model for the OCEAN_CONTIGUOUS Benchmark

Parameter	DirCC	EM^2
$cost_{L1}$	2 cycles	2 cycles
$cost_{L2}$ *(in addition to L1)*	5 cycles	5 cycles
$cost_{\$\ invalidate}$, $cost_{\$\ flush}$	7 cycles	—
cache line size	64 bytes	64 bytes
average network distance	12 hops	12 hops
$cost_{per\text{-}hop}$	2 cycles	2 cycles
$cost_{congestion}$	12 cycles	12 cycles
$cost_{DRAM\ latency}$	235 cycles	235 cycles
$cost_{DRAM\ serialization}$ (one cache line)	50 cycles	50 cycles
$cost_{dir\ lookup}$	10 cycles	—
flit size	256 bits	256 bits
execution context (32-bit x86)	—	1.5 Kbit
$rate_{L1\ miss}$ / $rate_{\$\ miss}$ (both L1 and L2 miss)	5.8% / 4.8%	2.4% / 0.8%
$rate_{core\ miss,\ EM^2}$	—	21%
$rate_{rdI}$, $rate_{rdS}$, $rate_{rdM}$, $rate_{wrI}$, $rate_{wrS}$, $rate_{wrM}$	31.5%, 21.4%, 12%, 22.4%, 12.6%, 0.1%	

a 64-byte cache line must be serialized over many cycles ((7.8) and (7.14)). Because there were dramatically fewer cache misses under EM^2, we observed relatively little DRAM queue contention (11 cycles), whereas the higher off-chip access rate of DirCC resulted in significantly more contention on average (43 cycles):

$$cost_{DRAM,\ EM^2} = 235 + 50 + 11 = 299,$$
$$cost_{DRAM,\ DirCC} = 235 + 50 + 43 = 331. \tag{7.19}$$

7.3.3 EM^2 Memory Access Latency

Given the network and DRAM costs, it is straightforward to compute the average memory latency (AML) for EM^2, which depends on the cache access cost and, for every cache miss, the cost of accessing off-chip RAM; under EM^2, we must also add the cost of migrations caused by core misses (7.11).

The cache access cost is incurred for every memory request and it depends on how many accesses hit the L1 cache: for EM^2,

$$cost_{\$\ access\ EM^2} = 2 + 2.4\% \times 5 = 2.12. \tag{7.20}$$

Each miss under EM^2 contacts the memory controller, retrieves a cache line

TABLE 7.2

System Configurations Used

Parameter	Settings
Cores	256 in-order, 5-stage pipeline, single issue cores
	2-way fine-grain multithreading
L1 instruction/L1 data/L2 cache per core	32/16/64 KB, 4/2/4-way set associative
Electrical network	2D Mesh, XY routing
	2 cycles per hop (+ contention), 256-bit flits
	1.5 Kbits execution context size
	(similar to Rangan, Wei, and Brooks 2009)
	Context load/unload latency:
	$\left\lceil \dfrac{\text{packet size}}{\text{flit size}} \right\rceil = 6$ cycles
	Context pipeline insertion latency $= 3$ cycles
Data Placement scheme	FIRST-TOUCH, 4 KB page size
Coherence protocol	Directory-based MOESI
	Full-map distributed directories $= 8$
	Entries per directory $= 32{,}768$
	16-way set associative
Memory	30 GB/s bandwidth, 75 ns latency

from DRAM, and sends it to the requesting core:

$$cost_{\text{\$ miss, }EM^2} = 37 + 299 + 38 = 374. \tag{7.21}$$

Finally, then, we arrive at the average memory latency:

$$AML_{EM^2} = 2.12 + 0.8\% \times 374 + 21\% \times 45 = 14.5. \tag{7.22}$$

7.3.4   Directory Coherence Memory Access Latency

Since DirCC does not need to migrate execution contexts, memory latency depends on the cache access and miss costs (7.1). Because fewer accesses hit the L1 cache, even the cache access cost itself is higher than under EM^2:

$$cost_{\text{\$ access}} = 2 + 5.8\% \times 5 = 2.29. \tag{7.23}$$

The cost of a cache miss is much more complex, as it depends on the kind of access (read or write, respectively, *rd* and *wr* below) and whether the line is cached nowhere (*I* below) or cached at some other node in shared (*S*) or modified (*M*) state:

- Non-invalidating requests (75.3% of L2 misses for OCEAN_CONTIGUOUS) — loads and stores with no other sharers, as well as loads when there are other read-only sharers — contact the directory and retrieve the cache line from DRAM:

$$cost_{rdI,wrI,rdS} = 37 + 10 + 331 + 38 + 7 = 423. \qquad (7.24)$$

- Stores to data cached in read-only state elsewhere (12.6%) must invalidate the remote copy before retrieving the data from DRAM: in the best case of only one remote sharer,

$$cost_{wrS} = 37 + 10 + 37 + 7 + 37 + 331 + 38 + 7 = 504. \qquad (7.25)$$

- Loads of data cached in modified state elsewhere (11.9%) must flush the modified remote cache line and write it back to DRAM before sending the data to the requesting core via a cache-to-cache transfer:

$$cost_{rdM} = 37 + 10 + 37 + 7 + 38 + 310 + 38 + 7 = 484. \qquad (7.26)$$

- Stores to data cached in modified state elsewhere (0.1%) must also flush the cache line but avoid a write-back to DRAM by sending the data to the requesting core via a cache-to-cache transfer:

$$cost_{wrM} = 37 + 10 + 37 + 7 + 38 + 38 + 7 = 174. \qquad (7.27)$$

Combining the cases with their respective observed rates (cf. Table 7.1), we arrive at the mean cost of a cache miss for DirCC:

$$
\begin{aligned}
cost_{\$\,miss,\,DirCC} = (31.5\% + 22.4\% + 21.4\%) &\times 423 \\
+12.6\% &\times 504 \\
+11.9\% &\times 484 \qquad (7.28) \\
+0.1\% &\times 174 \\
&= 440,
\end{aligned}
$$

and, finally, the average memory cost for DirCC:

$$AML_{DirCC} = 2.29 + 4.8\% \times 440 = 23.4, \qquad (7.29)$$

over **1.6×** greater than under EM².

Although the memory latency model does not account for some effects (for EM², the possibility of 2:1 serialization when both contexts on a given core are filled, and for DirCC, invalidations due to directory evictions or the extra delays associated with sending invalidations to *many* core caches and waiting for their responses), it gives a flavor for how EM² might scale as the number of cores grows. Centralized effects like off-chip memory contention and network congestion around directories, which limit performance in DirCC, will only increase as the ratio of core count to off-chip memory bandwidth increases. Performance-limiting costs under EM², on the other hand, are either decentralized (core migrations are distributed across the chip) or much smaller (contention for off-chip memory is much lower because cache miss rates are small compared to DirCC), and will scale more gracefully.

7.4   Methods

7.4.1   Architectural Simulation

We use Pin (Bach et al. 2010) and Graphite (Miller et al. 2010) to model the proposed execution migration (EM^2), remote-access (RA), and hybrid EM^2 architectures as well as the cache-coherent (DirCC) baseline. Pin enables runtime binary instrumentation of parallel programs, including the SPLASH-2 benchmarks we use here (Woo et al. 1995); Graphite implements a tile-based multicore, memory subsystem, and network, modeling performance and ensuring functional correctness.

The default settings used for the various system configuration parameters are summarized in Table 7.2; any deviations are noted when results are reported. In experiments comparing EM^2/RA architectures against DirCC, the parameters for both were identical, except for (a) the memory directories which are not needed for EM^2/RA and were set to sizes recommended by Graphite on the basis of the total cache capacity in the simulated system, and (b) the 2-way multithreaded cores which are not needed for cache-coherent baseline.

To exclude differences resulting from relative scheduling of Graphite threads, data were collected using a homogeneous cluster of machines.

7.4.2   On-Chip Interconnect Model

Experiments were performed using Graphite's model of an electrical mesh network with XY routing with 256-bit flits. Since modern network-on-chip routers are pipelined (Dally and Towles 2004), and 2- or even 1-cycle per hop router latencies (Amit Kumar et al. 2007) have been demonstrated, we model a 2-cycle per hop router delay; we also account for the appropriate pipeline latencies associated with loading and unloading a packet onto the network. In addition to the fixed per-hop latency, contention delays are modeled using a probabilistic model similar to the one proposed in Konstantakopulos et al. (2008).

7.4.3   Application Benchmarks

Our experiments used a set of SPLASH-2 benchmarks: FFT, LU_CONTIGUOUS, OCEAN_CONTIGUOUS, RADIX, RAYTRACE, and WATER-N^2. For the benchmarks for which versions optimized for cache coherence exist (LU and OCEAN (Woo, Singh, and Hennessy 1994; Woo et al. 1995)), we chose the versions that were most optimized for directory-based cache coherence. It is important to note that these benchmarks have been extensively optimized to remove false sharing and improve working set locality, fitting our requirement for the best-case loads for directory coherence.

Application benchmarks tend not to perform well in RA architectures with simple striped data placements (Hardavellas et al. 2009), and sophisticated data placement and replication algorithms like R-NUCA (Hardavellas et al. 2009) are required for fair comparisons. We therefore used the modified SPLASH-2 benchmarks presented in Shim et al. (2011) that represent a reference placement/replication scheme through source-level transformations that are limited to rearranging and replicating the main data structures. As such, the changes do not alter the algorithm used and do not affect the operation of the cache coherence protocol. In fact, the modified benchmarks are about 2% faster than the originals when run on the cache-coherent baseline.

Each application was run to completion using an input set that matched the number of cores used; e.g., we used 4,000,000 keys for RADIX sort benchmark, a 4× increase over the recommended input size. For each simulation run, we tracked the total application completion time, the parallel work completion time, the percentage of memory accesses causing cache hierarchy misses, and the percentage of memory accesses causing migrations. While the total application completion time (wall clock time from application start to finish) and parallel work completion time (wall clock time from the time the second thread is spawned until the time all threads re-join into one) show the same general trends, we focused on the parallel work completion time as a more accurate metric of average performance in a realistic multicore system with many applications.

7.4.4   Directory-Based Cache Coherence Baseline Selection

In order to choose a directory-based coherence (DirCC) baseline for comparison, we considered the textbook protocol with Modified/Shared/Invalid (MSI) states as well as two alternatives: on the one hand, data replication can be completely abandoned by only allowing modified or invalid states (MI); on the other hand, in the presence of data replication, off-chip access rates can be lowered via protocol extensions such as *owned* and *exclusive* states (MOESI) combined with cache-to-cache transfers whenever possible.

To evaluate the impact of these variations, we compared the performance for various SPLASH-2 benchmarks under MSI, MI, and MOESI (using parameters from Table 7.2). As shown in Figure 7.4, MI exhibits by far the worst memory latency: although it may at first blush seem that MI removes sharing and should thus improve cache utilization much like EM^2/RA, in actuality, eschewing the S state only spreads the sharing — and the cache pollution which leads to capacity misses — over time when compared with MSI and MOESI. At the same time, MI gives up the benefits of read-only sharing and suffers many more cache evictions: its cache miss rates were 2.3× greater than under MSI. The more complex MOESI protocol, meanwhile, stands to benefit from using cache-to-cache transfers more extensively to avoid writing back modified data to off-chip RAM, and take advantage of exclusive cache line ownership to speed up writes. Our analysis shows that, while cache-to-cache transfers

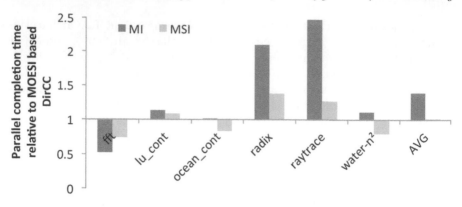

FIGURE 7.4

Parallel completion time under different DirCC protocols, normalized to MOESI. Although parallel completion time for different coherence protocols varies somewhat across the benchmarks (notably, the high directory eviction rate in FFT leads to rampant invalidations in MSI and MOESI and favors MI), generally MOESI was the most efficient protocol and MI performed worst.

result in many fewer DRAM accesses, they instead induce significantly more coherence traffic (even shared reads now take 4 messages); in addition, they come at a cost of significantly increased protocol, implementation, and validation complexity. Nevertheless, since our simulations indicate (Figure 7.4) that MOESI is the best-performing coherence protocol out of the three, we use it as a baseline for comparison in the remainder of this chapter.[5]

Finally, while we kept the number of memory controllers fixed at 8 for all architectures, for the cache-coherence baseline we also examined several ways of distributing the directory among the cores via Graphite simulations: central, one per memory controller, and fully distributed. On the one hand, the central directory version caused the highest queueing delays and most network congestion, and, while it would require the smallest total directory size, a single directory would still be so large that its power demands would put a significant strain on the 256-core chip.[6]At the other end of the spectrum, a fully distributed directory would spread congestion among the 256 cores, but each directory would have to be much larger to allow for imbalances in accesses to cache lines in each directory, and DRAM accesses would incur additional network latencies to contact the relatively few memory controllers. Finally, we considered the case of 8 directories (one for each of the 8 memory controllers), which removed the need for network messages to access DRAM and performed as well as the best-case fully distributed variant. Since the 8-

---

[5]We used MSI in our analytical model (Section 7.3) for simplicity.

[6]Power demands scale quadratically with SRAM size.

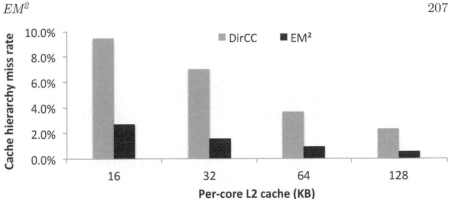

FIGURE 7.5
Cache hierarchy miss rates at various cache sizes show that, by eschewing replication, the EM^2/RA architecture achieves cache miss rates much lower than the DirCC baseline at all cache sizes. (Settings from Table 7.2.)

directory configuration offered best performance and a good tradeoff between directory size and contention, we used this design in our evaluation.

7.4.5 Remote-Access NUCA Baseline Selection

To compare against an RA architecture baseline, we considered two approaches: the traditional S-NUCA approach where the L1 and L2 caches are shared (that is, a local L1 or L2 may cache only a subset of the address space), and a hybrid NUCA/coherence approach where private L1 caches are maintained via a coherence protocol. Although the hybrid variant offers some relief from remote accesses to frequently used locations, the L1 caches must be sized to match the working set of the applications — if the L1 caches are too small, they will suffer frequent misses and the combined performance will revert toward a remote-access-only design. Based on these considerations we chose to compare our hybrid architecture to a fully shared L1/L2 remote-access-only baseline.

7.4.6 Cache Size Selection

We ran our SPLASH-2 simulations with a range of cache sizes under both an execution-migration design and the cache-coherent baseline. While adding cache capacity improves cache utilization and therefore performance for both architectures, cache miss rates are much lower for the migration-based approach and, with much smaller on-chip caches, EM^2/RA achieves significantly better results (Figure 7.5). When caches are very large, on the other hand, they tend to fit most of the working set of our SPLASH-2 benchmarks and both designs almost never miss the cache. This is, however, not a realistic scenario

TABLE 7.3
Area and Energy Estimates

Component	#	Area (mm^2)	Energy (nJ) Read	Write	Details
Register file	256	2.48	0.005	0.002	4-Read, 4-Write ports; 64×24 bits
EM^2 Router	256	15.67	0.022	0.007	5-Read, 5-Write ports; 256×20 bits
RA/DirCC Router	256	7.54	0.011	0.004	5-Read, 5-Write ports; 128×20 bits
Directory cache	8	9.06	1.12	1.23	1 MB cache (16-way assoc)
L2 Cache	256	26.65	0.086	0.074	64 KB (4-way assoc)
L1 Data Cache	256	6.44	0.034	0.017	16 KB cache (2-way assoc)
Off-chip DRAM	8	N/A	6.333	6.322	1 GB RAM

in a system concurrently running many applications: we empirically observed that as the input data set size increases, larger and larger caches are required for the cache-coherent baseline to keep up with the migration-based design. To avoid bias either way, we chose realistic 64 KB L2 data caches as our default configuration because it offers a reasonable performance tradeoff and, at the same time, results in 28 Mbytes of on-chip total cache (not including directories for DirCC).

7.4.7 Instruction Cache

Since the thread context transferred in an EM^2 architecture does not contain instruction cache entries, we reasoned that the target core might not contain the relevant instruction cache lines and a thread might incur an instruction cache miss immediately upon migration. To evaluate the potential impact of this phenomenon, we compared L1 instruction cache miss rates for EM^2 and the cache-coherent baseline in simulations of our SPLASH-2 multithreaded benchmarks.

Results indicated an average instruction cache miss rate of 0.19% in the EM^2 design as compared to 0.27% in the DirCC baseline. The slight improvement seen in EM^2 is due to the fact that non-memory instructions are always executed on the core where the last memory access was executed (since only another memory reference can cause a migration elsewhere), and so non-memory instructions that follow references to shared data are cached only on the core where the shared data resides.

7.4.8   Area and Energy Estimation

For area and energy, we assume 32 nm process technology and use CACTI
(Thoziyoor et al. 2008) to estimate the area requirements of the on-chip caches
and interconnect routers. To estimate the area overhead of extra hardware
context in the 2-way multithreaded core for EM², we used Synopsys Design
Compiler (Synopsys Inc. n.d.) to synthesize the extra logic and register-based
storage. We also use CACTI to estimate the dynamic energy consumption of
the caches, routers, register files, and DRAM. The area and dynamic energy
numbers used in this chapter are summarized in Table 7.3. We implemented
several energy counters (for example, the number of DRAM reads and writes)
in our simulation framework to estimate the total energy consumption of running SPLASH-2 benchmarks for both DirCC and EM². Note that DRAM only
models the energy consumption of the RAM and the I/O pads and pins will
only add to the energy cost of going off-chip.

7.5   Results and Analysis

7.5.1   Advantages over Directory-Based Cache Coherence

Since the main benefit of remote-access and migration architectures over
cache coherence protocols comes from improving on-chip cache utilization by
not replicating writable shared data and minimizing cache capacity/conflict
misses, we next investigated the impact of data sharing in cache-coherent
architectures. With this in mind, we created synthetic benchmarks that randomly access addresses with varying degrees of read-only sharing and read-
write sharing (see Table 7.4). The benchmarks vary along two axes: the fraction of instructions that access read-only data, and the degree of sharing of the
shared data: for example, for read-write shared data, degree d denotes that
this data can be read/written by up to d sharers or threads. We then simulated
the benchmarks using our cache-coherent (MOESI) baseline (Table 7.2), and
measured parallel application performance (which, unlike our memory latency
model above, includes effects not directly attributable to memory accesses like
serialization or cache/directory eviction costs).

Figure 7.6 shows that cache coherence performance worsens rapidly as
the degree of sharing increases. This is for two reasons: one is that a write
to shared data requires that all other copies be invalidated (this explains
the near-linear growth in parallel completion time when most accesses are
writes), and the other is that even read-only sharing causes one address to be
stored in many core-local caches, reducing the amount of cache left for other
data (this is responsible for the slower performance decay of the 100% read-
only benchmarks). These results neatly illustrate the increasing challenge of
data sharing with DirCC designs as the number of cores grows: even in the

TABLE 7.4

Synthetic Benchmark Settings

% of non-memory instructions	70%
% of memory instructions accessing shared data	10%
% of memory instructions accessing private data	20%
% of read-only data in shared data	{25%, 75%, 95%, 100%}
Load:store ratio	2:1
Private data per thread	16 KB
Total shared data	1 MB
Degree of sharing	{1, 2, 4, 8, 32, 64, 128, 256}
Number of instructions per thread	100,000

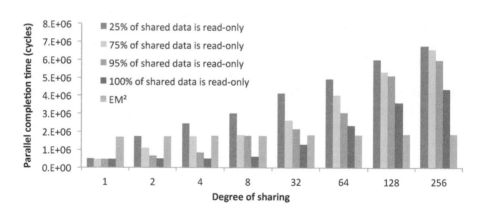

FIGURE 7.6

The performance of DirCC (under a MOESI protocol) degrades as the degree of sharing increases: for read-write sharing this is due to cache evictions, and for read-only sharing to reduced core-local cache effectiveness when multiple copies of the same data are cached. Under EM^2 performance degrades much more slowly.

unrealistic case where *all* shared data is read-only, the higher cache miss rates of cache coherence cause substantial performance degradation for degrees of sharing greater than 32; when more and more of the shared data is read-write, performance starts dropping at lower and lower degrees of sharing. With an EM^2 architecture, on the other hand, each address — even shared by multiple threads — is still assigned to only one cache, leaving more total cache capacity for other data and the performance of EM^2 degrades much more slowly.

Because the additional capacity arises from not storing addresses in many locations, cache miss rates naturally depend on the memory access pattern of specific applications; we therefore measured the differences in cache miss rates for several benchmarks between our EM^2/RA designs and the DirCC baseline. (Note that the cache miss rates are virtually identical for all our EM^2/RA designs). The miss rate differences in realistic benchmarks, shown in Figure 7.7, are attributable to two main causes. On the one extreme, the FFT benchmark does not exhibit much sharing and the high cache miss rate of 8% for MOESI is due mainly to significant directory evictions; since in the EM^2/RA design the caches are only subject to capacity misses, the cache miss rate falls to under 2%. At the other end of the spectrum, the OCEAN_CONTIGUOUS benchmark does not incur many directory evictions but exhibits significant read-write sharing, which, in directory-based cache coherence (DirCC), causes mass invalidations of cache lines actively used by the application; at the same time, replication of the same data in many per-core caches limits effective cache capacity. This combination of capacity and coherence misses results in a 5% miss rate under MOESI; the EM^2/RA architecture eliminates the coherence misses, increases effective cache capacity, and only incurs a 0.8% miss rate. The remaining benchmarks fall in between these two extremes, with a combination of directory evictions and read-write sharing patterns.

Cache miss rates illustrate the core potential advantage of EM^2/RA designs over DirCC: significantly lower off-chip access rates given the same cache sizes. Although miss rates in DirCC architectures can be reduced by increasing the per-core caches, our simulation results (not shown here) indicate that, overall, the DirCC design would need in excess of $2\times$ the L2 cache capacity to match the cache miss rates of EM^2/RA.

7.5.2   Advantages over Traditional NUCA (RA)

Although RA architectures eschew automatic sharing of writable data and significantly lower cache miss rates, their main weakness lies in not being able to take advantage of shared data locality: even if many consecutive accesses are made to data on the same remote core, sequential consistency requires that each be an independent round-trip access. To examine the extent of this problem, we measured the *run length* for non-local memory access: the number of consecutive accesses to memory cached in a non-local core not interrupted by any other memory accesses.

Figure 7.8 shows this metric for two of our benchmarks. Predictably, the

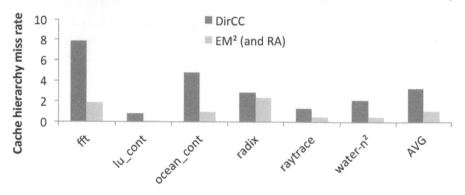

FIGURE 7.7

For our benchmarks, under our EM^2 and RA designs the cache miss rates are on the average 3.2× lower because storing each cache line in only one location eliminates many capacity- and coherence-related evictions and effectively increases the availability of cache lines.

number of remote accesses with a run length of one (a single access to a remote core followed by access to another remote core or the local core) is high; more significantly, however, a great portion of remote memory accesses in both of the benchmarks shown exhibit significant core locality and come in streaks of 40–50 accesses. Although core locality is not this dramatic in all applications, these examples show precisely where a migration-based architecture shines: the executing thread is migrated to a remote core and 40–50 now effectively 'local' memory accesses are made before incurring the cost of another migration.

To examine the real improvement potential offered by extending RA with efficient execution migrations, we next counted the *core miss* rates — the number of times a round-trip remote-access or a migration to a remote core must be made — for the RA baseline and our EM^2 architecture.

Figure 7.9 shows core misses across a range of benchmarks. As we would expect from the discussion above (Section 7.5.2), OCEAN_CONTIGUOUS and LU_CONTIGUOUS show that migrations significantly lower core miss rates, and most other benchmarks also improve. The outlier here is FFT: most of the accesses it makes are to each thread's private data, and shared accesses are infrequent and brief.

Figure 7.10 shows how many overall core misses were handled by remote accesses and migrations in several EM^2/RA variants. In the hybrid EM^2 scheme that performed best, namely, EM^2 (distance = 11) (see Figure 7.11), both migrations and remote access play a significant role, validating our intuition behind combining them into a hybrid architecture.

It is important to note that the cost of core misses is very different under RA and under EM^2: in the first, each core miss induces a *round-trip* remote

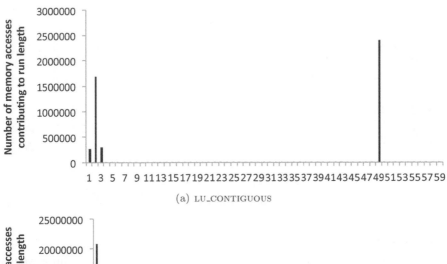

(a) LU_CONTIGUOUS

(b) OCEAN_CONTIGUOUS

FIGURE 7.8
Non-local memory accesses in our RA baseline binned by the number of surrounding contiguous accesses to the same remote core. Although, predictably,
many remote operations access just one address before accessing another core,
a surprisingly large number belong to streaks of 40–50 accesses to the same
remote core and indicate significant data locality.

access, while in the second it causes a *one-way* migration (the return migration, if any, is counted as another core miss). Adding efficient migrations
to an RA design therefore offers significant performance potential, which we
examine next.

7.5.3 Overall Area, Performance, and Energy

The EM^2/RA architectures do not require directories and as can be seen from
Table 7.3 are over $1\,\text{mm}^2$ smaller than the DirCC baseline.

Figure 7.11 shows the parallel completion time speedup relative to the
DirCC baseline for various EM^2/RA schemes: a remote-access-only variant,

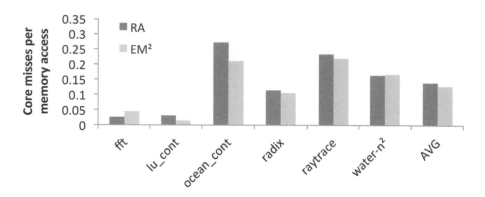

FIGURE 7.9

Per-benchmark core miss rates demonstrate the potential for improvement over the RA baseline. When efficient core-to-core thread migrations are allowed, the number of memory accesses requiring transition to another core (core misses) decreases.

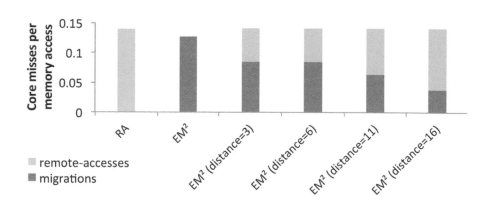

FIGURE 7.10

Core miss rates handled by remote accesses (light) and migrations (dark), averaged over all benchmarks. The fraction of core miss rates handled by remote accesses and migrations in various migration/remote-access hybrids shows that the best-performing scheme, EM^2 (distance = 11), has significant migration and remote access components.

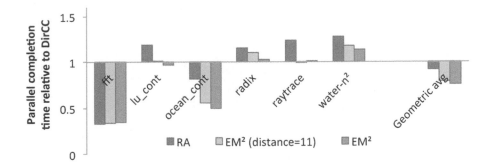

FIGURE 7.11

The performance of EM^2 and RA variants relative to DirCC. Although results vary significantly by benchmark, the best hybrid scheme, namely, EM^2 (distance = 11) outperforms DirCC by 1.25× on average.

a migrations-only variant, and a range of hybrid schemes where the remote-access versus migration decision is based on hop distance. Overall, performance is very competitive with the cache coherent baseline and the EM^2 and best hybrid EM^2 design (distance = 11) show an average 1.3× and 1.25× improvement over the DirCC baseline. Hybrid EM^2 with too small (large) a distance threshold does not perform as well because there are too many (few) migrations. For the chosen context size and hop latency, when distance ≥ 11, migrations become cheaper than data word round-trips, resulting in the best performance. Additionally, EM^2 has 1.2× higher performance than the RA baseline.

The benefits are naturally application-dependent: as might be expected from Figure 7.7, the benchmarks with the largest cache miss rate reductions (FFT and OCEAN_CONTIGUOUS) offer the most performance improvements. At the other extreme, the WATER benchmark combines fairly low cache miss rates under DirCC with significant read-only sharing and is very well suited for directory-based cache coherence; consequently, DirCC outperforms all EM^2/RA variants by a significant margin.

The result also shows the benefits of a combined EM^2/RA architecture: in some benchmarks (e.g., RADIX, LU_CONTIGUOUS, OCEAN_CONTIGUOUS), a migration-only design significantly outperforms remote accesses, while in others (e.g., WATER-N^2) the reverse is true. On average, the best distance-based hybrid EM^2 performs better than either EM^2 or RA and renders the EM^2/RA approach highly competitive with directory-based MOESI cache coherence.

Since energy dissipated per unit performance will be a critical factor in next-generation massive multicores, we employed an energy model (cf. Section 7.4) to estimate the dynamic energy consumed by the various EM^2/RA variants and DirCC. On the one hand, migrations incur significant dynamic

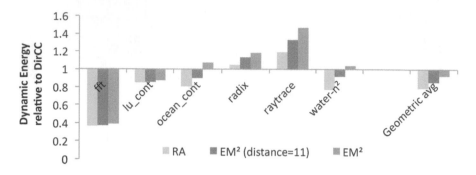

FIGURE 7.12
Dynamic energy usage for all EM^2 and RA variants improve compared to DirCC. Note that for energy consumption, DirCC and RA assume a 128-bit flit size, while EM^2 and the hybrid EM^2 variant utilize a higher bandwidth network with 256-bit flit sizes.

energy costs due to increased traffic in the on-chip network and the additional register file per core; on the other hand, dramatic reductions in off-chip accesses equate to very significant reductions in DRAM access energy.

As illustrated in Figure 7.12, energy consumption depends on each application's access patterns. For FFT, for example, which incurs crippling rates of eviction invalidations, the energy expended by the DirCC protocol messages and DRAM references far outweighs the cost of energy used by remote accesses and migrations. On the other extreme, the fairly random patterns of memory accesses in RAYTRACE, combined with a mostly private-data and read-only sharing paradigm, allows DirCC to efficiently keep data in the core caches and consume far less energy than EM^2/RA. The high cost of off-chip DRAM accesses is particularly highlighted in the WATER-N^2 benchmark: although the trend in cache miss rates between DirCC and EM^2/RA is similar for WATER-N^2 and RAYTRACE, the overall cache miss rate is markedly higher in WATER-N^2; combined with the associated protocol costs, the resulting off-chip DRAM accesses make the DirCC baseline consume more energy than the EM^2/RA architecture.

We note that our energy numbers for directory-based coherence are quite optimistic, since we did not include energy consumed by I/O pads and pins; this will result in higher energy for off-chip accesses of which DirCC makes more.

7.5.4 Performance Scaling Potential for EM^2 Designs

Finally, we investigated the scaling potential of the EM^2 architecture. We reasoned that, while directory-based coherence is limited by cache sizes and

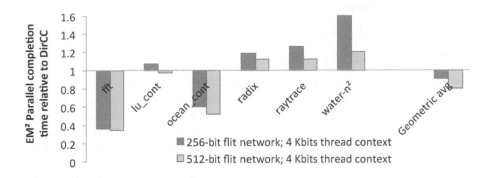

FIGURE 7.13
EM^2 performance scales with network bandwidth. Assuming a modern 64-bit core with 32 general purpose and 16 floating point registers, we calculated a thread context of 4 Kbits (register files plus TLB state). EM^2 still outperforms DirCC by 1.1× on a 256-bit flit network, but when the on-chip network bandwidth is scaled to 512-bit flit, EM^2 outperforms DirCC by 1.25×.

off-chip bandwidth and RA performance is restricted by interconnect *latencies*, EM^2 can be improved by increasing interconnect *bandwidth* (network link and router bandwidth can be scaled by widening data paths, allowing more bits to be transferred in a single cycle): with higher on-chip network bandwidth, the main effect is that messages carrying the thread's context consume fewer cycles.

With this in mind, we evaluated our 256-core system with a much larger context size of 4 Kbits and compared EM^2 against DirCC at our default 256-bit flit network as well as a higher bandwidth 512-bit flit network. As illustrated in Figure 7.13, the performance of EM^2 dropped from a 1.3× advantage over DirCC (using 1.5 Kbits context size) to a 1.1× advantage. When the network bandwidth was doubled to a 512-bit flit size, EM^2 outperformed DirCC by 1.25×. Since scaling of network bandwidth is easy — although buffers and crossbars must be made wider so area increases linearly — the fundamental design of the interconnect remains constant and the clock frequencies are not appreciably affected. Moreover, since the same amount of data must be transferred, dynamic energy consumption does not grow in tandem. Contrasted with the off-chip memory bandwidth wall and quadratically growing power requirements of large caches limiting directory-based architecture performance, and the difficulty in reducing electrical network hop counts limiting remote-access performance on the other hand, an EM^2 or hybrid EM^2 architecture offers an attractive way to significantly increase performance at sublinear impact on cost.

7.6   Related Work

7.6.1   Thread Migration

Migrating computation to the locus of the data is not itself a novel idea. Hector Garcia-Molina in 1984 suggested moving execution to data in memory bound architectures (Garcia-Molina, Lipton, and Valdes 1984). Nomadic threads reduce the number of messages needed to access data in multi-chip computers (Jenks and Gaudiot 1996), and Jenks and Gaudiot (2002) use migration to improve spatial locality of distributed array structures.

In recent years migrating execution context has re-emerged in the context of single-chip multicores. Michaud shows the benefits of using execution migration to improve the overall on-chip cache capacity and utilizes this for migrating selective sequential programs to improve performance (Michaud 2004). Computation spreading (Chakraborty, Wells, and Sohi 2006) splits thread code into segments and assigns cores responsible for different segments, and execution is migrated to improve code locality. Kandemir et al. (2008) present a data migration algorithm to address the data placement problem in the presence of non-uniform memory accesses within a traditional cache coherence protocol. This work attempts to find an optimal data placement for cache lines. A compile-time program transformation based migration scheme is proposed in Hsieh, Wang, and Weihl (1993) that attempts to improve remote data access. Migration is used to move part of the current thread to the processor where the data resides, thus making the thread portion local; this scheme allows the programmer to express when migration is desired. Dataflow machines (e.g., Papadopoulos and Culler 1990) — and, to some extent, out-of-order execution — are superficially similar as they allow an activated instruction to be claimed by any available execution unit, but cannot serve as a shared-memory abstraction. The J-machine (Noakes, Wallach, and Dally 1993) ties processors to on-chip memories, but relies on user-level messaging and does not address the challenge of off-chip memory bandwidth. Our proposed execution migration machine is unique among the previous works because we completely abandon data sharing (and therefore do away with cache coherence protocols). Instead, in this chapter we have proposed to rely on execution migration to provide coherence and consistency.

7.6.2   Remote-Access NUCA and Directory Coherence

Remote memory access is performed in S-NUCA (Kim, Burger, and Keckler 2002) and its variants (Fensch and Cintra 2008): these architectures unify the per-core caches into one large shared cache, in their pure form keeping only one copy of a given cache line on chip and thus steeply reducing off-chip access rates compared to directory-based coherence; in addition, because only one copy is ever present on chip, cache coherence is trivially ensured without a coherence protocol. This comes at the price of a potentially expensive two-message round-trip, as mentioned in the introduction. Various NUCA and hybrid proposals have therefore leveraged data migration and replication techniques previously explored in the NUMA context (e.g., Verghese et al. 1996) to move private data to its owner core and replicate read-only shared data among the sharers at OS level (Cho and Jin 2006; Hardavellas et al. 2009; Awasthi et al. 2009; Boyd-Wickizer, Morris, and Kaashoek 2009) or aided by hardware (Zhang and Asanović 2005; Chaudhuri 2009; Sudan et al. 2010), but while these schemes improve performance on some kinds of data, they still do not take full advantage of spatio-temporal locality and require either coherence protocols or repeated remote accesses to access read/write shared data. In this chapter, we use source-level transformations in conjunction with a first-touch scheme to obtain good data placements, and instruction-level thread migration to take advantage of locality.

We have also compared against MOESI directory-based coherence which is state-of-the-art and includes cache to cache transfers to minimize off-chip memory accesses (e.g., Brown, Kumar, and Tullsen 2007; Kurian et al. 2010). Some recent work has addressed the reduction of directory sizes required for cache coherence in large-scale multicores (e.g., Zebchuk et al. 2009; Zhao, Shriraman, and Dwarkadas 2010). These schemes typically trade off performance for reduced directory area.

7.7   Conclusion

In this chapter, we have extended the family of directoryless NUCA architectures by adding efficient, hardware-level core-to-core thread migrations as a way to maintain sequential consistency and memory coherence in a large multicore with per-core caches. Taking advantage of locality in shared data accesses exhibited by many applications, migrations amortize the cost of remote accesses that limit traditional NUCA performance. At the same time, an execution migration design retains the cache utilization benefits of a shared cache distributed among many cores and brings NUCA performance up to the level of directory-based cache-coherent designs.

We have demonstrated that appropriately designed execution migration (EM^2) and remote cache access (RA) hybrid designs do not cause deadlock. We have explored very straightforward hybrid EM^2 architectures in this chapter; future work involves the development of better-performing migration predictors. Perhaps most promisingly, we have shown that the performance of EM^2 designs is relatively easy to improve with low area cost, little power overhead, and virtually no verification cost, allowing chip designers to easily select the best compromise for their application space.

8

CAFÉ: Cache-Aware Fair and Efficient Scheduling for CMPs

Richard West

Department of Computer Science, Boston University, Boston, MA, USA

Puneet Zaroo

VMware Inc., Palo Alto, CA, USA

Carl A. Waldspurger

Formerly at VMware Inc., Palo Alto, CA, USA

Xiao Zhang

Google, Inc., Mountain View, CA, USA[1]

CONTENTS

Modern chip-level multiprocessors (CMPs) typically contain multiple processor cores sharing a common last-level cache, memory interconnects, and other

---

[1] Xiao Zhang was formerly at VMware Inc. for this work.

hardware resources. Workloads running on separate cores compete for these resources, often resulting in highly variable performance. Unfortunately, commodity processors manage shared hardware resources in a manner that is opaque to higher-level schedulers responsible for multiplexing these resources across workloads with varying demands and importance. As a result, it is extremely challenging to optimize for efficient resource utilization or enforce quality-of-service policies.

Effective cache management requires accurate measurement of per-thread cache occupancies and their impact on performance, often summarized by utility functions such as miss-ratio curves (MRCs). We introduce an efficient online technique for generating MRCs and other cache utility curves, requiring only performance counters available on commodity processors. Building on these monitoring and inference techniques, we also introduce novel methods to improve the fairness and efficiency of CMP scheduling decisions. *Vtime compensation* adjusts a thread's scheduling priority to account for cache and memory system interference from co-runners, and *cache divvying* estimates the performance impact of co-runner placements. We demonstrate the effectiveness of our monitoring and scheduling techniques with quantitative experiments, including both simulation results and a prototype implementation in the VMware ESX Server hypervisor.

8.1   Introduction

Advances in processor architecture have led to a proliferation of multi-core processors, commonly referred to as chip-level multiprocessors (CMPs). Commodity client and server platforms contain one or more CMPs, with each CMP consisting of multiple processor cores sharing a common last-level cache, memory interconnects, and other hardware resources (AMD 2009; Intel Corporation 2009). Workloads running on separate cores compete for these shared resources, often resulting in highly variable or unpredictable performance (Fedorova, Seltzer, and Smith 2006; Kim, Chandra, and Solihin 2004).

Operating systems and hypervisors are designed to multiplex hardware resources across multiple workloads with varying demands and importance. Unfortunately, commodity CMPs typically manage shared hardware resources, such as cache space and memory bandwidth, in a manner that is opaque to the software responsible for higher-level resource management. Without adequate visibility and control over performance-critical hardware resources, it is extremely difficult to optimize for efficient resource utilization or enforce quality-of-service policies.

Many hardware approaches have been proposed to address this problem, introducing low-level architectural mechanisms to support cache occupancy monitoring and/or the ability to partition cache space among multiple workloads (Albonesi 1999; Chang and Sohi 2007; Dybdahl, Stenström, and Natvig 2006; Iyer 2004; Kim, Chandra, and Solihin 2004; Liu, Sivasubramaniam, and Kandemir 2004; Rafique, Lim, and Thottethodi 2006; Ranganathan, Adve, and Jouppi 2000; Srikantaiah, Kandemir, and Irwin 2008; Suh, Rudolph, and Devadas 2004). To further understand the impact of shared caches on workload performance, methods have also been devised to construct cache utility functions, such as miss-ratio curves (MRCs), which capture miss ratios at different cache occupancies (Berg, Zeffer, and Hagersten 2006; Qureshi and Patt 2006; Suh, Devadas, and Rudolph 2001; Suh, Rudolph, and Devadas 2004; Tam et al. 2009). However, existing techniques for generating MRCs either require custom hardware support, or incur non-trivial software overheads.

Constructing cache utility curves is an important step toward effective cache management. To utilize caches more efficiently and provide differential quality of service for workloads, higher-level resource management policies are needed to leverage them. For example, schedulers can exploit cache performance information to make better co-runner placement decisions (Calandrino and Anderson 2008; Suh, Devadas, and Rudolph 2001; Tam, Azimi, and Stumm 2007), improving cache efficiency or fairness. Unfortunately, strict quality-of-service enforcement generally requires hardware support. While software-based page coloring techniques have been used to provide isolation (Cho and Jin 2006; Liedtke, Härtig, and Hohmuth 1997; Lin et al. 2008), such hard partitioning is inflexible and generally prevents efficient cache utilization. Moreover, without special hardware support (Sherwood, Calder, and Emer 1999), dynamically recoloring a page is expensive, requiring updates to page mappings and a full page copy, making this approach unattractive for dynamic workload mixes in general-purpose systems.

We offer an alternative for cache-aware fair and efficient scheduling in a system called *CAFÉ*. Unlike most previous approaches, CAFÉ requires no special hardware support, using only basic performance counters found on virtually all modern processors, including commodity x86 CMPs (AMD 2007; Intel Corporation 2009). Several new cache modeling and inference methods are introduced for accurate cache performance monitoring. Building on this basic monitoring capability, we also introduce new techniques for improving the fairness and efficiency of CMP scheduling decisions.

CAFÉ efficiently computes accurate per-workload cache occupancy estimates from per-core cache miss counts. Occupancy estimates are leveraged to support inexpensive construction of general cache utility curves. For example, miss-ratio and miss-rate curves can be generated by incorporating additional performance counter values for instructions retired and elapsed cycles, avoiding the need for special hardware or memory address traces.

We leverage CAFÉ's cache monitoring infrastructure to perform proper charging for resource consumption, accounting for dynamic interference between co-running workloads within a CMP. A new *vtime compensation* technique is introduced to compensate a workload for interference from co-runners. We also present CAFÉ's *cache divvying* policy for predicting approximate cache allocations during co-runner execution. Using estimated cache utility curves, we are able to determine good co-runner placements to maximize aggregate throughput.

The next section presents our cache occupancy estimation approach, including a detailed description of its mathematical basis, together with simulation results demonstrating its effectiveness. Section 8.3 builds on this foundation, explaining our method for online construction of cache utility curves. Using a prototype implementation in the VMware ESX Server hypervisor, we examine its accuracy by comparing CAFÉ's dynamically generated MRCs with MRCs for the same workloads collected via static page coloring. Section 8.4 introduces our cache-aware scheduling policies: vtime compensation for cache-fair scheduling, and our cache-divvying strategy for estimating the performance impact of co-runner placements. Quantitative experiments in the context of ESX Server show that these schemes are able to improve fairness and efficiency. Related work is examined in Section 8.5. Finally, we summarize our conclusions and highlight opportunities for future work in Section 8.6.

8.2   Cache Occupancy Estimation

In this section, we present our approach for estimating cache occupancy. We begin with a formal explanation of our basic model, which requires only cache miss counts for each co-running thread. We then examine the effects of pseudo-LRU[2] set-associativity as implemented in modern processors and extend our model to additionally incorporate cache hit counts to improve accuracy for such configurations.

We demonstrate the effectiveness of our cache occupancy estimation techniques with a series of experiments in which Standard Performance Evaluation Corporation (SPEC) benchmarks execute concurrently on multiple cores. Since real processors do not expose the contents of hardware caches to software,[3] we measure accuracy using the Intel CMPSched$im simulator (Moses et al. 2009) to compare the results of our model with actual cache occupancies in several different configurations.

For the purposes of our model, we consider a shared last-level cache that may be direct-mapped or n-way set associative. Our objective is to determine the current amount of cache space occupied by some thread, τ, at time t, given contention for cache lines by multiple threads running on all the cores that share that cache. At time t, thread τ may be descheduled, or it may be actively executing on one core while other threads are active on the remaining cores.

---

[2]Least Recently Used.

[3]Current processor families do not allow software to inspect cache tags, although the MIPS R4000 (Heinrich 1994) did provide a `cache` instruction with this capability.

8.2.1   Basic Cache Model

Since hardware caches reveal very little information to software, in order to derive quantitative information about their state, we must rely on inference techniques using features such as hardware performance counters. Virtually all modern processors provide performance counters through which information about various system events can be determined, such as instructions retired, cache misses, cache accesses, and cycle times for execution sequences. Using two events, namely, the *local* and *global* last-level cache misses, we estimate the number of cache lines, E, occupied by thread τ at time t. By global cache misses, we mean the cumulative number of such events across all cores that share the same last-level cache.

We assume that the shared cache is accessed uniformly at random. Results show this to be a reasonable assumption, given the unbiased nature of memory allocation, and the desire for all cache lines to be used effectively across multiple workloads and execution phases. Observe that for n-way set-associative caches, a cache set is selected by using a subset of bits in a memory address, and then a victim cache block within the set is typically chosen using an LRU-like algorithm. Our own observations suggest that n-way set-associative caches in modern multicore processors have some element of randomness to their line replacement policies within sets. In many cases, these policies use some form of binary decision tree as well as a degree of random selection to reduce the bitwise logic when approximating algorithms such as LRU. It is reasonable to assume that randomness will have a greater effect as the number of ways in cache sets is increased in future processors.

In this work, we also assume each cache line is allocated to a single thread at any point in time. Furthermore, we do not consider the effects of data sharing across threads, although this is an important topic for future work.

Cache occupancy is effectively dictated by the number of misses experienced by a thread because cache lines are allocated in response to such misses. Essentially, the current execution phase of a thread τ_i influences its cache investment, because any of its lines that it no longer accesses may be evicted by conflicting accesses to the same cache index by other threads. Evicted lines no longer relevant to the current execution phase of τ_i will not incur subsequent misses that would cause them to return to the cache. Hence, the cache occupancy of a thread is a function of its misses experienced over some interval of time. For subsequent discussion, we introduce the following notation:

- Let C represent the number of cache lines in a shared cache, accessed uniformly at random.

- Let m_l represent the number of misses experienced by the *local* thread, τ_l, under observation over some sampling interval. This term also represents the number of cache lines allocated due to misses.

- Let m_o represent the aggregate number of misses by every thread *other* than τ_l on all cores of a CMP that cause cache lines to be allocated in response to such misses. We use the notation τ_o to represent the aggregate behavior of all other threads, treating it as if it were a single thread.

Theorem 8.1 *Consider a cache of size C lines, with E cache lines belonging to τ_l and $C-E$ cache lines belonging to τ_o at some time, t. If, in some interval, δt, there are m_l misses corresponding to τ_l and m_o misses corresponding to τ_o, then the expected occupancy of τ_l at time $t + \delta t$ is approximately $E' = E + (1 - \frac{E}{C}) \cdot m_l - \frac{E}{C} \cdot m_o$.*

Proof 8.1 *First, at time t, it is assumed that τ_l and τ_o are sufficiently memory-intensive, and have executed for enough time, to collectively populate the entire cache. Now, considering any single cache line, i, at time $t + \delta t$ we have*

$$Pr\{i \text{ belongs to } \tau_l\} =$$
$$Pr\{i \text{ belongs to } \tau_l \mid i \text{ belonged to } \tau_l\} \cdot Pr\{i \text{ belonged to } \tau_l\} +$$
$$Pr\{i \text{ belongs to } \tau_l \mid i \text{ belonged to } \tau_o\} \cdot Pr\{i \text{ belonged to } \tau_o\}$$

This follows from the prior probabilities, at time t:

$$Pr\{i \text{ belonged to } \tau_l\} \quad = \quad \frac{E}{C} \tag{8.1}$$

$$Pr\{i \text{ belonged to } \tau_o\} \quad = \quad 1 - \frac{E}{C} \tag{8.2}$$

Additionally, after $m_l + m_o$ misses, the probability that τ_l replaces line i, previously occupied by τ_o, is one minus the probability that τ_l does not replace τ_o after $m_l + m_o$ misses. More formally,

$$Pr\{\tau_l \text{ replaces } \tau_o \text{ on line } i\} = 1 - \left[1 - \frac{m_l}{C(m_l + m_o)}\right]^{(m_l + m_o)} \tag{8.3}$$

In (8.3), $m_l/[C(m_l+m_o)]$ represents the probability that a miss by τ_l will result in an arbitrary line, i, being populated by contents for τ_l. We know that the probability of a particular line being replaced by a single miss is $1/C$, and the ratio $m_l/(m_l+m_o)$ corresponds to the probability of that miss being caused by one of τ_l's accesses. Note that here we make no assumptions about the order of interleaved memory accesses made by two or more co-running threads. Instead, the ratio $m_l/(m_l + m_o)$ is based on the probability that, among all possible interleaved misses from τ_l and τ_o, τ_l will have the last miss associated with a given cache line.

It follows from (8.3) that the probability of τ_o replacing τ_l on line i at the end of $m_l + m_o$ misses is

$$Pr\{\tau_o \text{ replaces } \tau_l \text{ on line } i\} = 1 - \left[1 - \frac{m_o}{C(m_l + m_o)}\right]^{(m_l + m_o)} \tag{8.4}$$

Therefore,

$$Pr\{i \text{ belongs to } \tau_l \mid i \text{ belonged to } \tau_l\} = 1 - Pr\{\tau_o \text{ replaces } \tau_l \text{ on line } i\}$$

$$= \left[1 - \frac{m_o}{C(m_l + m_o)}\right]^{(m_l + m_o)} \tag{8.5}$$

$$Pr\{i \text{ belongs to } \tau_l \mid i \text{ belonged to } \tau_o\} = Pr\{\tau_l \text{ replaces } \tau_o \text{ on line } i\}$$

$$= 1 - \left[1 - \frac{m_l}{C(m_l + m_o)}\right]^{(m_l + m_o)} \tag{8.6}$$

From (8.1), (8.2), (8.5), and (8.6), we have

$$Pr\{i \text{ belongs to } \tau_l\} = \frac{E}{C} \cdot \left[1 - \frac{m_o}{C(m_l + m_o)}\right]^{(m_l + m_o)}$$

$$+ \left(1 - \frac{E}{C}\right) \cdot \left[1 - \left[1 - \frac{m_l}{C(m_l + m_o)}\right]^{(m_l + m_o)}\right] \tag{8.7}$$

Ignoring the effects of quadratic and higher-degree terms, the first-degree linear approximation of (8.7) becomes

$$Pr\{i \text{ belongs to } \tau_l\} = \frac{E}{C} \cdot (1 - m_o/C) + \left(1 - \frac{E}{C}\right) m_l/C \tag{8.8}$$

This is a reasonable approximation given that $1/C$ is small. Consequently, the expected number of cache lines, E', belonging to τ_l at time $t + \delta t$ is

$$E' = E(1 - m_o/C) + \left(1 - \frac{E}{C}\right) m_l = E + \left(1 - \frac{E}{C}\right) \cdot m_l - \frac{E}{C} \cdot m_o \tag{8.9}$$

This follows from (8.8) by considering the state of each of the C cache lines as independent of all others.

Observe that the recurrence relation in (8.9) captures the changes in cache occupancy for some thread over a given interval of time, with known local and global misses. The terms $[1 - m_o/(C(m_l + m_o))]^{(m_l + m_o)}$ and $[1 - m_l/(C(m_l + m_o))]^{(m_l + m_o)}$ in (8.7) approximate to $e^{-m_o/C}$ and $e^{-m_l/C}$, respectively. Thus, for situations where $m_l + m_o \gg 1$, (8.9) becomes

$$E' = Ee^{-m_o/C} + C(1 - E/C)(1 - e^{-m_l/C}) \tag{8.10}$$

Equation (8.10) is significant in that it shows that the cache occupancy of a thread (here, τ_l) mimics the charge on an electrical capacitor. Given some initial occupancy, E, a growth rate proportional to $(1 - e^{-m_l/C})$ applies to lines currently unoccupied by τ_l. Similarly, the rate of reduction in occupancy (i.e., the equivalent discharge rate in a capacitor) is proportional to $e^{-m_o/C}$.

The linear model in (8.9) is practical for online occupancy estimation, since it consists of an inexpensive computation that requires only the ability to measure per-core and per-CMP cache misses, which is provided by most modern processor architectures. For example, in the Intel Core architecture (Intel Corporation 2009) used for our experiments in Section 8.3, the performance counter event L2_LINES_IN represents lines allocated in the L2 cache, in response to both on-demand and prefetch misses. A mask can be used to specify whether to count misses on a single core or on both cores sharing the cache.

8.2.2   Extended Cache Model for LRU Replacement Policies

So far, our analysis has assumed that each line of the cache is equally likely to be accessed. Over the lifetime of a large set of threads, this is a reasonable assumption. However, commodity CMP configurations feature n-way set-associative caches, and lines within sets are not usually replaced randomly. Rather, victim lines are typically selected using some approximation to a least recently used (LRU) replacement policy. We modified (8.9) to additionally incorporate cache *hit* information, modeling the reduced replacement probability due to LRU effects when lines are reused. Equation (8.9) can be rewritten as

$$E' = E(1 - m_o p_l) + (C - E)m_l p_o \qquad (8.11)$$

where p_l is the probability that a miss falls on a line belonging to τ_l, and p_o is the probability that a miss falls on a line belonging to τ_o. Since (8.9) does not model LRU effects, each line is equally likely to be replaced and $p_l = p_o = 1/C$. In order to model LRU effects, we calculate

$$r_l = (h_l + m_l)/E \qquad (8.12)$$
$$r_o = (h_o + m_o)/(C - E) \qquad (8.13)$$

to quantify the frequency of reuse of the cache lines of τ_l and τ_o, respectively. h_l and h_o represent the number of cache hits experienced by τ_l and τ_o, respectively, in the measurement interval. As with miss counts, these hit counts can be obtained using hardware performance counters available on most modern processors.

When the cache replacement policy is an LRU variant, r_o and r_l approximate the frequency of reuse of the cache lines belonging to τ_0 and τ_1, respectively, since we are unable to precisely know which line is the most recently accessed. Since the probability that a miss evicts a line belonging to a thread is inversely proportional to its reuse frequency, we assume the following relationship:

$$p_o/p_l = r_l/r_o \qquad (8.14)$$

Furthermore, since a miss must fall on some line in the cache with certainty:

$$p_l E + p_o(C - E) = 1 \qquad (8.15)$$

Solving (8.14) and (8.15), we obtain:

$$p_o = r_l/[r_o E + r_l(C - E)] \qquad (8.16)$$
$$p_l = r_o/[r_o E + r_l(C - E)] \qquad (8.17)$$

The values of p_o and p_l obtained from (8.16) and (8.17) can be substituted in (8.11) to obtain the hit-adjusted occupancy estimation model which handles LRU cache replacement effects.

8.2.3 Experiments

We evaluated the cache estimation models on Intel's CMPSched$im simulator (Moses et al. 2009), which supports binary execution and co-scheduling of multiple workloads. This enabled us to measure the accuracy of our cache occupancy models by comparing the estimated occupancy values with the actual values returned by the simulator. The ability to control scheduling allowed us to perform experiments in both under-committed and over-committed scenarios.

By default, the Intel simulator implements a CMP architecture using a pseudo-LRU policy used in modern processors, although it is also configurable to simulate random and other replacement policies. We configured the simulator to use a 3 GHz clock frequency, with private per-core 32 KB 4-way set-associative L1 caches, and a shared 4 MB 16-way set-associative L2 cache. All caches used a 64-byte line size. The number of hardware cores and software threads was varied across different experiments to test the effectiveness of our occupancy estimation models under diverse conditions.

During simulation, the per-core and per-CMP performance counters measuring L2 misses and hits were sampled once per millisecond, after which the occupancy estimates were updated for each software thread. Since cache occupancies exhibit rapid changes at this time scale, we averaged occupancies over 100 millisecond intervals. We plot one value per second for both the estimated and actual occupancy values, in order to display results more clearly over longer time scales. We refer to the miss-based occupancy estimation technique using the basic cache model presented in Section 8.2.1 as method *Estimate-M*. The extended cache model presented in Section 8.2.2 that also incorporates hit information to better model associativity is referred to as method *Estimate-MH*.

Our first experiment tests the effectiveness of the basic Estimate-M method in a dual-core configuration where a 16-way set-associative L2 cache is configured to use a simple random cache line replacement policy instead of pseudo-LRU. Figure 8.1 plots the estimated and actual cache occupancies over time when the two cores were running `mcf` and `omnetpp` from the SPEC CPU2006 benchmark suite. The estimated occupancy for each benchmark tracks its actual occupancy very closely, which is expected since the random replacement policy is consistent with our assumption of random cache access.

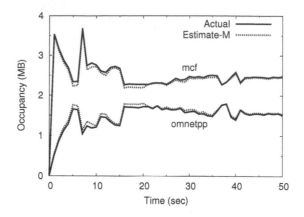

FIGURE 8.1
Accuracy of basic Estimate-M method on a dual-core system with random line replacement policy

Our next experiment evaluates the same workload with the default pseudo-LRU line replacement policy which is used by actual processor hardware. Figure 8.2(a) and (b) plot the estimated and actual cache occupancies over time, for `mcf` and `omnetpp`, respectively, using both the basic Estimate-M and extended Estimate-MH methods. Figure 8.2(c) and (d) present the absolute error between the actual and estimated values. The workloads in this experiment were selected to highlight the difference in accuracy between the two estimation methods, which generally agreed more closely for other workload pairings. In this case, the Estimate-M method is considerably less accurate, often showing a substantial discrepancy relative to the actual occupancies, especially during the interval between 8 and 18 seconds. On the other hand, the hit-adjusted Estimate-MH method, designed to better reflect LRU effects, is much more accurate and tracks the actual occupancies fairly closely.

The remaining experiments focus on the more accurate Estimate-MH method with various sets of co-running workloads. Figure 8.3 presents the results of two separate experiments with different co-running SPEC CPU2006 benchmarks with a dual-core configuration. Figure 8.3(a) and (b) show `mcf` running with `gcc` on the two cores; `omnetpp` and `perlbmk` are co-runners in Figure 8.3(c) and (d). The estimated occupancies match the actual values very closely.

Figure 8.4 shows the cache occupancy over time for four different co-running benchmarks from the SPEC CPU2006 suite in a quad-core configuration. Although not shown, we also conducted similar experiments with other benchmarks from the SPEC CPU2000 and 2006 suites, achieving similar levels of accuracy between estimated and actual values. As with the dual-core results, experiments on a quad-core platform are of similar precision.

We also evaluated the effectiveness of occupancy estimation in an over-

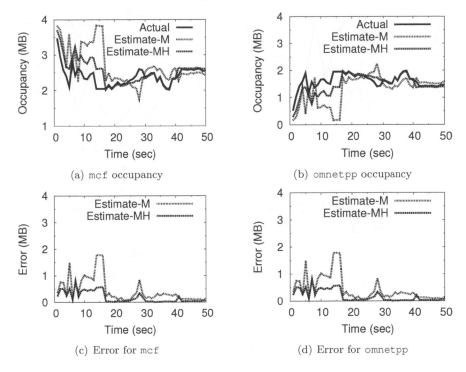

(a) mcf occupancy

(b) omnetpp occupancy

(c) Error for mcf

(d) Error for omnetpp

FIGURE 8.2
Occupancy and estimation error for the Estimate-M and Estimate-MH methods

committed system, in which many software threads are time-multiplexed onto a smaller number of hardware cores. In such a scenario, some threads will be descheduled at various points in time, waiting in a scheduler run queue to be dispatched onto a processor core. In our experiments, we used a 100 millisecond scheduling time quantum, with a simple round-robin scheduling policy selecting threads from a global run queue.

Figures 8.5 and 8.6 show plots of the actual and estimated occupancies over time for an over-committed quad-core system. Together, the two figures show ten software threads running various benchmarks from the SPEC CPU2000 and CPU2006 suites.[4] In the corresponding experiment, the ten threads are scheduled to run on the four cores sharing the L2 cache. The accuracy of occupancy estimation remains high, despite the time-sliced scheduling.

In order to look at the estimation accuracy over shorter time intervals, Figure 8.7 zooms in to examine the first three seconds of execution for the mcf and equake00 workloads from Figure 8.5(a) and (c), respectively. The actual

---

[4]Benchmarks with names ending in 00 are from SPEC CPU2000, while all others are from CPU2006.

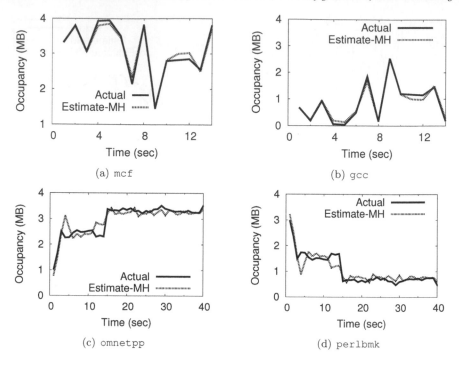

FIGURE 8.3

Two pairs of co-runners in dual-core systems: mcf versus gcc, and omnetpp versus perlbmk

and estimated occupancies are plotted every 100 ms. Estimated occupancy tracks actual occupancy very closely, even during periods when a thread is descheduled and its occupancy falls to zero. Although these fine-grained results are reported for only two of the ten workloads from Figures 8.5 and 8.6, we observed similar behavior for the remaining benchmarks.

8.3 Cache Utility Curves

Central to CAFÉ's resource management framework for fair and efficient scheduling is an understanding of workload-specific cache utility curves. These curves are presented with cache occupancy as the independent variable on the x-axis, and a dependent performance metric on the y-axis, such as the number of cache misses per reference, instruction, or cycle at different occupancies. In this section we explain our technique for lightweight online construction of

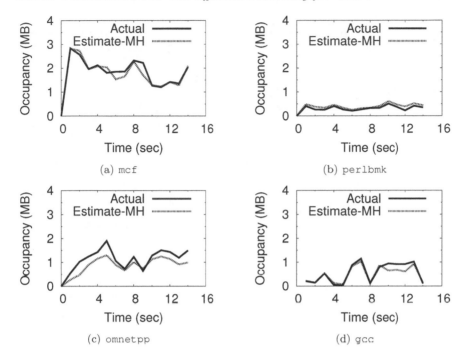

FIGURE 8.4

Cache occupancy over time for four co-runners in a quad-core system

cache utility curves, yielding information about the effect of cache size on expected performance for running workloads. We then present experimental MRC results for a series of benchmarks, using a prototype CAFÉ implementation, and compare them to MRCs collected for the same workloads using static page coloring.

All experiments were conducted on a Dell PowerEdge SC1430 host, configured with two 2.0 GHz Intel Xeon E5535 processors and 4 GB RAM. Each quad-core Xeon processor actually consists of two separate dual-core CMPs in a single physical package. The two cores in each CMP share a common 4 MB L2 cache. We implemented our CAFÉ prototype in the VMware ESX Server 4.0 hypervisor (VMware, Inc. 2009). Each benchmark application was deployed in a separate virtual machine, configured with a single CPU and 256 MB RAM, running an unmodified Red Hat Enterprise Linux 5 guest OS (Linux 2.6.18-8.e15 kernel).

8.3.1 Curve Types

Most work in this area has focused on per-thread *miss-ratio curves* that plot cache misses per memory reference at different cache occupancies (Berg, Zeffer,

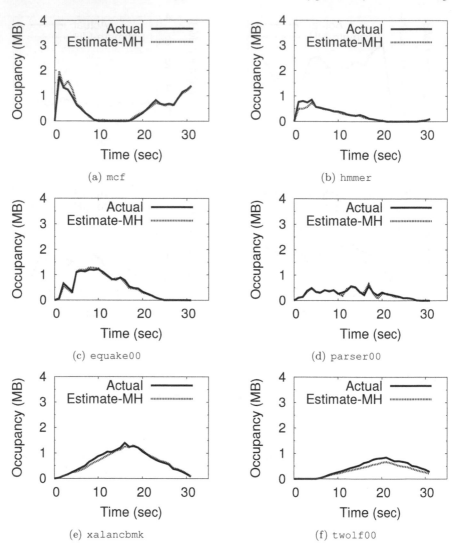

FIGURE 8.5

Occupancy estimation for an over-committed quad-core system (Part 1)

and Hagersten 2006; Qureshi and Patt 2006; Suh, Devadas, and Rudolph 2001; Suh, Rudolph, and Devadas 2004; Tam et al. 2009). Another type of miss-ratio curve plots cache misses per instruction retired at different cache occupancies. We refer to miss-ratio curves in units of misses per kilo-reference as *MPKR* curves and to those in units of misses per kilo-instruction as *MPKI* curves.

It is also possible to construct *miss-rate curves*, defined in terms of misses

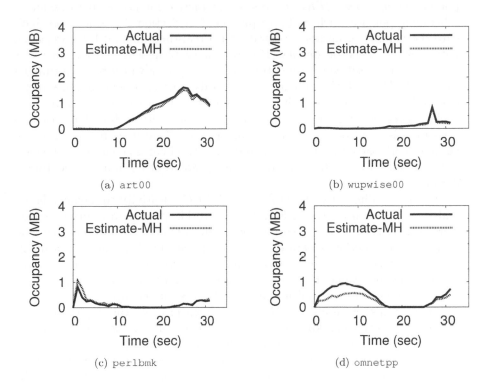

FIGURE 8.6

Occupancy estimation for an over-committed quad-core system (Part 2)

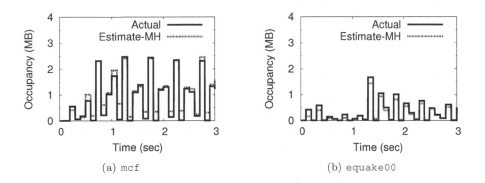

FIGURE 8.7

Fine-grained occupancy estimation in an over-committed quad-core system

per kilo-cycle. Such *MPKC* curves are attractive for use with cache-aware scheduling policies, such as those presented in Section 8.4, since they indicate the number of misses expected over a real-time interval for a workload with a given cache occupancy. However, a problem with MPKC curves is that they are sensitive to contention for memory bandwidth from co-running workloads. Under high contention, workloads start experiencing more memory stalls, throttling back their instruction issue rate, thereby decreasing their cache misses per unit time. Consequently, a cache utility function based on miss rates is dependent on dynamic memory bandwidth contention from co-running workloads. In contrast, MPKR and MPKI curves measure cache metrics that are intrinsic to a workload, independent of co-runners and timing details.

Figure 8.8 illustrates the problem of MPKC sensitivity to memory bandwidth contention using the SPEC2000 mcf workload. Miss-rate curves for mcf were collected using page coloring, but with different levels of memory read bandwidth contention generated by a micro-benchmark running on a different CMP sharing the same memory bus, but not the same cache. For a given cache occupancy value, the miss rates are higher when there is less memory bandwidth contention, resulting in variable miss-rate curves.

One can also generate *CPKI* curves, which measure the impact of cache size on the cycles per kilo-instruction efficiency of a workload. The CPKI metric has the advantage of directly showing the impact of cache size on a workload's performance, reflecting the effects of instruction-level parallelism that help tolerate cache miss latency. However, like MPKC curves, CPKI curves suffer from the problem of co-runner variability due to contention for memory bandwidth or other shared hardware resources.

Since MPKI and MPKR curves do not vary based on memory contention caused by co-runners, they are good candidates for determining a workload's intrinsic cache behavior. In some cases, however, it is also useful to infer the impact on workload performance due to the combined effects of cache and memory bandwidth contention. Therefore CAFÉ generates both MPKI and CPKI curves and utilizes them to guide its higher-level scheduling policies.

8.3.2   Curve Generation

We implemented CAFÉ's online cache-utility curve generation in ESX Server. Utilizing the occupancy estimation method described in Section 8.2, curve generation consists of two components at different time scales: fine-grained occupancy updates and coarse-grained curve construction.

8.3.2.1   Occupancy Updates

Each core updates the cache occupancy estimate for its currently running thread every two milliseconds, using the linear occupancy model in (8.9). A high-precision timer callback reads hardware performance counters to obtain

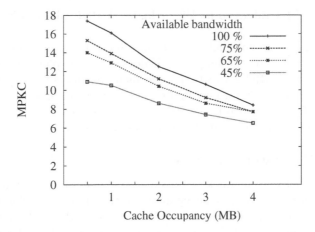

FIGURE 8.8
Effect of memory bandwidth contention on the MPKC miss-rate curve for the SPEC CPU2000 `mcf` workload

the number of cache misses for both the local core and the whole CMP since the last update. In addition to this periodic update, occupancy estimates are also updated whenever a thread is rescheduled, based on the number of intervening cache misses since it last ran.

Our current implementation tracks cache occupancy in discrete units equal to one-eighth of the total cache size. We construct discrete curves to bound the space and time complexity of their generation, while providing sufficient accuracy to be useful in cache-aware CPU scheduling enhancements. During each cache occupancy update for a thread, several performance metrics are associated with its current occupancy level, including accumulated cache misses, instructions retired, and elapsed CPU cycles. Since occupancy updates are invoked very frequently, we tuned the timer callback carefully and measured its cost as approximately 320 cycles on our experimental platform.

8.3.2.2 Generating Miss-Ratio Curves

Miss-ratio curves are generated after a configurable time period, typically several seconds spanning thousands of fine-grained occupancy updates. For each discrete occupancy point, an MPKI value is computed by dividing the accumulated cache misses by the accumulated retired instructions at that occupancy.

MPKI values are expected to be monotonically decreasing with increasing cache occupancy; i.e., more cache leads to fewer misses per instruction. CAFÉ enforces this monotonicity property explicitly by adjusting MPKI values. Preference is given to those occupancy points which have the most updates, since we have more confidence in the performance metrics corresponding to these points. Starting with the most-updated occupancy point with MPKI value m, any lower MPKI values to its left or higher MPKI values to its right are set to m.

Interestingly, monotonicity violations are good indicators of phase changes in workload behavior, although CAFÉ does not yet exploit such hints. We instrumented our MRC generation code, including monotonicity enforcement, and found that it takes approximately 2850 cycles to execute on our experimental platform. The overheads for occupancy estimation and MRC construction are sufficiently low that they can remain enabled at all times.

8.3.2.3 Generating Other Curves

The basic CAFÉ framework is extremely flexible. By recording appropriate statistics with each discrete occupancy point, a variety of different cache performance curves can be constructed. By default, CAFÉ collects cache misses, instructions retired, and elapsed cycles, enabling generation of MPKI, MPKC, and CPKI curves.

We could not experiment with generating MPKR curves, due to limitations of our experimental platform. The Intel Core architecture provides only two programmable counters, which were used to obtain core and whole-CMP cache misses, respectively. MPKI, MPKC, and CPKI curves can be generated by CAFÉ, since retired instructions and elapsed cycles are available as additional fixed hardware counters.

8.3.2.4 Obtaining Full Curves

A key challenge with CAFÉ's approach is obtaining performance metrics at all discrete occupancy points. In the steady state, a group of threads co-running on a shared cache achieve equilibrium occupancies. As a result, the cache performance curve for each thread has performance metrics concentrated around its equilibrium occupancy, leading to inaccuracies in the full cache performance curves.

In addition to passive monitoring, we have explored ways to actively perturb the execution of co-running threads to alter their relative cache occupancies temporarily. For example, varying the group of co-runners scheduled with a thread typically causes it to visit a wider range of occupancy points. An alternative approach is to dynamically throttle the execution of some cores, allowing threads on other cores to increase their occupancies. CAFÉ cannot use frequency and voltage scaling to throttle cores, since in commodity CMPs, all cores must operate at the same frequency (Naveh et al. 2006). However, we did have some success with duty-cycle modulation techniques (Intel Corporation 2009; Zhang, Dwarkadas, and Shen 2009) to slow down specific cores dynamically.

For thermal management, Intel processors allow system code to specify a multiplier (in discrete units of 12.5%) specifying the fraction of regular cycles during which a core should be halted. When a core is slowed down, its co-runners get an opportunity to increase their cache occupancy, while the occupancy of the thread running on the throttled core is decreased. To limit any potential performance impact, we enable duty-cycle modulation during less than 2% of execution time. Experiments with SPEC CPU2000 benchmarks did not reveal any observable performance impact due to cache performance curve generation with duty-cycle modulation.

8.3.3 Experiments

We evaluated CAFÉ's cache curve construction techniques using our ESX Server implementation. We first collected the miss-ratio curves for various SPEC CPU2000 benchmarks (`mcf`, `swim`, `twolf`, `equake`, `gzip`, and `perlbmk`), by running them to completion with access to an increasing number of page colors in each successive run. We then ran all six benchmarks together on a single CMP of the Dell system, with CAFÉ generating the miss-ratio curves, configured to construct the curves at benchmark completion time.

Figure 8.9 compares the miss-ratio curves of the benchmarks obtained by CAFÉ with those obtained by page coloring. In most cases, the MRC shapes and absolute MPKI values match reasonably well. However, in Figure 8.9(a), the MRC generated by CAFÉ for `mcf` is flat at lower occupancy points, differing significantly from the page-coloring results. Even with duty-cycle modulation there is insufficient interference from co-runners to push `mcf` into lower occupancy points. Since there are no updates for these points, the miss-ratio values for higher occupancy points are used as the best estimate due to monotonocity enforcement.

To analyze this further, Figure 8.10 shows separate MRCs generated by CAFÉ for `mcf` with different co-runners, `swim` and `gzip`. The MRC generated when `mcf` is running with `gzip` is flat because `mcf` only has updates at the highest occupancy point. The miss ratio of `mcf` at the highest occupancy point is a factor of sixty more than the miss ratio of `gzip`, which renders duty-cycle modulation ineffective, since it can throttle a core by at most a factor of eight. In contrast, the MRC generated with co-runner `swim` matches the MRC obtained by page coloring closely.

8.3.4 Discussion

Our online technique for MRC construction builds upon our cache occupancy estimation model. While the MRCs generated for a working system in Section 8.3.3 are encouraging, there remain several open issues. By using only commodity hardware features, our MRCs may not always yield data points across the full spectrum of cache occupancies. Duty cycle modulation addresses this problem to some degree, but some sensitivity to co-runner selection may still remain. Although an MPKI curve is intrinsic to a workload, and does not vary based on contention from co-runners, the workload may be prevented from visiting certain occupancy levels due to co-runner interference,

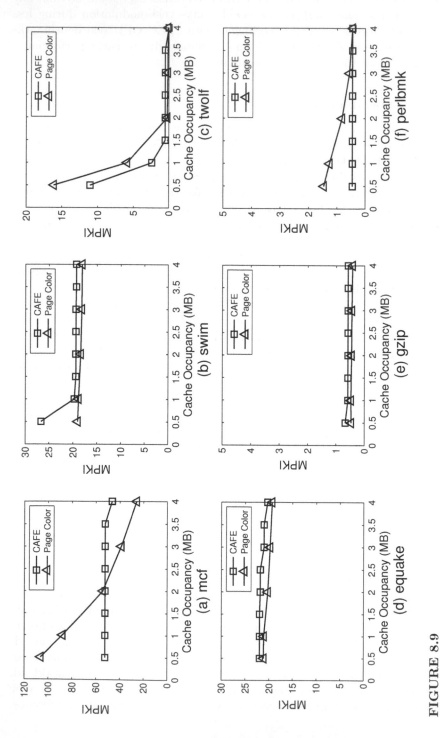

FIGURE 8.9

Miss-ratio curves (MRCs) for various SPEC CPU workloads, obtained online by CAFÉ versus offline by page coloring

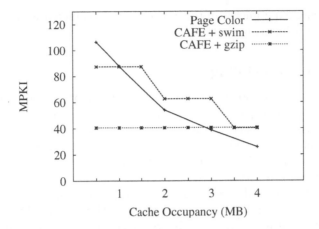

FIGURE 8.10
MRC for `mcf` with different co-runners

as observed in Figure 8.10. In practice, it may be necessary to vary co-runners selectively during some execution intervals, in order to allow a workload to reach high cache occupancies, or alternatively, to force a workload into low occupancy states, depending on the memory demands of the co-runners.

While the experiments in Section 8.3.3 compare offline MRCs with our online approach, they are produced at the time of benchmark completion. This introduces some potential differences between the online and offline curves, since online we plot MPKI values based on the time *during workload execution* at which a given occupancy is reached. We are currently investigating MRCs at different time granularities. Early investigations yield curves that remain stable for an execution phase, but which fluctuate while changing phases. We intend to study how MRCs can be used to identify phase changes as part of future work.

8.4 Cache-Aware Scheduling

In this section, we present higher-level scheduling policies that leverage CAFÉ's low-level methods for estimating cache occupancies and generating cache utility curves. We first examine the issue of fairness in CMPs, and present a new *vtime compensation* technique for improving CMP fairness in proportional-share schedulers. Next, we show how to use cache utility functions for estimating the impact of co-runner placements via a novel *cache divvying* approach. The scheduler considers new co-runner placements periodically, in order to maximize aggregate throughput. Unless otherwise stated, all scheduling experiments in this section were conducted using the same system configuration as in Section 8.3.

8.4.1   Fair Scheduling

Operating systems and hypervisors are designed to multiplex hardware resources across multiple workloads with varying demands and importance. Administrators and users influence resource allocation policies by specifying settings such as priorities, reservations, or proportional-share weights. Such controls are commonly used to provide differential quality of service, or to enforce guaranteed service rates.

When all workloads are assigned equal allocations, *fairness* implies that each workload should receive equal service. More generally, a scheduler is considered fair if it accurately delivers resources to each workload consistent with specified allocation parameters.

Fair scheduling requires accurate accounting of resource consumption, although few systems implement this properly (Zhang and West 2006). For example, if a hardware interrupt occurs in the context of one workload, but performs work on behalf of a different workload, then the interrupt processing cost must be subtracted from the interrupted context and added to the workload that benefited. The VMware ESX Server scheduler (VMware, Inc. 2009), used for our experiments, implements proper accounting for interrupts, bottom halves, and other system processing; we extended this with cache-miss accounting for CAFÉ.

8.4.1.1   Proportional-Share Scheduling

In this work, we focus on *proportional-share* scheduling. Resource allocations are specified by numeric *shares* (or, equivalently, *weights*), which are assigned to threads that consume processor resources.[5] A thread is entitled to consume resources proportional to its share allocation, which specifies its importance relative to other threads.

Most proportional-share scheduling algorithms (Bennett and Zhang 1996; Parekh 1992; Stoica et al. 1996; Waldspurger and Weihl 1995; Zhang and Keshav 1991; Goyal, Vin, and Cheng 1996) use a notion of *virtual time* to represent per-thread progress. Each thread τ_i has an associated virtual time v_i, which advances at a rate that is directly proportional to its resource consumption q_i, and inversely proportional to its share allocation w_i:

$$v_i' = v_i + q_i/w_i \qquad (8.18)$$

The scheduler chooses the thread with the minimum virtual time to execute next. For example, consider threads τ_i and τ_j with share allocations $w_i = 2$

---

[5]Although we use the term *thread* to be concrete, the same proportional-share framework can accommodate other abstractions of resource consumers, such as processes, applications, or VMs.

and $w_j = 1$. Thread τ_i is entitled to execute twice as quickly as τ_j; this 2:1 ratio is implemented by advancing v_i at half the rate of v_j for the same execution quantum q.

Some proportional-share schedulers differ significantly in their treatment of virtual time for threads blocked waiting on I/O or synchronization objects. For example, some algorithms partially credit a thread for time when it was blocked, while others do not. Here, we focus on CPU-bound threads, so these differences are not important; time spent blocking will be addressed in future work.

8.4.1.2   Fair Scheduling for CMPs

How should fairness be defined in the context of a CMP, where multiple processor cores may share last-level cache space, memory bandwidth, and other hardware resources? Accounting based solely on the amount of real time a thread has executed is clearly inadequate, since the amount of useful computation performed by a thread varies significantly with resource contention from co-runners.

One option is to define *cache-fair* as equal sharing of CMP cache space among co-running threads (Fedorova, Seltzer, and Smith 2006). However, this definition does not reflect the marginal utility of additional cache space, which typically differs across threads. For efficiency, we want to allocate more cache space to those threads which can utilize it most productively. Moreover, this definition of cache-fair does not facilitate our goal of proportional-share fairness, where different threads may be entitled to unequal amounts of shared resources.

We instead assume that a thread is entitled to consume *all* shared CMP resources while it is executing, including the entire last-level cache, in the absence of competition from co-running threads. At runtime, we dynamically estimate the actual performance degradation experienced by a thread due to co-runner interference and compensate it appropriately. Since most threads are negatively impacted to some degree by co-runners, this means that most threads will receive at least some compensation.

To quantify fairness, we first define the *weighted slowdown* for each thread to be the ratio of its actual execution time (in the presence of co-running threads) to its ideal execution time when running alone without co-runners, scaled by the thread's relative share allocation. The relative share allocation is, itself, the ratio of the local thread's weight to total weights of all competing threads. We then use the coefficient of variation of these per-thread weighted slowdowns as an unfairness metric; with perfect fairness, all weighted slowdowns are identical.

8.4.1.3   Virtual-Time Compensation

In a proportional-share scheduler, a convenient way to compensate threads for co-runner interference is to adjust the virtual time update in (8.18). In particular, when a thread τ_i is charged for consuming its timeslice, we reduce its consumption q_i to account for the time it was stalled due to contention for shared resources. We call this virtual-time adjustment technique *vtime compensation.*[6] We present two different vtime compensation methods – an initial approach that compensates for conflict misses and an improved method that compensates for negative impacts on cycles per instruction (CPI).

Compensating for Conflict Misses

Our initial attempt at vtime compensation was designed to compensate a thread for conflict misses that it incurred while executing with co-runners and while on a ready queue waiting to be dispatched. We first estimate the cache occupancy that a thread τ_i would achieve without interference from other threads. Starting with (8.9), this reduces to

$$E_{i,NI} = E_i + \left(1 - \frac{E_i}{C}\right) m_i \tag{8.19}$$

where $E_{i,NI}$ represents the expected occupancy of thread τ_i with *no interference* from other threads.

We then use the *miss-rate* curve for τ_i to obtain two values: $M(E_i)$, the miss rate at E_i, and $M(E_{i,NI})$, the miss rate at $E_{i,NI}$, according to (8.9) and (8.19), respectively. Given our monotonicity enforcement for miss-rate curves, it must be the case that $M(E_i) \geq M(E_{i,NI})$.

Taking the difference between these two miss rates over τ_i's most recent timeslice, q_i, provides a measure of the *conflict misses* experienced by the thread. In practice, the latency of a cache miss is not constant, depending on several factors, including prefetching and contention for memory bandwidth. However, if we assume the average latency of a single last-level cache (LLC) miss is L, then we can approximate the stall cycles due to conflict misses, denoted by S_i, as

$$S_i = (M(E_i) - M(E_{i,NI})) \cdot L \tag{8.20}$$

Given this measure of the conflict stall cycles experienced by a thread, we modify the virtual time update from (8.18) accordingly:

$$v'_i = v_i + (q_i - S_i)/w_i \tag{8.21}$$

In (8.21), the updated virtual timestamp, v'_i factors in the amount of time τ_i stalls during its use of a CPU due to conflict misses with other threads. The number of conflict misses considers both the time during which τ_i executes *and* the time it waits for the CPU, since during this time its cache state may be evicted by other threads. This method of virtual time compensation attempts

---

[6]Similar compensation approaches could be used in proportional-share schedulers that are not based on virtual time. For example, in probabilistic lottery scheduling (Waldspurger and Weihl 1994), the concept of 'compensation tickets' introduced to support non-uniform quanta could be extended to reflect co-runner interference.

to benefit those threads that are affected by cache interference by reducing their effective resource consumption, which increases their scheduling priority.

Unfortunately, this approach requires miss-rate curves, which, as explained in Section 8.3, are difficult to derive accurately in the presence of co-runners competing for limited memory bandwidth. A related problem is modeling the average cache miss latency L, which may vary due to contention for memory bandwidth.

Compensating for Increased CPI

To address these issues, we revised our vtime compensation strategy to simply determine the *actual* cycles per instruction, CPI_{actual}, at the current occupancy, as well as the *ideal* cycles per instruction CPI_{ideal} if the thread were to experience no resource contention from other threads. We obtain CPI_{ideal} from the value at full occupancy in the CPKI curve. This is more robust than simply measuring the minimum observed CPI value, because the CPKI curve captures the average value over an interval, reducing sensitivity to phase transitions. Hence, our revised virtual time adjustment for τ_i becomes

$$v'_i = v_i + \frac{CPI_{ideal}}{CPI_{actual}} \cdot q_i/w_i \qquad (8.22)$$

This approach effectively replaces the use of miss-ratio curves with cache performance curves that provide CPI values at different cache occupancies (i.e., CPKI instead of MPKI). As a result, it reflects contention for *all* shared CMP resources, including memory interconnect bandwidth. Thus, compensating for negative impacts on CPI is simpler and more accurate than compensating only for cache conflict misses.

8.4.1.4 Vtime Compensation Experiments

We implemented vtime compensation in the VMware ESX Server hypervisor. ESX Server implements a proportional-share scheduler that employs a virtual-time algorithm similar to those described in Section 8.4.1.1. Our experiments ran two instances each of four different SPEC2000 benchmark applications: `mcf`, `swim`, `twolf`, and `equake`. In this case, we restricted all software threads to run on one package of the Dell PowerEdge machine, as described in Section 8.3. This meant that four cores were overcommitted with eight threads that were scheduled by the ESX Server hypervisor. The hypervisor was responsible for the assignment of threads to cores.

In Figure 8.11(a), all benchmark instances had equal share allocations, while in Figure 8.11(b), a 2:1 share ratio was specified for the two instances of each application. To evaluate the efficacy of vtime compensation, we measured per-application weighted slowdown, as defined in Section 8.4.1.2. The overall slowdown was calculated as the arithmetic mean of the weighted slowdowns of all the applications. Although CAFÉ only slightly reduces the average slowdown, it significantly reduces the variation in slowdowns experienced by all

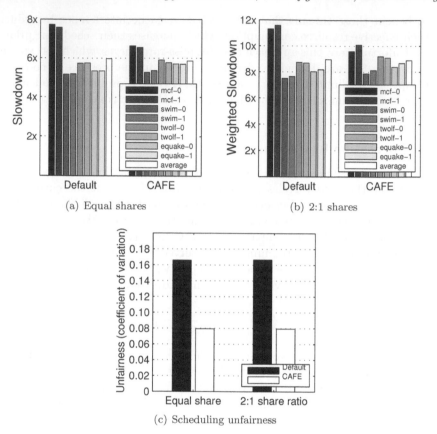

(a) Equal shares

(b) 2:1 shares

(c) Scheduling unfairness

FIGURE 8.11
Vtime compensation

workloads. For both Figures 8.11(a) and (b), the slowdown experienced by mcf is much less when using CAFÉ compared to the default ESX Server scheduler.

Figure 8.11(c) plots the unfairness measured for the equal-share and 2:1-share ratio experiments. The unfairness metric is the coefficient of variation of the per-application weighted slowdowns, and vtime compensation improves it by approximately 50%. Overall, vtime compensation provides a slight increase in performance while reducing unfairness significantly.

8.4.2   Efficient Scheduling

Now we describe how CAFÉ's cache monitoring infrastructure can be leveraged to improve the performance of co-running workloads. We start by introducing the concept of *cache pressure*, which represents how aggressively a thread competes for additional cache space. We then present a *cache divvying* algorithm, based on cache pressure, for approximating the steady-state cache occupancies of co-running threads. Using cache divvying to determine the performance impact of various co-runner placements, we demonstrate simple scheduler modifications for selecting good co-runner placements to maximize aggregate system throughput.

8.4.2.1   Cache Pressure

To understand cache pressure, recall that CAFÉ estimates the cache occupancy for a single thread using (8.9), which defines a recurrence relation between its previous and current occupancies. Since $(1 - E/C) \cdot m_l$ specifies the increase in occupancy, we define the cache pressure P_i exerted by thread τ_i as

$$P_i = (1 - E_i/C) \cdot M(E_i) \tag{8.23}$$

where C is the total number of cache lines in the shared cache, and $M(E_i)$ is the miss rate of τ_i at its current occupancy, E_i. In short, cache pressure reflects how aggressively a thread tends to increase its cache occupancy.

A key insight is that at equilibrium occupancies, the cache pressures exerted by co-running threads are either equal or zero. If the cache pressures are not equal, then the thread with the highest cache pressure increases its cache occupancy. We have observed that in most cases, co-running threads do not converge at equilibrium occupancies, but instead cycle through a series of occupancies with oscillating cache pressures.

Calculating a thread's cache pressure requires $M(E_i)$, which is obtained from its miss-rate curve. As explained earlier, since miss-rate curves are sensitive to contention for memory bandwidth and other dynamic interference, we instead construct miss-ratio curves, despite our desire to examine time-varying behavior. To translate MRCs that track MPKI values into misses per cycle, we normalize each point on the discrete curve by the ideal CPI for the corresponding thread. While this is not completely accurate, it nonetheless provides a practical way to generate approximate miss-rate curves that are not sensitive to interference from co-runners.

8.4.2.2   Cache Divvying

Using the insight above that cache pressures of co-running threads should match at equilibrium occupancies, we are able to estimate their average occupancies, enabling us to predict how the cache will be divided among them. Our *cache divvying* technique does not control how cache lines are actually allocated to threads, but rather serves to predict how cache lines would be allocated given their current occupancy and working-set demands. It also captures the average occupancies of co-running threads that cycle through a series of occupancy values at equilibrium.

Algorithm 8.1 summarizes the cache divvying strategy, assuming the cache is initially empty. In reality, each thread, τ_i, will have a potentially non-zero

---

Algorithm 8.1: Cache Divvying

```
// initialize surplus cache lines S
S = C;
foreach τi do
    Ei = 0; // initial occupancy
end
repeat
    // reset max pressure
    Pmax = 0; foreach τi do
        // pressure at current occupancy
        Pi = (1 − Ei/C) · M(Ei);
        if Pi > Pmax then
            // record thread with max pressure
            Pmax = Pi;
            max = i;
        end
    end
    // greedily assume chunk of size B
    // allocated to thread with max pressure
    Emax = Emax + B;
    S = S − B;
until S = 0 or ∀Pi = 0;
```

---

current occupancy, E_i. The algorithm compares the pressures of each thread at their initial occupancies, by using miss-rate information obtained from MRC data. The thread with the highest pressure is assumed to be granted a chunk of cache. The allocatable chunk size, B, is configurable, but serves to limit the number of iterations of the algorithm required to predict steady-state occupancies for the competing threads. In practice we have found that setting B to one-eighth or one-sixteenth of the total cache size works well with our MRCs, which are also quantized using discrete cache occupancy values.

During each iteration, the thread with the highest pressure increases its hypothetical cache occupancy. This in turn affects its current miss rate, $M(E_i)$, and hence its current pressure, P_i, for its new occupancy. As the pressure from a thread subsides, its competition for additional cache lines diminishes. When the entire cache is divvied, or when all pressures reach zero, the algorithm terminates, yielding a prediction of cache occupancies for each co-runner.

Figure 8.12 shows Algorithm 8.1 used with our simulator for a dual-core system as described in Section 8.2.2. Cache divvying is used to predict the occupancies for six pairs of co-runners, separated by vertical dashed lines in the figure. In each case, the chunk size, B, is set to one-sixteenth of the cache size (i.e., 256 KB). Each co-runner generates 10 million interleaved cache references from a Valgrind trace. While this is insufficient to lead to full cache occupancy in all cases, results show that predicted and actual occupancies are almost always within one chunk size of the actual occupancy. This suggests

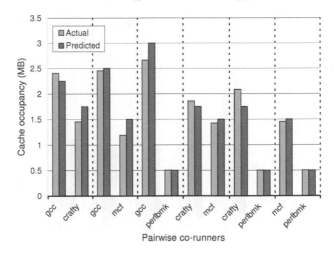

FIGURE 8.12
Cache divvying occupancy prediction

cache divvying is an accurate method of determining cache shares among co-runners. We are investigating its accuracy on architectures with higher core counts.

8.4.2.3   Co-Runner Selection

Cache divvying provides the ability to predict the equilibrium occupancies achieved by workloads co-running on a shared cache. This information can be used in CPU scheduling decisions to enhance overall system throughput.

We extended the VMware ESX Server scheduler with a simple heuristic. A user-level thread periodically snapshots the miss-ratio curves generated by CAFÉ and evaluates various co-runner pairings using *cache divvying* to predict their associated equilibrium occupancies. Based on a workload's estimated occupancy, we predict its miss ratio by consulting the workload's miss-ratio curve. We employ a simple approximation to convert the predicted miss ratio into a time-based miss rate, multiplying the workload's miss ratio by $1/CPI_{ideal}$, its instructions-per-cycle metric at full occupancy. The pairing which achieves the smallest aggregate *conflict miss rate* is chosen and communicated to the scheduler, which migrates threads to implement the improved placements.

The conflict miss rate is the miss rate in excess of what a thread experiences at full cache occupancy. By selecting pairings which reduce aggregate conflict misses, CAFÉ tries to improve performance as well as fairness. While we have demonstrated one practical heuristic incorporating cache divvying predictions, many other scheduler optimizations could benefit from this information.

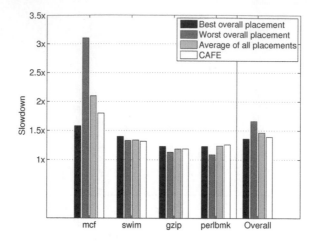

FIGURE 8.13
Co-runner placement

8.4.2.4 Co-Runner Selection Experiments

To evaluate our implementation of the co-runner placement heuristic in the ESX Server scheduler, we used the SPEC2000 benchmarks `mcf`, `swim`, `gzip`, and `perlbmk`, each running on a separate core. To focus on the effectiveness of CAFÉ at finding good co-runner placements, we restricted the workloads to execute on a single package containing two dual-core CMPs, each with its own last-level cache.

As before, we use the average of the per-application slowdowns as the metric for overall efficiency and their coefficient of variation as the metric for unfairness. At the start of the experiment, the co-runner pairings were manually selected to be the pairing that was determined to result in the worst overall performance (`mcf` paired with `swim` and `perlbmk` paired with `gzip`).

Note that in Figure 8.13, the 'Worst overall placement' column for each separate workload shows the slowdown of that benchmark when running in the worst overall configuration. As can be seen, some benchmarks do not suffer as much as others in this worst-case configuration, but `mcf` was the one that incurred significant slowdown. Notwithstanding, the rightmost 'Overall' column shows that when `mcf` experiences its worst slowdown that is when we have the worst overall slowdown across all workloads.

As Figure 8.13 shows, CAFÉ was able to achieve performance close to the best overall placement by adjusting the workload assignments to better cores. CAFÉ co-runner placement reduces unfairness by 24% and improves performance by 5% compared to the average of all placements. Compared to the worst overall placement, CAFÉ reduces unfairness by 64% and improves performance by 16%.

8.5   Related Work

The focus of this chapter encompasses several areas of related work, from shared-cache resource management to co-scheduling of threads on parallel or multi-core architectures. In the area of shared-cache resource management, there is a significant literature on cache partitioning, using either hardware or software techniques (Albonesi 1999; Chang and Sohi 2007; Dybdahl, Stenström, and Natvig 2006; Iyer 2004; Kim, Chandra, and Solihin 2004; Liu, Sivasubramaniam, and Kandemir 2004; Rafique, Lim, and Thottethodi 2006; Ranganathan, Adve, and Jouppi 2000; Srikantaiah, Kandemir, and Irwin 2008; Suh, Rudolph, and Devadas 2004). This has been prompted by the observation that multiple workloads sharing a cache may experience interference in the form of conflict misses and memory bus bandwidth contention, resulting in significant performance degradation. For example, Kim, Chandra, and Solihin (2004) showed significant variation in execution times of SPEC benchmarks, depending on co-runners competing for shared resources.

Cache partitioning has the potential to eliminate conflict misses and improve fairness or overall performance. While hardware-based approaches are typically faster and more efficient than those implemented by software, they are not commonly available on current processors (Suh, Devadas, and Rudolph 2001; Suh, Rudolph, and Devadas 2004). Software techniques such as those based on page coloring require careful coordination with the memory management subsystem of the underlying OS or hypervisor and are generally too expensive for workloads with dynamically varying memory demands (Cho and Jin 2006; Liedtke, Härtig, and Hohmuth 1997; Lin et al. 2008).

A significant challenge with cache partitioning is deriving the optimal allocation size for a workload. One way to tackle this problem is to construct cache utility functions, or performance curves, that associate workload benefits (e.g., in terms of miss ratios, miss rates, or CPI) with different cache sizes. In particular, methods to construct miss-ratio curves (MRCs) have been proposed that capture workload performance impacts at different cache occupancies, but either require special hardware (Qureshi and Patt 2006; Suh, Devadas, and Rudolph 2001; Suh, Rudolph, and Devadas 2004) or incur high overhead (Berg, Zeffer, and Hagersten 2006; Tam et al. 2009).

The Mattson Stack Algorithm (Mattson et al. 1970) can derive MRCs by maintaining an LRU-ordered stack of memory addresses. RapidMRC uses this algorithm as the basis for its online MRC construction (Tam et al. 2009). This requires hardware support in the form of a *Sampled Data Address Register* (SDAR) in the IBM POWER5 performance monitoring unit to obtain a stream of memory addresses that match a pre-specified selection criterion. The total cost of online MRC construction is several hundred milliseconds, with more than 80 ms of workload stall time due to the high overhead of trace collection. This overhead is mitigated by triggering MRC construction only when phase transitions are detected, based on changes in the overall cache miss rate. However, since changes in cache miss rates can be triggered by cache contention caused by co-runners and not necessarily phase changes; the phase transition detection in RapidMRC does not seem robust in overcommitted environments.

In contrast, we deploy an online method to construct MRCs and other cache-performance curves efficiently, requiring only commonly available performance counters. Due to the low overhead of our cache-performance curve construction, it can remain enabled at all times, providing up-to-date information pertaining to the most recent phase. As a result, CAFÉ does not require an offline reference point to account for vertical shifts in the online curves due to phase transitions and is also robust in the presence of cache contention from co-runners. We do, however, suffer from the problem of obtaining enough occupancy data points to construct full curves. Using duty-cycle modulation to temporarily reduce the rate of memory access by competing workloads is one technique that has the potential to alleviate this problem.

Other researchers have inferred cache usage and utility of different cache sizes. In CacheScouts (Zhao et al. 2007), for example, hardware support for monitoring IDs and set sampling are used to associate cache lines with different workloads, enabling cache occupancy measurements. However, the use of special IDs differs from our occupancy estimation approach, which only requires currently available performance monitoring events common to modern CMPs.

Given cache utility curves, we attempt to perform fair and efficient scheduling of workloads on multiple cores. Fedorova, Seltzer, and Smith (2006) devised a cache-fair thread scheduler that redistributes CPU time to threads to account for unequal cache sharing. This work assumes that different workloads competing for shared resources should receive equal cache shares to be fair, regardless of different memory demands from workloads. A two-phase procedure is employed, first computing the fair cache miss rate of each thread, followed by adjustments to CPU allocations. Computing fair cache miss rates requires sampling a subset of co-runners followed by a linear regression and is potentially expensive. In contrast, we derive a workload's current and fair CPI values inexpensively and then perform vtime compensation to improve fairness.

8.6   Conclusions and Future Work

This chapter introduces several novel techniques for chip-level multiprocessor resource management. In particular, we focus on the management of shared last-level caches and their impact on fair and efficient scheduling of workloads. Toward this end, our first contribution is the online estimation of cache occupancies for different threads, using only performance counters commonly available on commodity processors. Simulation results verify the accuracy of our mathematical model for cache occupancy estimation.

Building on occupancy estimation, we demonstrate how to dynamically generate cache performance curves, such as MRCs, that capture the utility of cache space on workload performance. Empirical results using the VMware ESX Server hypervisor show that we are able to construct per-thread MRCs online with low overhead, in the presence of interference from co-runners. We show how duty cycle modulation can be used to help a thread increase its cache occupancy by reducing interference from co-runners. This approach facilitates obtaining a wide range of occupancy data points for MRCs.

Our fast online MRC construction technique is used as part of a cache divvying heuristic to predict the average occupancies of a set of co-running workloads. Simulation results show this to be an effective method of using MRCs to estimate the expected occupancies if two or more workloads were to co-execute and compete for cache space. Cache divvying forms the basis of our co-runner selection strategy, which partitions threads across separate CMPs. By carefully partitioning threads, we avoid potentially bad groupings of co-runners that could negatively impact the shared last-level cache on the same CMP. Experiments show that for a group of SPEC CPU workloads, we are able to reduce slowdown by as much as 5% in the average case and 16% in the best case.

Finally, we attempt to improve fairness by compensating a workload for the resource conflicts it experiences when co-running with other workloads. Our vtime compensation technique accounts for the time a thread is stalled contending for resources, including the stall cycles caused by last-level cache conflict misses and memory bus access. Estimates of performance degradation experienced by a thread due to co-runner interference are calculated online. Results show as much as 50% improvement in fairness using vtime compensation.

While we have presented several new online techniques for CMP resource management, a variety of interesting research opportunities remain. We are exploring various approaches for improving CAFÉ's ability to generate accurate cache performance curves at all occupancy points. We continue to investigate new scheduling heuristics that leverage our cache monitoring capabilities, and we are examining applications of vtime compensation to other problems, such as NUMA locality management. We also plan to extend our modeling techniques to address the impact of threads that block waiting for events such as I/O completion, and to incorporate the effects of data sharing and constructive interference between threads. Finally, we are actively exploring ways to extend and integrate our software techniques with future hardware, such as architectural support for cache quality of service (QoS) monitoring and enforcement, and large-scale CMPs containing tens to hundreds of cores.

Part IV

Debugging

9

Software Debugging Infrastructure for Multicore Systems-on-Chip

Bojan Mihajlović, Warren J. Gross, and Željko Žilić

Department of Electrical and Computer Engineering, McGill University, Montreal, Canada

CONTENTS

9.1   Introduction

In recent years, increasing numbers of processor cores have begun to be integrated into typical systems-on-chip (SoCs). From high-performance server CPUs, to embedded systems such as mobile phones, multicore processors are now present in virtually every domain of computing. In certain domains, multicore hardware has been ubiquitous for some time, while the availability of software that fully exploits such hardware has not kept pace. This trend can partly be explained by the fact that creating multithreaded software can involve a substantial rise in complexity over a single-threaded version. In addition to necessitating the use of a different programming model, multithreaded software is not suited to the same debugging methods as its single-threaded counterpart. For many decades, a standard way to debug a single software thread has been to halt its execution at an opportune point and to observe its internal data structures before resuming its execution. However, this method can cause the failure of a multithreaded program that relies on timely data exchange with other threads. A parallel can be drawn to the *observer effect*, which holds that the mere act of observing something will alter the subject of the observation.

While developers are starting to embrace multithreaded programming models, traditional software debugging methods still dominate in the realm of development. Ironically, resolving certain types of bugs common in multithreaded software can be an intractable problem using entrenched debugging methods. Evidence of this disconnect can be seen in statistics that show 50–75% of the total cost of software development is spent in verification, testing, and debugging (Hailpern and Santhanam 2002).

In the past, a major debugging obstacle has been the lack of adequate software debugging methods that address the complexities of a multicore environment. Established methods of debugging multithreaded software have traditionally been highly specialized, slow, and expensive to use. A method known as *trace generation* has been used for decades to debug real-time software, by generating logs of debug data in real time without interrupting software execution. The method has recently seen renewed interest as a potentially good solution to the debugging of many other types of multithreaded software. Recent advances are increasingly solving outstanding problems, bringing new debug methods closer to mainstream use.

Many complex SoCs are also starting to add basic support for trace generation into their designs. At the moment, the method commonly known as *tracing* is positioned as a complementary method to traditional debugging, to be used on the cumbersome bugs that cannot easily be resolved by traditional means. There are both hardware and software challenges to be overcome before tracing can further displace traditional debug methods. However, as the infrastructure becomes available, there is a need for software developers to start embracing the tools that will allow more efficient resolution of multi-threaded bugs.

In this chapter we introduce the traditional approaches to software debugging, give an assessment of how future systems will be debugged, and outline work, including our own, currently being done to realize that goal.

9.2 Software Debugging

Software faults (or *bugs*) are unintended segments of code in a program which can lead to *errors* in the intended functionality of software. In formal terms, that means that the software is not conforming to its requirements or specifications. The process by which such faults are identified and fixed is called software debugging.

In general, the process of software debugging aims to identify bugs and trace their manifestations back to an offending segment of source code. Tracking the bug to a certain point within a program's execution requires that the system allows the program's internal data structures to be observed in the midst of execution, known as *observability*. Observable structures can include variables, registers, memory addresses, or other internal data that may help to identify the fault, both in terms of location and time. Once a fault is identified, its cause can be determined if the program's *symbol table* can be used to link its executable machine code back to its source code.

The program can then be re-executed to verify that the fault has been corrected. This typically means modifying and recompiling source code and re-executing the program, but if either execution or recompilation will take a long time, it can also mean utilizing other methods. Without recompilation, *controllability* can be exercised on the program's data structures in the midst of execution. Variables, registers, or memory locations can be changed to their expected or known-good values, and execution can be resumed.

9.2.1 Debugger Programs

A program known as a *software debugger* is most often used to centralize and facilitate the debugging process, including observing and tracing faults, and controlling data structures. A software debugger and program-being-debugged can either run *locally* on the same machine or *remotely* on different machines. In a remote debug scenario, an example of which is seen in Figure 9.1, the machine running the debugger is known as the *host*, which tends to be a performance-oriented workstation that software developers use to run various development tools. The machine running the program-being-debugged is known as the *target* and can range from high-performance CPUs to the smallest embedded processors. Local debugging is commonly used when the

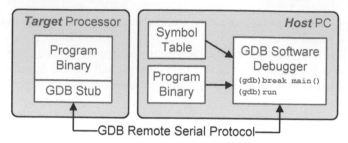

FIGURE 9.1

Remote debugging scenario – software view

debugger and program-being-debugged can be run on the same machine, such as for native application development. Remote debugging is sometimes the only choice when the target system lacks debugging resources or interfaces, such as for embedded software. It is also useful when cross-compiling software for a different target Instruction Set Architecture (ISA).

Using the multitarget software debugger GDB (GNU Project 2010) in a remote debugging scenario involves executing a small piece of interface software on the target system. Known as *gdbserver*, or the smaller *gdbstub*, these interfaces enable access to the target from a host system running the full GDB debugger. This keeps the complexity and memory footprint of the target relatively small while allowing the same debugging capabilities.

Many software debuggers support debugging both with and without an underlying operating system (OS), and some have support for multiple threads. When debugging a single-threaded program under an OS, the thread will effectively be executed under full control of the debugger. The OS will schedule the debugger thread as it would any other, allowing the debugger to be oblivious to any OS thread scheduling or control mechanisms. However, this is not the case when debugging multithreaded programs. Since other threads may be instantiated at any time, the debugger must be aware of the application-programming interface (API) under which threads are instantiated and terminated in order to take control of them.

9.2.2   Fault Types

Software faults may take several different forms. From the standpoint of the software developer, it may be of benefit to differentiate between bugs based upon their apparent manifestations, as it can allow for different approaches to uncovering and resolving them. Such a categorization was first made by J. Gray (1986), so we will summarize the two main types of bugs:

1. Bohrbugs: As in the Bohr model of the atom, these bugs are detectable and 'manifest consistently under a well-defined set of conditions' (Grottke and Trivedi 2005). These easily reproducible bugs can be defined as Bohrbugs.

2. Mandelbugs: As in the Mandelbrot set, 'a fault whose activation and/or error propagation are complex' (Grottke and Trivedi 2005), where the complexity is derived from many dependencies or is related to the time required for the bug to manifest. These bugs can be much more difficult to reproduce.

The ease of *reproducibility* is a primary distinction between the two types of bugs. Using a software debugger, Bohrbugs may be identified and resolved by reproducing the conditions under which they appear. However, Mandelbugs first require the developer to identify the conditions under which they are manifest. In a complex system with ever-changing sets of conditions, the most cumbersome Mandelbugs may appear only infrequently over long periods of execution. In this case, periodically recording the conditions of the system may help determine the combination that led to the bug. Using controllability to recreate these same conditions can help to reproduce the bug more consistently and can help uncover its underlying cause.

9.2.3   Multithreaded Software

There are unique challenges to debugging some multithreaded software. In single-threaded software, instructions will be retired in the same order as they are encountered in the instruction stream. This property of *determinism* allows the results of multiple executions to be compared with each other, aiding in the identification of software faults. If multiple software threads are time scheduled to share a single core, the execution order of threads can change depending on the thread-scheduling algorithm and the presence of other active threads. In this case, access to the OS and thread-scheduling routine is needed to determine which executed instructions belong to which thread. The same is true for multicore SoCs when debugging multiple software threads executing on multiple cores simultaneously.

In an effort to extract maximum performance from a chip, the different processing elements of multicore SoCs are often clocked in different clock domains. This means that each clock is asynchronous with respect to the others found on-chip. While the order of executed instructions is maintained within each thread, execution order with respect to other threads is non-deterministic. In other words, there is no way to know when each instruction will execute with respect to instructions in other threads. In fact, this order can change with each execution of the program, which makes it much more difficult to reproduce both Bohrbugs and Mandelbugs.

Even when thread-enabled software debuggers are used, a concurrently executing multithreaded program can have a different flow of execution every time it is run. This is because of the physical limitations of non-determinism, and not a problem of the software debugger itself. Both of the fault types described above can appear to be Mandelbugs because of decreased reproducibility due to non-deterministic execution. Clearly, this increases the effort needed to debug multithreaded software above that of a single-threaded version of the same software. For this reason, multithreaded software development typically begins as a single thread, where most bugs are resolved before creating a fully multithreaded application.

9.3 Traditional Debugging Methods

Over the years, a multitude of debugging methods have been employed to identify and eliminate software faults. All methods require some amount of supporting hardware, software, or some combination thereof. If the method changes the software behavior or timing in any way, it is known as *intrusive*. Intrusive debugging methods have enjoyed wide popularity for decades because they are effective in identifying most bugs in many types of programs. If single-threaded software does not have real-time scheduling requirements, it can have its timing altered without major consequences to its functionality. *Transparent* debugging methods have traditionally been used only when intrusive methods could not identify bugs adequately. This is the case with real-time software, where altered timing could cause additional faults in the system, such as by missing deadlines. This is also the case with modern interdependent, concurrently executing threads of the same program. In the following discussion we examine traditional debugging methods and the implications of using them with multithreaded programs.

9.3.1 Software Instrumentation

It can be said that the simplest type of debugging uses a purely software-based approach. Most software developers are familiar with an informal method of adding *print* statements to a program in order to observe a variable at an opportunistic point during execution. This form of *software instrumentation* allows a limited degree of observability to be achieved at the expense of much manual effort, including a re-compilation and re-execution of code for every change that is made. Software instrumentation can also be done with a *monitor* program, which is an environment under which the program being debugged can be executed. Monitors can instrument software by modifying it so that, for example, execution proceeds until a predetermined memory address is reached, known as a *software breakpoint*. Data structures can then be observed or changed within the monitor environment, and execution resumed. Monitor programs are useful when a processor lacks debug support hardware, or access to it is limited. While monitor programs have mostly been superseded by software debuggers in recent years, if debug support hardware is unavailable, some software debuggers will also fall back on software instrumentation as a means of performing debugging tasks.

One of the drawbacks of using software instrumentation is that a limited number of locations can be instrumented, which exclude protected (or privileged) memory locations such as interrupt service routines. In all cases, the original 'release' version of the software is modified because of the added software-based instrumentation. The resulting 'debug' version of the software may cause additional faults in multithreaded or real-time applications, primarily due to the added time needed to achieve the desired controllability and observability. Software instrumentation methods can therefore be categorized as intrusive.

9.3.2   Scan-Chain Methods

Unlike software instrumentation methods, the addition of on-chip debug support hardware, known as *hardware instrumentation*, allows unaltered software to be executed on-chip. This alleviates some of the problems associated with software-based debug approaches, namely, the functionality changes between debug and release versions of the same software.

In designing SoCs, some Design-for-Test (DFT) hardware is often added to aid in hardware verification and manufacturer testing of products. The IEEE 1149.1 standard (*IEEE Standard Test Access Port and Boundary-Scan Architecture* 2001), also known as JTAG, describes a Test Access Port (TAP) that links the flip-flops of a chip together in one or more *scan-chains*. This allows data I/O to all flip-flops connected to the scan-chain through the TAP. The feature is useful for hardware testing purposes, but can also be repurposed for software debugging. Many processor cores also include a 'debug' mode which allows data, such as registers, to be output to a host computer by using a scan-chain. In the simplest scenario, this mode also allows the processor clock to be supplied by an external host computer, which is able to scan-out data between executed instructions. In these methods the host is exercising *execution control* over the target as to when to start and stop execution, in order to achieve the desired controllability and observability.

In a multicore SoC, the scan-chains of each core could either be linked serially, as seen in Figure 9.2, or provided in parallel as multiple individual scan-chains. Taking either of these approaches poses several additional problems. Multiple scan-chains require many external pins, which are a scarce commodity in many SoCs. A long serial scan-chain would require the least number of external pins, but would incur a latency proportional to its length. There have been some efforts to reduce the I/O latency of such long scan-chains by bypassing cores that are not under observation, this being one of the improvements of the IEEE 1149.7 standard (*IEEE Standard for Reduced-Pin and Enhanced-Functionality Test Access Port and Boundary-Scan Architecture* 2009).

While debugging, the use of a synchronous, externally supplied clock can change the behavior of software that normally executes on cores in asynchronous clock domains. In such a scenario, timing bugs may not manifest

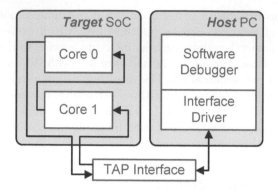

FIGURE 9.2
Debugging multiple cores through IEEE 1149.1 (JTAG)

when they are related to data synchronization between clock domains. This is a significant change in software behavior for some applications, and as a result scan-chain methods can be considered intrusive.

However, the biggest impediment to the use of scan-chain methods, even for single-threaded debugging, is that external clocks can be an order of magnitude slower than their on-chip equivalents. Many modern SoC cores are clocked above 2 GHz, while the maximum scan-chain clock speeds can be said to be in the 100 MHz range. Using this method alone, scan-chain driven execution times may rise dramatically, and with them, the time spent both debugging and waiting for a bug to manifest. Since the cost of debugging software is proportional to the amount of time spent debugging, this method would incur an increasing cost as SoC clock frequencies rose over time.

9.3.3   In-Circuit Emulation

Using DFT hardware to exercise fine-grained processor control may be an advantage during hardware testing, but it can be excessively slow for software debugging. For this reason, some commonly used software debugging control logic can be moved from the host to the target to enhance overall debugging speed and functionality. These methods can roughly be grouped into a category known as In-Circuit Emulation (ICE), many of which aim to speed up what is already possible using scan-chain methods at the expense of additional on-chip hardware instrumentation. The enhancements are often complementary and augment the execution control provided by scan-chains. For example, in Figure 9.3, the addition of hardware breakpoint registers and their corresponding comparators (not shown) allows the software to execute in real time using an on-chip clock while waiting for the desired point of execution to arrive. In the simplest scenario, when a comparator determines that the program counter address is equal to the address stored in the breakpoint

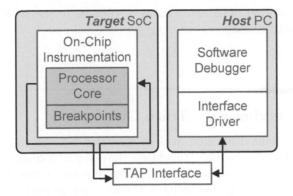

FIGURE 9.3
Debugging a single-core SoC through In-Circuit Emulation (ICE)

register, execution is stopped and control is transferred to the host. Systems with expanded ICE capabilities also allow hardware breakpoints to be loaded with data values in addition to addresses, which can then be compared to general-purpose registers or memory locations. They are then known as *hardware watchpoints* and allow execution to be stopped when a desired data value has been observed. In either case, after execution is stopped and control transferred to the host, a scan-chain can then be used to enable controllability and observability at precisely the desired point of execution.

Applying execution control through ICE can help avoid the problems of using a slow external clock in scan-chain-based methods. For single-threaded programs, it allows the execution of individual cores to be paused in order to control or observe them. However, applying this method to one thread in a multithreaded program can lead the paused thread to induce timing faults in its dependent threads. Likewise, exercising execution control over an entire group of concurrently executing threads can lead to non-deterministic behavior. For example, a breakpoint on one core could be 'cross-triggered' to stop the execution of all other cores, but the delay in propagating such a signal to asynchronous clock domains leads the cores to stop at non-deterministic points (at unpredictable locations). Some SoCs with homogeneous cores that use this imprecise approach (Intel Corporation 2010b) report hundreds of cycles of 'skid' before all cores can be stopped or started. When debugging heterogeneous cores, some of which can lack cross-triggering hardware, the time needed for execution control to be performed through software is reported to be on the order of seconds (ARM Ltd. 2010b). This method clearly does not allow for the desired points of execution to be observed in a multicore system. As the complexity and number of cores per chip rises, the latency of performing centralized execution control will also continue to grow. While execution control, as performed through ICE, will continue to be essential in software debugging, it is not seen as a solution to the debugging of future multithreaded software.

9.4   Debugging with Trace Generation

From the previous discussion, it is evident that a transparent, non-intrusive debugging method is required to address the needs of debugging multithreaded software. Fortunately, multithreaded software is not unique in its intolerance to intrusive debugging. Real-time software must respect deadlines that require data processing within a specified period of time, at the risk of a fault or failure. A method known as *trace generation* has been used for decades for real-time software debugging (Plattner 1984) and is now increasingly being suggested as most suited to the debugging of multithreaded software (Hopkins and McDonald-Maier 2006; Mihajlovic et al. 2009). In fact, trace generation capabilities are now being integrated into many modern SoC designs and are an integral part of many commercial multicore debugging strategies, including ARM's Coresight (ARM Ltd. 2010a) and Infineon's Multi-Core Debug Solution (MCDS) (Infineon Technologies 2010).

Trace generation involves transparently generating a log (or *trace*) of on-chip data structures as a program is being executed. By observing the desired registers, variables, or memory locations passively without changing software behavior, an actual record of the events leading to a fault can be kept. The simplest type of trace is known as an *execution trace* (or *address trace*) and is a list of instruction addresses which have been executed, indicating the flow of software execution.

In modern SoCs, trace generation is typically accomplished by adding on-chip hardware dedicated to the task. By distributing such hardware to each SoC clock domain, traces can be generated synchronously with respect to each core. Prior to program execution, an on-chip hardware structure known as a *trigger* must be user configured to select both when tracing should occur and what data should be output. Useful trace data can include processor registers, memory values, or any other data which will help the software debugging effort. When the trigger is 'hit' during execution, the trace data can be stored in an on-chip trace memory (also known as a *trace buffer*), as seen in Figure 9.4. Alternatively, trace data can be streamed directly off-chip in real time to the host machine, seen as part of the debugging scenario incorporating a trace compression scheme (see Figure 9.6).

To preserve the global order of events that lead to a fault, an implicit or explicit mechanism is also needed for determining the order that traces of multiple cores were generated. Analyzing a chronologically ordered trace *off-line* (after execution is complete) with the help of a software tool can help determine the cause of the fault.

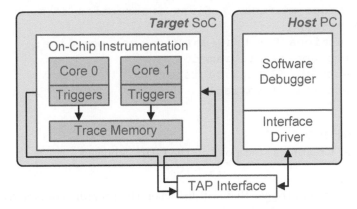

FIGURE 9.4
Debugging through a combination of trace generation and ICE

While many modern SoCs have recently added tracing hardware to their designs, few have the capability of being controlled through standardized debug interfaces. As trace generation is still a developing field, most trace generation capabilities are currently extended through proprietary software tools. Software debuggers such as GDB are starting to add support for the running of *trace experiments*, though few designs currently embrace such interfaces. There have also been recent efforts to standardize the hardware interface and protocols that will be needed for both trace generation and other debugging methods. The Nexus 5001 standard (Nexus 5001™ Forum 2012) is a good candidate and has been progressing into a superset of debug interfaces to ensure backward compatibility with older designs.

The following sections will discuss in greater detail various considerations in the field of trace generation-based debugging.

9.4.1 Triggers

Triggers used in trace generation can be compared to the hardware breakpoints and watchpoints most commonly used in ICE methods. In fact, the hardware structure of a trigger can be similar to that of a breakpoint, consisting of comparators and/or counters. Like breakpoints, triggers can be used to compare instruction addresses or data values with a preset trigger value, or may use counters to follow events such as loop iterations. A hardware trigger unit can contain more than one comparator and counter, which allows for conditional trace generation using compound boolean conditions. When trigger conditions are met, it is said that the trigger has 'hit'. Rather than breaking the execution of the program, a *trace* of some desired data is output instead.

In the example shown in Figure 9.5, the trigger of tracepoint 1 is loaded with the address of the first instruction of function my_func, creating

```
(gdb) trace my_func // first instruction of my_func
(gdb) actions
Enter actions for tracepoint 1, one per line:
>collect $pc // program counter
>collect $r1 // data register
>collect var // variable
end
```

FIGURE 9.5
Example: Creating a tracepoint in the GDB debugger

an unconditional trigger that will *hit* when the program counter reaches that
address. If we assume that the trigger hardware supports the outputting of
three data values simultaneously, it can be configured to output the program
counter and r1 registers, as well as the variable var. For transparent tracing
to occur, var must be located on-chip at the time of the *hit*, and the debugger
must use the program symbol table to link it to a register location. If any of
these conditions are not satisfied, the debugger may create a *software tracepoint* instead.[1] This would instead force execution to be halted and a memory
read issued on the variable, in effect acting as a breakpoint with automated
data collection.

To maintain trigger flexibility, multiple triggers can also be compounded to
either create compound conditional statements or to output more data than
any one trigger unit supports (though in practice, this is difficult for reasons
discussed in Section 9.4.4).

The inclusion of cross-triggers, where the trigger *action* on one core acts as
the trigger *condition* on another core, can be useful for multithreaded debugging. Even though there is trigger-signal-proportionate delay between cores,
this type of transparent multicore trace generation would allow flexible debugging of multithreaded software.

9.4.2   Trace Ordering

Since the main goal of trace generation is the time-accurate observation of
events, the relative order in which traces were generated must be preserved.
This is especially important in an environment where traces are generated by
multiple cores simultaneously.

One of the most constrained SoC resources is the number of pins available
to interface its various functions with the outside world, including debugging
functions. For this reason, debug interfaces are usually restricted to a bare
minimum number of pins, including pins needed for transferring trace data.

---

[1]While some processors have hardware support for on-the-fly memory I/O that doesn't
halt software execution, a large number of operations could lead to an appreciable change
in software behavior. For this reason, the method is omitted from the discussion.

Before either transferring traces to the host in real time or storing them in a trace buffer, the traces of all cores would first need to be aggregated in a way that preserves their generated order.

In a simple multicore architecture where the trace data path from each core to the aggregator is the same length, trace ordering can be accomplished by implicitly processing the data that arrives first. However, complex interconnection schemes such as those employed by Networks-on-Chip (NoCs) pose a much greater challenge. In a typical NoC mesh interconnect, network congestion and routing delays may lead to out-of-order traces being collected by the aggregator. While this is far from a solved problem, one solution may be to arrange for accurate time-keeping on each core and to time-stamp each trace before it is transmitted. This would allow traces to be accurately reordered off-line.

9.4.3 Debug Interface

Several TAP standards, including IEEE 1149.1 (JTAG) (*IEEE Standard Test Access Port and Boundary-Scan Architecture* 2001) and IEEE 1149.7 (*IEEE Standard for Reduced-Pin and Enhanced-Functionality Test Access Port and Boundary-Scan Architecture* 2009), have served as debug interfaces for execution control over the years. However, there have been efforts to replace them with a standardized TAP interface that has been designed with next-generation debugging in mind, including trace generation. The most promising successor is the Nexus 5001™ standard (Nexus 5001™ Forum 2012), which aims to be backward-compatible with both of the older IEEE standards that are now predominantly used for debugging. In terms of trace generation, one of the main advantages of Nexus is an optional *auxiliary* port of up to 16 bits which can be used to output trace data. The standard also includes a mechanism to reduce trace data volume that is robust yet lacking in performance compared to current research methods.

It is expected that widespread adoption of the Nexus standard would ensure interoperability between SoCs, debug adapters, and debugger software. This could decrease the cost of providing debugging infrastructure and, in turn, streamline software debugging itself.

9.4.4 Data Volume

Fundamentally, the use of trace generation involves moving on-chip data to an off-chip host machine for the purposes of debugging. Since traces are most useful if recorded in real time as software is executing, one option is to move traces off-chip in real time as they are generated. However, as the example below demonstrates, a considerable amount of bandwidth could be needed to accomplish this task at current SoC clock speeds.

Example: In order to find a software fault, a real-time execution trace is required of software running on a single-core 32-bit processor clocked at 2 GHz, with a performance of 1 cycle-per-instruction (CPI). A trace placed on the Program Counter (PC) register, collected on every cycle, will then generate 7.45 GB/s of trace data. The bandwidth needed to collect this data in real time rules out all but the most expensive communication links, while tracing an additional register or increasing the SoC clock speed would easily overwhelm even those links.

From the above scenario, real-time trace output would require either clocking the SoC at a slower speed or reducing the amount of trace data generated. In the designs of Chung and Hoe (2010) and Chung et al. (2008), slowing down SoC speed resulted in execution times an order of magnitude greater. As with scan-chain methods, reducing SoC clock speed would increase debugging time and cost, and is not a scalable solution for future debugging needs.

For this reason, many trace-capable SoCs implement an on-chip trace buffer into which trace data is written. After a trace experiment is complete, the contents of the trace buffer are then moved to the host machine. The drawback to this approach is that the small trace buffers that can typically fit on-chip will be quickly filled at typical SoC clock speeds. This means that only a small window of execution may be observed before the trace experiment is either forced to stop or before collected trace data must be overwritten. This is especially problematic for Mandelbugs that manifest only intermittently or after long periods of execution. Though recent research has explored improvements in trace buffer utilization and efficiency (Yang, Chiang, and Huang 2010), this method is not expected to scale as both clock speeds and the number of cores per chip increase.

On the other hand, real-time trace output allows trace experiments to be performed for indefinite periods of time and is suitable for all types of bugs. Current research, including that of our own team, has been attempting to aggressively reduce the volume of data that needs to be streamed off-chip in real time by developing on-chip trace compression techniques.

9.5 Generalized Debugging Procedure

This chapter discusses various debugging methods and techniques for uncovering software bugs, in both single-threaded and multithreaded applications. To understand how these methods relate to each other in a typical debugging scenario, an example is provided of the generalized debugging procedure that can be followed to resolve a bug. Using a software debugger that supports trace generation, a typical debug session would take the following form:

1. *Identify that the software is not behaving as expected.*

   Unintended software behavior can manifest in a number of ways. The spectrum can range from a catastrophic failure causing a fatal exception in hardware, to subtle behavioral changes that do not necessarily cause errors.

2. *Attempt to replicate the faulty behavior.*

   With the aid of a software debugger, executing the application under controlled conditions will allow Bohrbugs to manifest predictably, while Mandelbugs will appear to manifest in non-deterministic ways, and perhaps after longer periods of time. It is also possible that faulty hardware or transient environmental factors, like temperature or radiation, may be the cause of the fault. These external causes can be identified by executing the software on known-good hardware and placing the system in a controlled environment, respectively.

3. *Isolate the location of the bug in the source code.*

   The approach to isolating a bug depends upon the nature of the application. Software that is tolerant to intrusive debugging can be debugged by using an execution control method such as ICE. A faulty code segment can quickly be identified by placing breakpoints at various points of execution and observing internal data structures when execution is halted.

   Software intolerant of intrusive debugging, such as real-time software, or certain multithreaded software, can instead be the subject of a trace experiment. Assuming trace data is transferred off-chip in real time, instead of stored in a trace buffer, the following steps are required:

   (a) Determine the execution point(s) where traces should be collected;

   (b) Configure triggers to select all relevant data structures;

   (c) Enable trace generation and execute the application;

   (d) Collect trace data until the trace experiment is over;

   (e) Decompress the traces and forward them to a trace analysis tool.

   Similar to execution control methods, trace collection occurs at predetermined instruction addresses. At these addresses, triggers select the data structures to observe, often based upon one or more boolean conditions. Once traces have been collected, they are decompressed by the host machine and forwarded to a software tool for analysis. Such tools can be integrated with software debuggers to maintain a link with the program's symbol table. They are responsible for navigating through large amounts of trace data in order to connect the manifestation of a bug with its cause.

   Since more than a single trace experiment may be needed to find a bug, there are efficient ways to perform consecutive trace experiments. An initial experiment may trace a small amount of data over a long time period to identify the approximate location of errors. In each subsequent trace experiment, the time window being traced can be narrowed while the amount of data being traced can be increased. This iterative refinement finally allows a full picture of the software's data structures to be observed closest to the location of the fault.

4. *Correct the faulty segment of code and recompile the application.*

5. *Re-execute the software to observe whether the bug has been resolved.*

The software can be executed with a range of inputs to verify that the fault has been corrected. Regression testing can also be done to ensure that no new bugs have been introduced into the application.

9.6   Trace Compression Scheme

Even though there is no simple solution to the difficulties of moving large volumes of data off-chip, much attention has recently been given to the reduction of trace data volumes. One of the most promising ways to achieve this is through the use of on-chip trace compression techniques, which are now being used in commercial trace generation hardware such as ARM Coresight. While commercial tools generally apply simple compression methods, many recent research efforts have focused on applying both novel and well-known data compression techniques (Lelewer and Hirschberg 1987; Burtscher et al. 2005) to achieve greater compression performance.

Execution traces are one of the most useful types of trace data that can be collected and have also been shown to be highly compressible. Recent on-chip hardware trace compression schemes (Mihajlovic and Zilic 2011; Milenkovic et al. 2011; Uzelac et al. 2010; Uzelac and Milenkovic 2009; Kao, Huang, and Huang 2007; Milenkovic and Burtscher 2007) have shown that lossless data compression techniques can reduce execution traces by three orders of magnitude from their original sizes. Even with the hardware area and memory limitations of implementing on-chip compression hardware, most schemes achieve compression results superior to even well-known general-purpose algorithms. This is done by exploiting the generally predictable nature of processor instruction execution.

Data traces are less predictable and as such generally less compressible. Nevertheless, recent work has allowed moderate compression performance to be achieved (Anis and Nicolici 2007), with some data types achieving compression ratios of up to an order of magnitude (Uzelac and Milenkovic 2010).

In all cases, trace decompression must be performed on the host machine to restore the original trace data. Depending upon the complexity of the compression scheme, either software or specialized hardware can be used for the task. The sections below describe one of our team's recent contributions (Mihajlovic and Zilic 2011) to the area of execution trace compression, which follows past work in the compression of certain Reduced Instruction Set Computer (RISC) machine code (Mihajlovic, Zilic, and Radecka 2010, 2007).

9.6.1 Overview

In the hardware development of SoCs, it is common for a developer to prototype a design on a Field-Programmable Gate Array (FPGA) in order to perform functional verification. This is commonly referred to as emulation. Once verified, some low-volume designs may eventually even be released as emulated implementations on an FPGA, while higher-volume designs can eventually be manufactured as Application-Specific Integrated Circuits (ASICs).

Many modern FPGAs contain on-chip embedded SRAM that may be harnessed by a developer to use in the design. It is difficult (and sometimes not preferable) to match the requirements of a design with an FPGA whose resources will be fully utilized by that design. Thus, an FPGA will often contain additional logic and memory that remains unused by the design.

Our team has developed an address trace compression scheme (Mihajlovic and Zilic 2011) that harnesses the unused SRAM found on many FPGAs to compress address traces and allow them to be streamed off-chip to a host in real time. Since many ASIC processors already contain trace buffers for storing trace data, the method can also be used by ASICs to transform the memory ordinarily used as a trace buffer into a *compression dictionary*, enabling real-time streaming of address traces.

The role of the compression scheme within a trace-based debugging scenario is shown in Figure 9.6, applied to a single emulated or real SoC core. After a developer configures the tracing triggers (not seen) and initiates the trace experiment through the TAP interface, traces are streamed in real time to the host after passing through the *trace compression module*. Trace data is stored on the host until the trace experiment is complete, at which time it is decompressed off-line by host software. It is then passed to the software debugger, which is able to identify the execution path taken by the software and to presumably locate the bug. Trace data may optionally be provided to performance profiling software so that analysis and optimization of the software can be performed.

Software trace compression algorithms (Burtscher et al. 2005) are known to be computationally intense and to use a vast amount of memory, while hardware trace compression schemes must meet several constraints. Traces must be compressed in real time and are typically given limited amounts of hardware area and memory for their task. The trace compression scheme seen in Figure 9.7 is designed with those constraints in mind, and additionally targets fixed instruction-width (RISC) processors due to their greater prevalence in embedded computing and the increased potential of compressing their instruction addresses. The scheme is pipelined (stages denoted by dashed lines)

TABLE 9.1

Example of CAE with 16-Bit Addresses

Pre-CAE		Post-CAE	
Address	Instruction	BTA	Length
0xc130	add	0xc130	2
0xc132	sub		
0xc134	jmp 0xd1a4		
0xd1a4	load	0xd1a4	2
0xd1a6	sub	OR	
0xd1a8	bnz 0xe6d8	0xd1a4	4
0xd1ac	xor		
0xd1b0	jmp 0xe6d8		
0xe6d8	jmp 0xf632	0xe6d8	0

and progressively harnesses a number of transformations and a custom encoding to achieve its results. The sections below describe its functionality in greater detail.

9.6.2   Consecutive Address Elimination

Certain patterns can be noticed in the order of a typical sequence of executed instruction addresses. As can be seen in the leftmost column of Table 9.1, a series of sequential instruction addresses is repeatedly followed by a non-sequential address. These interruptions in the instruction sequence occur at branch instructions and represent that a branch has been taken. This property can be used to reduce the amount of address trace data that needs to be collected and is the basis of this first compression stage. In a fixed instruction-width processor, intermediate instructions can be inferred if only the instruction addresses of *taken* branches are collected, along with a count of instructions between them, denoted *length*. The reasoning behind the length value is demonstrated in Table 9.1. In that example, the address 0xe6d8 can be reached through two separate paths of execution. To determine the path taken, it is necessary to store the address with a *length* that indicates the end of a sequence of instructions. Together they are referred to as address-length (AL) combinations in later sections.

Fixed instruction-width processors also have the property that some of the least significant bits (LSBits) of the instruction addresses will always be zero. The Consecutive Address Elimination (CAE) stage uses this property to forward only the useful bits of an instruction address. In the case of 16-bit processors, only the leading 15 most significant bits (MSBits) are needed, while our design targets 32-bit processors, requiring only the leading 30 MSBits.

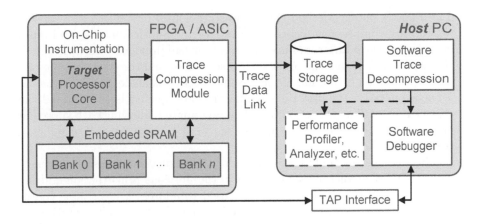

FIGURE 9.6
Trace-based debugging scenario with address trace compression

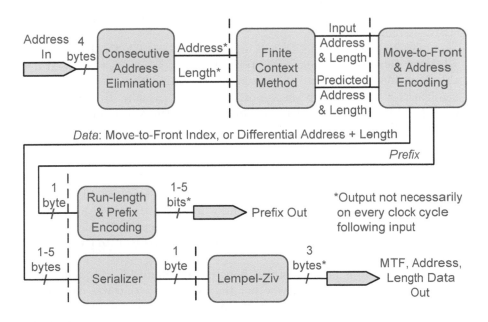

FIGURE 9.7
Trace compression scheme from Mihajlovic and Zilic (2011)

9.6.3   Finite Context Method

The Finite Context Method (FCM) was previously introduced in a software-based trace compression technique (Sazeides and Smith 1997) to avoid repeatedly storing the addresses of instructions that had previously been executed in the same order. Using the previous n instructions and the order in which they were executed, a prediction of the next instruction is made. Similar to the way a cache operates, the first time an instruction sequence is encountered there will be a *miss*. The prediction will then be updated, allowing each subsequent encounter with the same instruction sequence to be a *hit* if it matches the prediction.

Just as a cache would use portions of an instruction address to generate a *tag*, the FCM predictor uses portions of the previous n instruction addresses to construct a *search key*. The key is used to address a hash table containing a prediction of the 'next' AL combination. In the implementation seen in Figure 9.8, a search key is created by performing an exclusive-OR (XOR) on the 4 LSBits of the previous instruction's length term (L) together with staggered 16-bit portions of the previous 4 instruction addresses (h1–h4). The key is constructed is this particular way to obtain a unique value for each sequence of ALs, minimizing *aliasing* between different sequences of instructions. Each hash table entry contains the 30 MSBits of the predicted address concatenated with an 8-bit length. The size of the table can be catered to the amount of unused memory in an FPGA or to hardware area limitations of an ASIC design.

On each cycle, a prediction will be read from the table for the newest AL, while the table is simultaneously updated with a new prediction based upon the 4 previous ALs. Due to the nature of this transaction, the table must be implemented in at least a single-ported memory (usually SRAM). The predicted and input ALs are then both forwarded to the Move-to-Front/Address Encoding (MTF/AE) stage.

9.6.4   Move-to-Front and Address Encoding

This compression stage is responsible for performing two distinct operations: the first is a search of an AL dictionary implementing the Move-to-Front (MTF) transform (Bentley et al. 1986), and the second is the encoding of an AL as the smallest available representation before passing it to later compression stages.

A comparison is first performed between the input AL and predicted AL from the FCM stage to find whether the FCM prediction was correct. If it was not, the MTF array, which contains the last m previously encountered AL combinations, is searched for the input AL. The m parameter represents the length of the dictionary and can vary depending upon the number of available logic blocks on an FPGA or available hardware area on an ASIC. Using parallel comparators, each array element is compared with the input AL

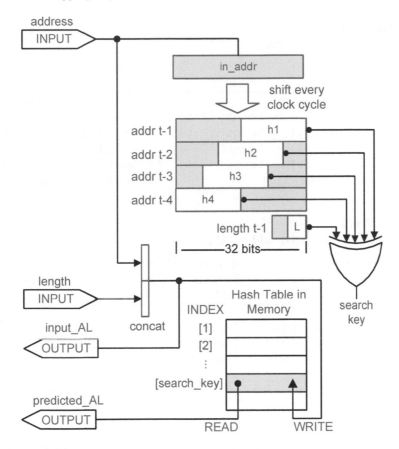

FIGURE 9.8
Finite context method

in a single clock cycle. If there is a match, the matching element's array index is noted. The element containing the matching AL is then moved to the top of the array, while other elements are shifted down to take up its old position. If a match cannot be found, the input AL is placed at the top of the array and the remaining array elements shifted down, in the process eliminating the array element which was least recently used (LRU).

Second, the encoding of ALs is based upon the scheme in Table 9.2, which separates encoded ALs into *prefix* and *data* streams. This is done to allow the two streams to be compressed differently in later stages, extracting additional redundancy from the data to achieve better compression performance.

The encoding works by outputting a single 0x00 prefix byte when a correct AL prediction is made by the FCM stage. If the AL was found in the MTF array instead, a prefix byte of 0x05 will be output along with a data byte representing the MTF array index. Failing both of the previous possibil-

TABLE 9.2
Address Encoding Scheme

Prefix Byte	Data Byte(s)	Meaning
0x00	–	correct FCM prediction
0x01	addr[9:2] + length[7:0]	1 byte Δ from previous address
0x02	addr[17:2] + length[7:0]	2 bytes Δ from previous address
0x03	addr[25:2] + length[7:0]	3 bytes Δ from previous address
0x04	addr[31:2] + length[7:0]	4 bytes Δ from previous address
0x05	mtf_index[7:0]	AL found in MTF array

TABLE 9.3
Example of Differential Address Encoding – 16-Bit Addresses

Effective Address[a] (binary)	Δ from Previous Address	Output Prefix	Differential Address
1010 0001 0010 000X	–	–	–
1010 0001 0100 010X	1 byte	0x01	1010 0010
1100 0010 1001 100X	2 bytes	0x02	0110 0001 0100 1100
1100 0011 0011 001X	1 byte	0x01	1001 1001

[a]After ignoring the redundant least significant bit of each 16-bit address, marked as X in the table.

ities, the differential address encoding scheme depicted in Table 9.3 is used to encode the AL. The scheme allows only the portion of the AL that differs from the previously encountered AL to be output. The table provides an example using 16-bit addresses for clarity, though 32-bit addresses are used in our implementation.

The differential encoding scheme also takes advantage of the fact that 32-bit fixed instruction-width processors have two redundant LSBits. After ignoring these bits, the instruction address is compared with the previously encountered instruction address in a byte-wise fashion. The resulting difference between the addresses could then be between 1 and 4 bytes, which are represented by a prefix of between 0x01 and 0x04, respectively. The prefix represents the number of difference bytes in the *data* stream, which are in every case followed by a single-byte length term.

9.6.5 Data Stream Serializer

The *data* stream output of the MTF/AE stage can be between 1 and 5 bytes in a single cycle, and it must be serialized before being passed to the Lempel–

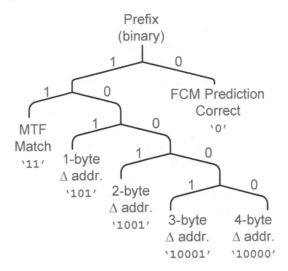

FIGURE 9.9
Huffman tree for prefix encoding

Ziv (LZ) stage. A simple solution is to implement a small buffer, the minimum size of which can be determined by profiling the instruction code to ensure there are no buffer overflows. A more robust solution would be to clock the serializer at 5× the clock rate of the other stages, which would ensure that it is able to continuously process the worst-case data flow.

9.6.6 Run-Length and Prefix Encoding

The input to this stage is any of 5 different prefix bytes seen in Table 9.2. If the prefixes were to be output in a simple binary encoding, their encodings would be the optimal if each prefix was encountered an equal number of times. In practice, however, the compression scheme will generate some prefixes much more frequently than others. To take advantage of this property, binary codes of differing lengths may be assigned to each prefix, such that they are inversely proportional to their frequencies.

Either a static or dynamic encoding may be used to generate codes. Static methods presume that frequencies will not change over time, while dynamic schemes adapt to changing frequencies over time by altering their encodings. Dynamic methods are more complex than static ones and will generally require more hardware area and memory to store a relative frequency count of the inputs. An example of the static Huffman Tree used in Mihajlovic and Zilic (2011) can be seen in Figure 9.9, where the shortest single-bit code is given to correct FCM predictions (prefix byte 0x00), which are found to occur with greatest frequency.

In addition to prefixes occurring with differing frequencies over time, it can be observed that the FCM predictor delivers correct predictions in bursts. Even though each of the predictions maps to a single zero-bit in the Huffman Tree, applying an additional Run-Length Encoding (RLE) allows long sequences of these bits to be additionally compressed.

The RLE replaces sequences of up to 256 zero-bits with a single byte. Since it should only capture bursts of predictions, there is no change to the encoding of Figure 9.9 if there are three consecutive zero-bits or less. However, four or more will result in the prefix output being paused and the zero-bits that follow being counted. When either 256 zero-bits are counted or the first non-prediction prefix is encountered, a byte containing the count will be output. This scheme encodes sequences of between 4 and 260 zero-bits as a 12-bit value, consisting of 4 initial zero-bits followed by a 1-byte count.

9.6.7   Lempel–Ziv Encoding

Just as prefix encoding and run-length encoding are used as a final compression step for the *prefix* stream, a final compression stage can be applied to the *data* stream to extract any vestigial redundancies that may persist. The input to this stage is the *data* stream, composed of bytes that represent both MTF indices as well as differential AL combinations of 2–5 bytes in length. This stage is particularly useful at identifying non-obvious redundancies in the input data, such as when there are repeating combinations inside the *data* stream. To allow this, the input data is processed at byte-level granularity by the Lempel–Ziv (LZ77) algorithm (Ziv and Lempel 1977). In part, this algorithm is chosen due to its simplicity and suitability to hardware implementation. The performance of LZ77 depends upon the amount of redundancy left in the input data and in theory does not guarantee compression. However, in practice it has been found that there is enough redundancy in the *data* stream to achieve up to 5× compression consistently.

The LZ77 implementation holds a dictionary array of 1-byte elements, the size of which can be adjusted to the number of available logic blocks on the FPGA, or hardware area on the ASIC. Upon receiving an input byte, a parallel comparator searches the array for matching elements and flags them as 'found' by writing to a *tag* array consisting of 1 bit per element. The dictionary array is then shifted by 1 element, and the input byte added to the top of the array. If there are any matching elements in the *tag* array, only they are searched upon receiving the next input. This allows a sequence of matching bytes to be identified by incrementally searching the array. If no match is initially found, or a sequence of inputs can no longer be matched, an output is issued that includes the initial matching dictionary array index, the number of elements matched, and the new input byte.

9.7 Conclusions

This chapter presents an overview of the software debugging challenges that are being faced by multicore systems-on-chip. Traditional debugging methods are no longer suitable for debugging concurrently executing multithreaded software and will become less tenable in the future as the sole solutions to debugging this type of software. The use of trace generation to complement traditional debug methods is gaining traction in both academic and commercial spheres and is expected to take an increased role in the debugging of future multithreaded software. However, there are some obstacles that must be overcome before trace generation can take an increased role, and prime among them is the issue of how to transfer massive amounts of trace data off-chip for analysis. We have presented our recent work in developing an instruction-address trace compression scheme that aims to mitigate this problem for what is arguably the most commonly traced data type. We have also identified three further impediments to the proliferation of tracing as a mainstream debugging tool:

1. *Dedicated debug hardware.* To reduce the amount of time and money spent by developers in debugging software, hardware designers will need to dedicate more on-chip area to debugging infrastructure. This includes trace generation, compression, triggering, and control.

2. *Moving real-time data traces off-chip.* The compression of traces containing more unpredictable data types have not been matched by recent advances in execution trace compression. Strategies are needed to address the volume of data typically generated when observing other types of on-chip data.

3. *Streamlined software debugging tools.* Prominent software debuggers such as GDB, and its many front-ends, will need to include more support for trace collection and analysis. Just as software development tools have increasingly reduced the manual effort needed to create software, so too is there a need for comprehensive tools to isolate and identify software bugs.

In the multicore era, there have already been widespread changes in the way software has been traditionally developed. As software complexity increases, it is essential that the ability to create software with few, or zero, bugs is preserved. The growing cost of software debugging is ample evidence that debugging methods must also adapt to the new software development paradigm. Trace generation is one tool that can help bridge the gap, allowing the potential of multithreaded software development to be realized.

9.8   Glossary

Breakpoint: A user-configurable instruction address at which execution is halted. Implemented either through hardware or software instrumentation.

Controllability: The ability to directly control the internal states of hardware.

Data Trace: A log of registers (excluding the program counter) or other data within a processor that are collected during a trace experiment.

Execution Trace: A log of executed instruction addresses that are collected during a trace experiment.

Field Programmable Gate Array (FPGA): A chip composed of logic, memory, and other cores that can be user-configured into a custom hardware circuit.

Host: The computer that is used to perform debugging on the *target*, usually with the aid of a software debugger.

Observability: The ability to directly observe the internal states of hardware.

Skid: The number of clock cycles that elapse between the time that a processor core is sent a signal to halt its execution and the time that its execution is actually halted.

Target: The processor core the contains the program-being-debugged.

Trace generation: A non-intrusive debugging method that logs on-chip data for later analysis.

Trigger: An on-chip, user-configurable hardware structure that is responsible for generating trace data when a desired event occurs.

Watchpoint: The user-configurable value of a register or other data structure that will cause execution to be halted. Implemented either through hardware or software instrumentation.

Part V

Networks-on-Chip

10

On Chip Interconnects for Multicore Architectures

Prasun Ghosal

Department of Information Technology, Bengal Engineering and Science University, Shibpur, India

Soumyajit Poddar

School of VLSI Technology, Bengal Engineering and Science University, Shibpur, India

CONTENTS

10.1 Introduction

How do we exploit current technology to harness the full power of parallel computing architectures within the constraints of minimum energy, die-area, and complexity? Finding a viable answer to the above question needs a paradigm shift from the 1990s philosophy that *a better processor has a functionally better and faster arithmetic logic unit (ALU) than its predecessor*. Today's process technologies are producing transistors in the Deep Sub-Micron (DSM) regime. These devices operate at GHz (gigahertz) frequencies at pico Watt power levels. Using these devices, it is possible to integrate several different types of functionalities into the processor itself. Also, Moore's Law has been followed quite well over the past few years. However, the current level of progress has reached a communication bottleneck in terms of interconnect and intra-chip communication resources. The latter problem is solved by putting emphasis on designing a reliable and efficient Network-on-Chip (NoC). Interconnects using metal wires have several issues for distributing both signals and clocks across the chip, e.g., power consumption of wires and use of repeaters. The remainder of this chapter discusses, from a communication viewpoint, the role of current and future interconnects used in multicore architectures.

10.2   Evolution of Interconnects for Multicore Architectures

There are various driving forces behind today's growth of the semiconductor industry. One of these forces is the continuing demand from the electronics and embedded market to pack more and more devices onto the same chip. According to Moore's Law, the number of components that can be incorporated per integrated circuit increases exponentially over time (Moore 1998). As the component count increases per chip, so does the connection infrastructure. At least, for the last decade, scaling down transistor size went hand-in-hand with scaling down of wire width. However, there is a limit to the minimum width a metal wire can be scaled down, below which fringing effects are exacerbated.

Interconnects provide two important functions, namely, supply rails and signal lines. Supply rails carry DC and signal lines carry AC (for digital circuits) or varying DC (for mixed signal circuits). In addition, for digital circuits, clock lines need careful design to minimize clock skew. Maintaining global synchrony is an important issue for multicore chips designed today. Providing a global clock signal across the chip is difficult due to the capacitance of such a network and its complexity (which is further increased by gated clocks). Applications such as multimedia and communication demand heterogeneous cores to be placed on the same chip. These cores may not be regular in physical size and/or function. This would cause the interconnect infrastructure also to be heterogeneous.

Mainly, wires are classified into three groups, based on their length. They are the local (short), intermediate (medium), and global (long) wires. According to *The International Technology Roadmap for Semiconductors 2009 for Interconnects* (2009), signal delay time for local wires is typically much smaller than a clock cycle and decreases with scaling for technology nodes greater than 45 nm. Below 45 nm, gate delay begins to dominate the local wire delay and offsets scaling benefits. The delay for intermediate wires will grow only slightly with technology scaling.

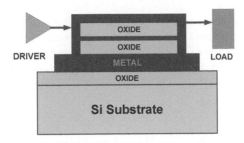

FIGURE 10.1
Side view of multipath interconnect

The global wire delay increases with scaling due to the increased resistance and increased length (capacitance and inductance effects). The power distribution network for the chip needs very low resistance global wires, mainly due to the low values of the supply voltage and high values of current. Alternate packaging techniques such as ball grid array are used to distribute the power across the chip and eliminate lateral power feeds that face voltage drop problems. In the absence of repeaters, a global wire suffers from a delay proportional to the square of its wire length. For this reason, a longer wire is usually broken into several segments (to reduce the delay quadratically), and repeaters are inserted between segments to boost the signal level (Rabaey, Chandrakasan, and Nikolic 2003).

Multipath interconnects are now used for high-performance interconnects carrying high currents. The concept is shown in Figure 10.1. It is a type of parallel interconnection (Goel 2001).

10.2.1   More than Moore Trends: A New Perspective

So far we have seen that scaling down transistors constitutes the main theme of *Moore's Law*. The reason for scaling down transistor size is to increase performance while keeping power dissipation low. New techniques and materials were introduced for *more Moore* performance in the digital-processing and storage domain. However, one more perspective about VLSI technology and innovation is the *More than Moore* trend (Arden et al. 2010). Judging from the present level of technological innovation, mere increase in digital processing capability is not enough. What is needed is the diversification of functionality (not necessarily digital functionality). This increase in diverse functionality (for example, mixed signal and radio frequency) integrated into a single chip constitutes the *More than Moore* trend. It is so named because it covers more applications than *Moore's Law* does. *More than Moore* trends are not a kind of competition to *Moore's Law*. The difference is in the type of functionality: *More than Moore* refers to interfacing the computationally intelligent devices of the *more Moore* trend.

For the case of interconnects this implies that emerging technologies need new types of interconnect technology, architecture, and physical design algorithms. Different kinds of emerging interconnect technology will be discussed in the next section. A brief overview of interconnect architecture is given in the next sub-section. However routing and other physical design issues are beyond the scope of this chapter.

10.2.2   From Single Bus Based to Network-on-Chip Architectures

In earlier microprocessors and microcontrollers, various component blocks including the CPU were attached to a bus. A bus is a large synchronous global interconnect. There are a few issues with buses, namely, the fact that all peripherals attached to the bus cannot transfer data through it at the same time. Moreover, these peripherals may not have a uniform interface with the global bus, resulting in arbitration problems. The wafer size is growing, so is the total chip area, resulting in large communication overhead (in the form of buses). This high communication cost cannot be fulfilled by single-bus architectures alone. So the next most viable choice was to use multiple buses and use sophisticated arbitration schemes to ensure communication reliability.

As semiconductor transistor dimensions shrink and increasing quantities of IP block functions are added to a chip, the physical infrastructure that carries data on the chip and guarantees quality of service fails to meet performance requirements. Many of today's systems-on-chip are too complex to utilize a traditional hierarchical bus or crossbar interconnect approach. The current trend for connecting large numbers of cores together is using Network-on-Chip architectures. The first of its kind was the *Arteris NoC* (Wein 2007). This NoC was based on a scalable Globally Asynchronous Locally Synchronous (GALS) architecture (Meincke et al. 1999). The main feature of Arteris was a flexible and user-defined network topology.

An NoC solves many problems that bus based systems on chips do not. For example, communication occurs in three distinct abstraction layers:

Physical Link Layer Deals with the issues of interconnection devices and signaling.

Transport Layer Deals with transporting packets across the physical network.

Transaction Layer Provides a network interface between the IP cores and transport layer.

This approach is shown in Figure 10.2. The regular mesh based architecture is advantageous from a design perspective because of regular electrical properties and reliable performance. The Arteris Network-on-Chip platform is highly developed today and provides the three layers mentioned above as follows:

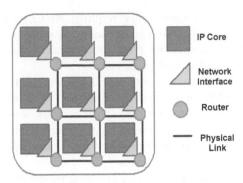

FIGURE 10.2
Network-on-chip concept

Transaction Layer With Arteris NoC technology, network interface units (NIU) manage communication with a connected IP core. NIUs convert traditional Advanced Microcontroller Bus Architecture (AMBA), Open Core Protocol (OCP), and proprietary protocol load and store transactions into packets for transport across the network.

Transport Layer This layer deals exclusively with packets. Only a limited amount of information in packet headers needs to be examined in order to determine the required transport operations. The transport layer can safely ignore the specifics of the transactions being managed at its own level. This simplifies the hardware required for switching and routing functions and enables higher operating frequencies.

Physical Layer This layer defines how packets are actually transmitted between NoC units. Various link types with different capabilities, such as transport links, globally asynchronous/locally synchronous (GALS) links for longer distances, or Chip-to-Chip links, can be employed. Separate layers make it possible to change links, or their characteristics, without affecting the transport or transaction layers. Because all connections are point-to-point, high fan-out nets are prevented, thus providing better performance and easier routing.

With today's platform based SoC design methodologies, it is important for design teams to be able to quickly generate derivatives based on a single SoC platform. These derivatives usually require changing IP blocks that use a different protocol configuration, or even an entirely different protocol, than the original IP blocks.

In traditional bus and crossbar interconnects, changing an IP block means not only dealing with a different transaction protocol, but also making changes to the path widths and physical bus topology to accommodate the new IP. At a minimum, bridging is usually required. In comparison, changing an IP block on an NoC interconnect only requires changing the configuration of the network interface unit (NIU) to accommodate the protocol and bit width changes for that IP block. The NoC interconnect's packet-based transport allows mixing data widths and clock rates.

Currently there are several successful implementations of Networks-on-Chip. As today's chips may contain hundreds of cores, the network must provide algorithms to route the data packets from source to destination in an efficient and reliable way. Several router architectures and routing algorithms (both adaptive and deterministic) have been proposed to handle the communication load of modern NoCs.

Network flow control, referred to as routing mode, determines how packets are transmitted inside a network. The mode is not directly dependent on routing algorithm. Many algorithms are designed to use some given mode, but most of them do not define which mode should be used.

Store-and-forward is the simplest routing mode. Packets move in one piece, and an entire packet has to be stored in the router's memory before it can be forwarded to the next router. So the buffer memory has to be as large as the largest packet in the network. The latency is the combined time of receiving a packet and sending it ahead. Sending cannot be started before the whole packet is received and stored in the memory of the router.

Virtual cut-through is an improved version of store-and-forward mode. A router can begin to send a packet to the next router as soon as the next router gives a permission. The packet is stored in the router until the forwarding begins. Forwarding can be started before the whole packet is received and stored in the router. The mode needs as much buffer memory as store-and-forward mode, but latencies are lower.

In wormhole routing, packets are divided into small and equal sized flits (flow control digit or flow control unit). A first flit of a packet is routed similarly to packets in virtual cut-through routing. After the first flit, the route is reserved to route the remaining flits of the packet. This route is called a wormhole. Wormhole mode requires less memory than the two other modes because only one flit has to be stored at once.

The interested reader is referred to Gebali, Elmiligi, and El-Kharashi (2009) and Nurmi et al. (2004) where NoC architecture and routing algorithms are discussed in detail.

10.2.3   On Chip Applications

As mentioned previously, due to the *More than Moore* trend, the versatility of chips is increasing rapidly. Apart from pure digital processing applications, communication resources and memory are being embedded into the chip. Analog and mixed signal peripherals are also needed in some sensing and signal processing applications. Power management cores like dynamic supply voltage controllers are needed for power handling in large many-core architectures. Also, radio frequency functionality and reconfigurable cores are now needed in some chips (apart from FPGAs). On the whole these applications demand new and non-conventional interconnect and packaging techniques which require much research.

10.3 Emerging Technologies for Interconnections

Recently researchers have been looking toward upcoming interconnect technologies and architectures to solve multicore communication problems. Some solutions have already found their way into the semiconductor ecosystem. These are as follows:

- 3D technology

- Wireless technology

- RF waveguide technology

- Photonic interconnect

- Nano materials

The above mentioned technologies will be discussed in detail in this section.

10.3.1 Three Dimensional Interconnects

In order to cope with the ever decreasing device sizes and increasing wiring complexity, a new technology came to the forefront, namely, the *3D IC* (3-dimensional integrated circuit) technology. This name was chosen because the wafers containing the devices were *stacked* one on top of another and each layer/plane communicates with its top or bottom layer using through silicon vias (TSV). These TSVs are built by making perforations in the wafer by deep etching techniques and then filling up the hole thus made with a conductive compound.

The primary advantage of using 3D IC technology is the reduced global interconnect length. This is because cores that were far apart in planar 2D wafers can now find alternate shorter 3D paths (combination of horizontal and vertical paths) to communicate due to the *close proximity* in their positions. Some other advantages of 3D interconnects may be listed as follows:

Heterogeneous integration Several different process technology wafers can be stacked up. Thus more than Moore functions (e.g., RF, mixed signal, optical, and wireless) can be performed by a single chip.

On chip interconnect power reduction Since signals need not travel from wafer to wafer via off chip interconnects, power consumption and RC delay is minimized.

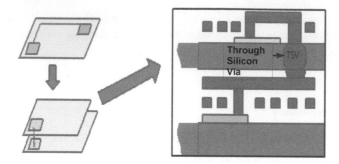

FIGURE 10.3
Reduction of interconnect length from 2D ICs to 3D ICs

Bandwidth increase There is a large number of TSVs that can be used to communicate data between two layers in a parallel fashion (e.g., between memory caches and cores).

On the right side of Figure 10.3 the cross section of a typical 3D IC is shown. The through silicon via connects the bottom device layer with the top device layer. This is contrary to the 2D case where long global interconnects are used to connect two cores placed far apart. There are several technologies used to fabricate 3D ICs (*The International Technology Roadmap for Semiconductors 2009 for Interconnects* 2009), as shown in Figure 10.4. Some of the major challenges are as follows.

EMI (Electromagnetic Interference) Coupling between the top layer of metal of the first active layer and the device on the second active layer increases interconnect coupling capacitance and crosstalk. Interconnect inductance is reduced in 3D ICs as shorter wire lengths help to reduce the inductance.

Reliability issues Electrothermal and thermomechanical effects between various active layers can influence electromigration. Die yield issues due to mismatches between die yields in different layers affect the net yield of 3D chips.

10.3.2   Photonic Interconnects

Silicon is transparent to light of wavelengths close to $1.5\,\mu$m. Silicon waveguides are now being used to transmit/receive large amounts of data through photonics. The capability for a single optical path to accommodate multiple wavelengths increases the data-carrying capacity by providing higher bandwidths not achievable by electrical means (with wires, for example). The optical path shown in Figure 10.5 is built using standard silicon photonic components (Li et al. 2009). An off chip laser source which has a broad bandwidth (of

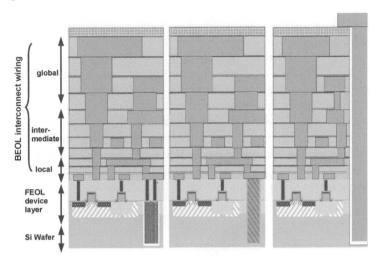

FIGURE 10.4
Schematic representation of TSV first, middle, and last processes (*The International Technology Roadmap for Semiconductors 2009 for Interconnects* 2009), (BEOL stands for Back End of Line and FEOL stands for Front End of Line).

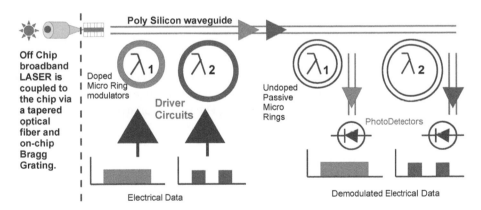

FIGURE 10.5
Schematic of photonic interconnect using micro ring resonators

wavelengths) is coupled to the chip by an optical fiber using standard coupling techniques (Kopp et al. 2011). Since the fiber diameter is around $100\,\mu$m and the on chip polysilicon waveguide has a cross-sectional area of $8\,\mu$m $\times$ $8\,\mu$m, the fiber is tapered and a lens is placed at the end where the fiber injects light into the chip. An on chip Bragg grating couples the injected light into the polysilicon waveguide that carries the coupled light throughout the chip. Another kind of architecture is the on-chip light source architecture, where laser sources like VCSELs (Vertical Cavity Surface Emitting Lasers) are present on the same die. Although such devices have been built that consume very little power, heat dissipation and area overhead, coupled with heterogeneous die integration, is still a hurdle for on chip light source architecture.

For optical interconnect architectures, it is possible to define a critical length above which optical interconnects are faster than their metal-wire counterpart. The critical length, which depends on the quality of the optical components, has been assessed to be on the order of millimeters. Optical interconnects have the potential to simplify design and do not suffer from undesirable crosstalk in metal-dielectric interconnects.

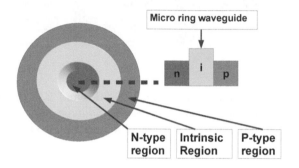

FIGURE 10.6
A simple schematic of a micro ring resonator

The heart of the photonic interconnect system is the micro ring resonator, as shown in Figure 10.6. Detailed operation of the micro ring device can be found in Mookherjea and Melloni (2008). The ring resonator is in fact a poly-Si waveguide bent onto itself. It has a particular resonant frequency that depends on the optical path length of the ring. Depending on its physical structure and material properties, the rings can function as filters, modulators, switches, or detectors. The key performance parameters include coupling efficiency, operating voltage, switching time, waveguide loss, overall power, modulation depth/extinction ratio, and area.

The main purpose of a modulator is the control of the flow of light into a particular waveguide with a standard digital signal. A frequency-dependent filter can be used to introduce multiplexing that enables the transmission of multiple signals in a single waveguide. A large variety of CMOS compatible modulators can be used, including resonators and Mach Zehnder interferometers.

Waveguides provide the means for light propagation on the chip with minimum loses. A large refractive index contrast between the waveguide and the surrounding materials enables tight turn radii and small pitches, but at the expense of lower speed that decreases with the inverse of the effective refractive

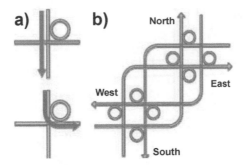

FIGURE 10.7
A photonic switch (left) and a non-blocking photonic router (right) (Wang et al. 2008)

index. Reported on-die waveguides using materials that are already common in the industry include Si, Si_3N_4, or $Si_3O_xN_y$ cores with SiO_2 cladding.

There are a few disadvantages with optical interconnects, for example, to maintain the frequency response constant, the optoelectronic devices (like rings) must be kept within a constant temperature range. This requires heating elements which in turn require more power.

The existing interconnect solutions for multicore architectures include optical routers and switches. There is a restriction on the interconnect topologies, namely, they must be non-blocking. This is because photonic signals cannot be stored in memory elements. The photonic ring can be used to switch the direction of light flow; in effect it can stop the flow of light in one direction, as shown on the left side of Figure 10.7. On the right side of Figure 10.7 is shown a router that can be used to steer the light in any of the four directions. Some of the topologies that are used generally are torus (Figure 10.8) and ring.

10.3.3   Wireless Interconnect Technology

An on chip wireless interconnect consists of metallic antennae on top of the silicon substrate. Sometimes a dielectric layer is placed above the Si substrate for better signal propagation (lower loss/attenuation due to better electromagnetic energy confinement). This provides a parallel channel to conventional interconnect layers as well. The doped Si substrate generates high losses and associated signal attenuation. The silicon substrate can be thinned below $100\,\mu$m to reduce this attenuation (*The International Technology Roadmap for Semiconductors 2009 for Interconnects* 2009). The antenna size scales down linearly with the operating frequency. The bandwidth of wireless interconnects is limited by the bandwidth of the circuits that can be realized in a given process technology. The bandwidth is 25–50% of transistor frequency f_T. With present CMOS technology, it should be possible to support a data rate of 100–200 Mbps.

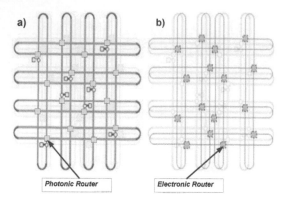

FIGURE 10.8

Torus topology: (a) the photonic network (with the routers shown as small rectangles), and (b) the electrical network (with the gateways shown as small boxes) (Wang et al. 2008)

Wireless interconnects are suited for global signals with large fan-outs such as global clock, reset, sleep, and other moderate bandwidth signals requiring multiple long metal lines. However, some multicore chips are also using wireless interconnect for local and intermediate links. An example is the scalable wireless interconnect (Lee et al. 2009) where the authors have used a concentrated mesh topology. The cores and L2 caches are connected as groups of four units. Sixteen such units form a cluster called a *WCube*. There is a dedicated wireless router for each *WCube*. Frequency Division Multiple Access (FDMA) is used for communicating data either as single-cast or multi-cast. The above schemes are shown in Figure 10.9. Basically, the architecture is a recursive one.

10.3.4   RF Waveguide Interconnects

Radio frequency transmission line type interconnections, also known as superconducting interconnects, are used today because of their very high signal propagation speeds as compared to metallic interconnects. A major contributor to the use of these interconnects is high critical temperature superconductors on silicon as well as gallium arsenide substrates. Apart from speed, these interconnects have higher packing density than metal interconnects and there is virtually no signal dispersion for frequencies up to many tens of gigahertz. The structure of these interconnects resembles micro-strip lines.

10.3.5   Carbon Nanotubes (CNT)

Carbon Nanotubes (CNTs) are a very interesting research topic today. A CNT is an allotrope of carbon. Carbon nanotubes are thin, hollow cylinders

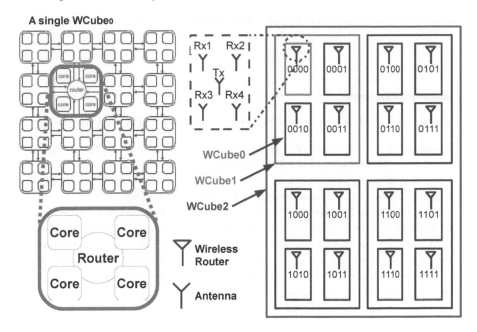

FIGURE 10.9
Concentrated mesh topology and wireless routers shown from left to right

made entirely of carbon. The walls of an NT are formed from graphene — a honey-comb lattice of carbon atoms.

CNTs have several advantages compared to the previous interconnect solutions (Goel 2007). They may be listed as follows:

- The carrier transport in CNTs is one dimensional, resulting in ballistic transport, no scattering, and lesser power dissipation.

- The carbon–carbon covalent bonds are very strong, hence CNTs are mechanically very strong and thermally stable.

- Due to their mechanical strength, CNTs have good resistance to electromigration.

- They can conduct currents 1000 times greater than copper wires.

There are two main types of CNTs: *single-walled nanotubes (SWCNTs)* and *multiwalled nanotubes (MWCNTs)* (Banerjee, Im, and Srivastava 2006). In SWCNTs the cylindrical structure consists of a single layer of graphene, while MWCNTs consist of multiple concentric cylinders or the graphene sheet is simply rolled in around itself, resembling a scroll of parchment. Integration of CNTs with silicon is now possible by two approaches, namely, the bottom-up approach and the buried-catalyst approach (Kreupl et al. 2002). The interested reader is referred to Banerjee, Im, and Srivastava (2006) for further details on CNTs.

10.4   Conclusion and Future Research Directions

Various types of interconnect have been studied throughout the chapter. Interconnect advantages and fabrication procedures have been very briefly discussed. A few applications, like Network-on-Chip, also have been given where required. More-than-Moore trends also have been stated.

An insight to be gained from this chapter is that much research is directed at finding the best solution toward an interconnect fabric and architecture that optimizes performance, power, and reliability. The emerging technologies for interconnects discussed previously need cheaper and simpler fabrication processes to enter into commercialization. Thus this field provides ample opportunity to device physicists, interconnect architects, and process engineers for research and development.

The future awaits many promising possible interconnect paradigms. Apart from the CNT based approach, which holds great scope, there are others like Graphene Nano Ribbons (GNRs) and spin based interconnects. GNRs can be considered as unrolled CNTs and have many electronic properties in common with CNTs. The bandgap in a GNR is determined by its width and edge geometry and can therefore be controlled through proper patterning.

In spin based devices, the spin of an electron or group of electrons or an electron-hole pair (called an *exciton*) is used to denote state variables (like low or high in electronic circuits). However, the difference is that, unlike electronic circuits, which suffer from energy dissipation (due to charging/discharging of capacitors), spin-tronic circuits can work at very low power, are fast, are nonvolatile, and can achieve high integration densities.

11

Routing in Multicore NoCs

Prasun Ghosal and Tuhin Subhra Das

Department of Information Technology, Bengal Engineering and Science University, Shibpur, India

CONTENTS

11.1   Introduction

Chip design has four distinct primary aspects, viz., computing, communication, memory, and I/O (input/output) or interfacing. For modern high-end System-On-Chip (SoC) designs, conventional bus based interconnection architectures fail to address the problem of on-chip communication due to increased integration density and communication complexity. Networks-on-Chip (NoC) is a viable alternative as well as a reliable solution for on-chip communication among multiple cores (Rantala, Lehtonen, and Plosila 2006). A Network-on-chip (NoC) is a pre-designed network fabric consisting of routers and links connected in a definite topology, used to communicate between different IP cores. An NoC can provide separation between computation and communication, it can support modularity and IP reuse via standard interfaces, handle synchronization issues, and hence increase system productivity by reducing the complexity of interconnects, consumed power, network latency, noise, and all other important metrics which are desired for a high performance system-on-chip. The performance of an NoC architecture mainly depends on its topology, switching mechanism, routing algorithms, and flow control (see Figure 11.1. This chapter will review these factors in detail.

In an NoC, network topology and the related routing policy play an important role in the improvement of overall performance. A structural scalable interconnection architecture may significantly increase the performance of an NoC (Kundu et al. 2008) in terms of scalability, modularity, transport latency, parallelism, and all other important metrics. Also, a well-constructed network architecture may balance the load on the network and try to reduce the chance of causing a hot spot. Therefore, from a design point of view, a very good on-chip network should have both low latency and a high level of parallelism that can effectively support a diverse variety of workloads. The interconnection architecture should also have a small diameter as well as a small node degree and large bisection width (Kundu et al. 2008). Network switches should be small, energy efficient, and fast. Generally it has been observed that the performance of a system decreases with increasing size. Scalability and performance issues are of primary concern in designing new architectures. The larger the architecture becomes (with a large number of IP cores), different issues like communication latency, congestion, fault tolerance, deadlock prevention policy, etc., become predominant factors in optimization of the design. As discussed earlier, the efficiency of the network largely depends on the underlying topology as this affects latency, bandwidth, throughput, overall area, and power consumption. Moreover, topology plays an important role in developing routing techniques. A routing technique is the strategy for moving data through the NoC. Routing is the intelligence that determines the path of data transport. In designing a topology that can give effective performance improvement, different types of regular and irregular topologies have

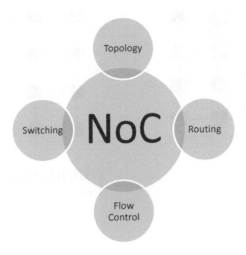

FIGURE 11.1
Factors affecting the performance of an NoC

been designed keeping in mind the diameter, bisection width, and bandwidth requirements. To map the IP cores into a topology structure, a number of approaches have been proposed so that the resources of the topology structure can be utilized efficiently. Besides the topology, many different routing algorithms, considering many aspects of NoC design, are discussed throughout the chapter.

11.2 Routing Topologies in NoC

Network topology is the primary factor that determines the ability of a network to communicate data. It affects the network bandwidth, latency, throughput, overall area, fault tolerance, and power consumption. Moreover, it plays an important role in designing the routing strategy and mapping the IP cores in the network.

A topology may be regular or irregular. But, with the advancement of technology, additional choices are to switch from 2D to 3D topology, as well as from electrical interconnection to optical or wireless interconnection. To perfectly suit the technology in use as well as to get the maximum benefit under that environment, previous 2D topologies are being modified by researchers. Depending upon the technology used, we can classify the topologies used in an NoC as follows:

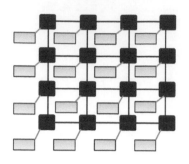

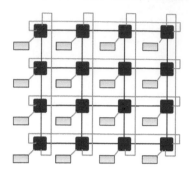

FIGURE 11.2
Mesh topology

FIGURE 11.3
Torus

- Topologies in 2D NoCs

- Topologies in 3D NoCs

- Topologies in Optical NoCs

- Topologies in Wireless NoCs

11.2.1   Topologies in 2D NoCs

A brief description of various topologies used in 2D NoCs follows.

11.2.1.1   2D Regular Topologies

Among different regular topologies *mesh* or *Manhattan street network* is the most common and simplest. The network is arranged in a matrix-like arrangement with m columns and n rows. At each intersection, routers are placed and cores (computational resources) are placed near routers. With a Cartesian XY coordinate system we can designate the address of a router or resource in this topology. Figure 11.2 shows the structure.

Torus and *folded torus* are improved versions of mesh. A torus network is formed from a mesh by connecting the heads of columns with their tails and the right-hand edges of the rows with their left-hand edges in a mesh. When a torus is folded, the long end-around connections are avoided. This avoids excessive delay due to long wires. These topologies are shown in Figures 11.3 and 11.4, respectively.

Another regular topology is the *octagon*, where each node is associated with an IP core and a switch, as shown in Figure 11.5. The network is extended to multidimensional space for a system consisting of more than eight nodes. In the case of a *star* (see Figure 11.6), the computational resources lie on the points of the star with the central router in the middle. The primary disadvantage of this topology is that the central router must be large enough

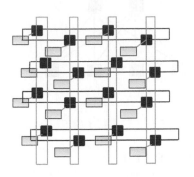

FIGURE 11.4
Folded torus

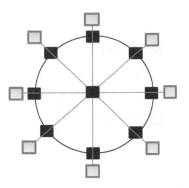

FIGURE 11.5
Octagon

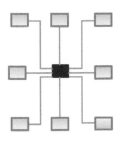

FIGURE 11.6
Star

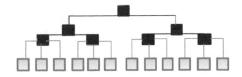

FIGURE 11.7
Binary tree

to handle the heavy traffic passing through it. Another important topology is the *tree*. A *binary tree* topology is shown in Figure 11.7. Here, three IP cores are connected at each of the leaf level nodes. Unlike the binary tree, the *butterfly fat tree* (see Figure 11.9) has IP cores at the leaves but switches are placed at the vertices. But in a simple *butterfly* topology (Figure 11.8), leaf nodes may be present on both sides. In a *polygon*, packets travel in a loop from routers to other nodes in a circular network. The addition of chords makes the network more diverse. This is called a *spidergon*, when there are chords only between opposite routers.

A *honeycomb* is another regular 2D topology (Li and Gu 2009) where regular hexagonal structures are tessellated, forming a regular 3-degree planar graph. The topological structure is shown in Figure 11.10.

11.2.1.2   2D Irregular Topologies

Irregular topologies are formed by mixing different regular topologies to get the advantages of each. Construction of this new efficient topology is usually based on the concept of clustering. Sometimes irregular topologies are formed

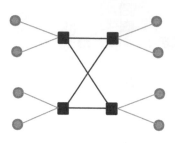

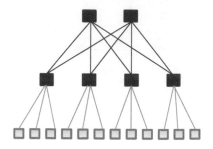

FIGURE 11.8
Butterfly

FIGURE 11.9
Butterfly fat tree

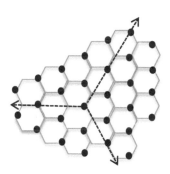

FIGURE 11.10
Honeycomb

when the sizes of the IP cores are unequal or the difference in size among the cores is large.

M-o-T (mesh of tree) An $(M \times N)$ MoT (Manna, Chattopadhyay, and Gupta 2010), where both M and N are powers of 2, and M and N denote the number of row trees and column trees, respectively, is a recursive structure. Figure 11.11 illustrates this. Filled nodes are column trees and blank nodes are row trees. At the leaf node two cores are connected.

Diametrical mesh In a diametrical 2D mesh (Reshadi et al. n.d.; Ghosal and Das 2013) the network is a 2D mesh with 8 extra bypass links (see Figure 11.12). Every router is connected to one IP core.

Diametrical 2D mesh of tree In a diametrical 2D mesh of tree (Ghosal and Karmakar 2012) an effort has been made to combine a diametrical mesh as well as a mesh of tree structure to get the benefits of both. This is shown in Figure 11.13.

Structural diametrical 2D mesh (SD2D) A structural diametrical 2D mesh architecture (Ghosal and Das 2012a, 2012b) is very similar to a 2D

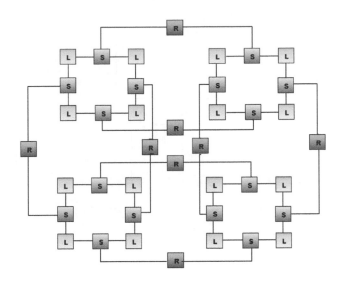

FIGURE 11.11
Mesh-of-tree

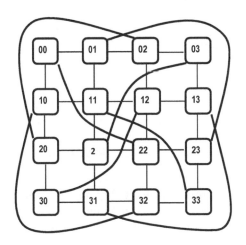

FIGURE 11.12
Diametric 2D mesh

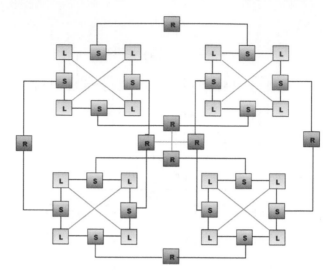

FIGURE 11.13
Diametric 2D mesh of tree

mesh topology but it has some additional diagonal connections compared to a simple 2D mesh. The network may be considered as a collection of three similar types of subnetworks following a specific pattern, as shown in Figure 11.14, where at most three types of square subsets (each subset is of size 3 × 3, as shown by different shades in Figure 11.14) may be considered in a network and each of these subnetworks has two diagonal links, connecting the opposite corners of that subset. Network size may grow in both rows and columns as required.

Star topology This topology (Chen, Peng, and Lai 2010) may be considered as a collection of subnetworks, where a central node of each subnetwork is connected to all of its neighbor nodes directly by links. The central nodes of all subnetworks are connected by a level-2 link (XY type) to reduce the network latency considerably. The size of the network may grow in both rows and columns as per the requirement by a multiple factor of 3, as shown in Figure 11.15.

Irregular mesh custom topology Depending upon the requirement of the application, the mesh structure may be changed so that it can accommodate any size of IP core, as is done in Bolotin et al. (2004), and shown in Figure 11.16.

SPIN Every node has four children and the parent is replicated four times at any level of the tree. The functional IP blocks reside at the leaves and the switches reside at the vertices.

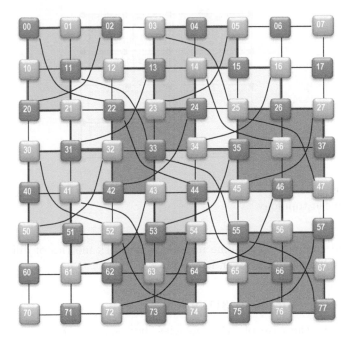

FIGURE 11.14
A 9 × 9 structural diametrical 2D mesh

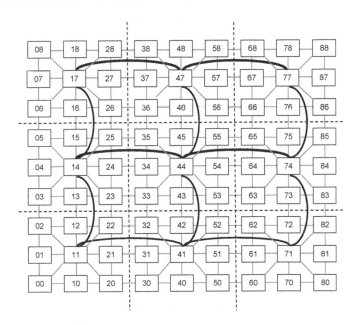

FIGURE 11.15
A 9 × 9 star type topology

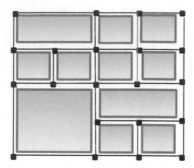

FIGURE 11.16
Custom mesh topology

A detailed comparison of some important network metrics of different of 2D irregular technologies is given in Table 11.1.

11.2.2   Topologies in 3D NoCs

Generally, in the case of 3D topologies, nodes in a 3D interconnect represent either cores or cache banks which are typically tiled in 2D and stacked regularly in 3D. Current fabrication technology only supports vertical links connecting nodes directly above or below between different layers. Presently, it is not feasible to directly connect nodes on different layers with a non-vertical link. This restriction eliminates a large number of topologies containing angled links such as trees. In fact, the only freedom in placing links lies within each layer (Xu et al. 2009). Topologies used in 3D NoCs can be described as follows.

11.2.2.1   Vertical Interconnection Topologies

In the case of vertical interconnection topologies, the connections are supported only among neighboring routers aligned to the same grid as well as in adjacent layers. For a 3D NoC with each grid having dimensions $X \times Y$ and $K\%$ of the routers having vertical connectivity, different possible methods of placement adopted are *uniform, center, periphery, full custom,* and *odd.* Interested readers are referred to Bartzas et al. (2007) for details.

11.2.2.2   Layer-Multiplexed 3D

In this type of topology (Bartzas, Papadopoulos, and Soudris 2009) simple multiplexing and demultiplexing techniques are used for one-layer-per-hop communication. The structure used is similar to a 3D mesh. During packet injection, flits are demultiplexed across different layers and routed across a

TABLE 11.1
Relative Comparison of 2D Irregular Topologies

Topology	Number of Nodes	Bisection Width	Node Degree	Network Diameter
MoT with M row trees and N column trees	$3MN - (M+N)$	$\min(M,N)$	Max 4, Min 2	$2\log_2 M + 2\log_2 N$
Diametrical mesh, N dimensional	N^2	$N+8$	Max 6, Min 4	$N+1$
Diametrical mesh of tree, $2MN$ IP blocks: (Ghosal and Karmakar 2012)	$3MN - (M+N)$	$\min(M,N)+1$	Max 5, Min 2	
SD2D Structural diametrical 2D mesh, $M \times N$ (Ghosal and Das 2012a, 2012b)	$M \times N$	$\leq 2 \times \min(M,N)$	Max 7, Min 3	$\max(M,N)$
STAR, $M \times N$ dimensional (M, N multiples of 3) (Chen, Peng, and Lai 2010)	$M \times N$	$\min(M,N)+\min(M,N)/3$	Max 9, Min 3	$(\max(M,N)-3)/3 + \\ (\min(M,N)-3)/3 + 2$
SPIN, N IP blocks, number of routers: $(N\log_2 N)/8$		$N/2$	8 (non-root), 4 (root)	$\log_2 N$

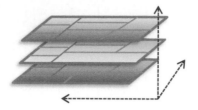

FIGURE 11.17
3D irregular mesh

particular layer. At their destination, during packet ejection, flits from all layers are multiplexed.

11.2.2.3   Hybrid 3D

Various approaches have been tried to integrate several topologies to make a hybrid for combining the advantages of several topologies. A large number of nodes may be accommodated for better performance using such architectures.

As reported by Ramanujam and Lin (2009), in a 3D NoC, the total number of nodes is $N = n_1 \times n_2 \times n_3$, where, n_3 is the number of nodes in the third dimension. The number of PEs (processing elements) on each of the n_3 physical planes is $n_1 \times n_2$ (as a PE can be implemented on only one of the n_3 physical planes of the system). This structure is known as a 2D IC–3D NoC structure.

Another structure may be possible when an interconnect network is contained within one physical plane (i.e., $n_3 = 1$), while each PE is integrated in multiple planes. This is called a 3D IC–2D NoC structure. In the hybrid 3D NoC mentioned above, two topologies have been merged to form a new topology with a 3D IC–3D NoC structure, where both the interconnect network and the PEs can span more than one physical plane of the stack. The structure of a 3D irregular mesh topology is shown in Figure 11.17.

11.2.3   Topologies in Optical or Photonic NoCs

Photonic or optical NoCs are being explored by present day researchers as a viable solution toward metal-interconnect problems such as delay and power consumption. The design of a photonic NoC, due to the high area and energy requirements of a photonic interconnect, is a challenging issue. Currently, photonic networks-on-chip utilize dense wavelength division multiplexing by modulating multiple high bandwidth wavelength channels from an off-chip broadband laser source. Different topologies used in designing optical NoCs are described as follows.

11.2.3.1   Optical Mesh

Due to its successful application in previous generations of NoCs, the mesh has been chosen as an obvious primary choice initially (Wang et al. 2008) for optical or photonic NoCs too. The primary advantage of this topology lies in the planar distribution of the network nodes. The organization of nodes is done in such a way that two nodes will occupy a given column or row in the network at most. Message injection is done by the gateway switches and upon reaching the destination or desired gateway switch the packet is ejected from the network. Due to its bidirectional propagation support through all paths, no message will block any other message.

11.2.3.2   Fat-Tree Based Optical NoC (FONoC)

In a FONoC topology the structure is of a fat tree that represents a hierarchical multi-stage network (Gu, Xu, and Zhang 2009). It uses both types of switching, i.e., circuit (for payload data) as well as packet (for network control data). It is a non-blocking network and provides path diversity to improve performance.

11.2.3.3   Crossbar

Another widely used topology for optical NoCs is the crossbar. For an N-core chip multiprocessor (CMP), switches are organized in an $N/2 \times N/2$ matrix structure and connected by bidirectional links. As reported by Pavlidis and Friedman (2007), each pair of facing gateways on a column shares a row for injection and a column for ejection. It eventually exploits the bi-directionality of the 4×4 switches. However, this topology suffers from limited scalability.

11.2.3.4   Clos

This topology (Joshi et al. 2009) has the primary advantage of very low optical power, thermal tuning power (i.e., the thermal power required to tune the refractive index), and area compared to global photonic crossbars over a range of photonic device parameters.

11.2.3.5   Dragonfly Topology

Dragonfly (Pan et al. 2009) has been primarily deployed for large-scale, off-chip networks to exploit the availability of economical, optical signaling technology and high-radix routers to create a cost-efficient topology. The structure of this topology is shown in Figure 11.18.

11.2.3.6   2D-Hert

The advantage of this topology (Koohi, Abdollahi, and Hessabi 2011) lies in a high degree of connectivity along with small node degree. Here, the clusters of processing cores are locally connected by optical 1D rings and global optical 2D rings are used for other clusters.

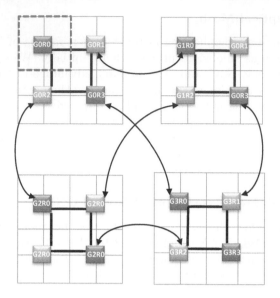

FIGURE 11.18
Dragonfly topology

11.2.3.7   Wavelength Routed Optical Network (WRON)

Another interesting optical topology is the generalized WRON. Primary components of this network are I/O nodes and optical switches across multiple stages. This may be of two types, viz., WRON type I and WRON type II. Interested readers are referred to Zhang et al. (2006) for details.

Table 11.2 compares various optical technologies, considering several performance parameters.

11.2.4   Topologies in Wireless NoCs

Research in wireless NoC design has progressed well in the past few years. Several topologies are also being explored for designing wireless NoCs. Generally, a wireless channel is more complex than a traditional wired transmission channel. But rapid development of wireless communication as well as the possibility of nano-scale fabrication of wireless antennas have worked as catalysts. In wireless NoCs, the whole system is divided into multiple small clusters of neighboring units called *subnets*. Being smaller, intra-subnet communication requires a shorter average path length than a single NoC encompassing the whole system. Wireless links are used to provide this inter-subnet communication, whereas intra-subnet communication is wired.

Popular topologies used for wireless NoCs are as follows. The *mesh* (Wang and Zhao 2007) is the most popular one and first explored due to its sim-

TABLE 11.2

Comparison of Optical Network Topologies

Topology		Buses & Channels				Routers		Latency				
		N_C	N_{BC}	b_C	$N_{BC}.b_C$	N_R	Radix	H_R	T_R	T_C	T_S	T_0
Crossbar	64 * 64	64	64	128	8,192	1	64 * 64	1	10	n/a	4	14
Butterfly	8-ary 2-stage	64	32	128	4,096	16	8 * 8	2	2	2-10	4	10-18
Clos	(8, 8, 8)	128	64	128	8,192	24	8 * 8	3	2	2-10	4	14-32
Torus	8-ary 2-dim	256	32	128	4,096	64	5 * 5	2-9	2	2	4	10-38
Mesh	8-ary 2-dim	224	16	256	4,096	64	5 * 5	2-15	2	1	2	7-46
CMesh	4-ary 2-dim	48	8	512	4,096	16	8 * 8	1-7	2	2	1	3-25

N_C = number of channels or buses;

N_{BC} = number of bisection channels or buses;

b_C = bits/channel or bits/bus;

N_R = number of routers;

H_R = number of routers along minimal routes;

T_R = router latency;

T_C = channel latency;

T_S = serialization latency;

T_0 = zero load latency (Ganguly et al. 2008).

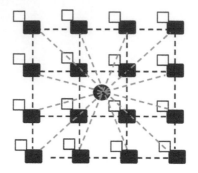

FIGURE 11.19
Wireless mesh

plicity. As mentioned previously, intra-subnet communication among cores is through wireline mesh and inter-subnet communication is through wireless (see Figure 11.19). Each subnet is connected through NoC switches and links as in a standard mesh based NoC. Moreover, all the individual switches have direct wireline links to a Wireless Base station (WB). Dimension order routing techniques are used for packet routing within the subnet.

11.2.4.1   MORFIC (Mesh Overlaid with RF Inter Connect)

This topology is also a hybrid, with the integration of conventional mesh topology augmented with RF-I enabled short-cuts (Ganguly et al. 2009). This can eventually reduce the communication latency. The flexibility of RF-I (due to the ability to allocate different frequencies for different transmission lines) combined with the extremely low latency of the RF interconnect made it possible to use a shared pool of transmission lines that can physically span the NoC. For the purpose of understanding the system it can be assumed to be analogous to road concepts where RF-I may be treated as highways and standard 2D mesh links as small city streets. The scheme is illustrated in Figure 11.20.

11.2.4.2   Hybrid Ring, Star, and Tree

Subnets are formed by connecting all the cores on a ring to two nearest neighbors on either side by wireline (see Figure 11.21). There is a central switch connected with all the cores as well as WB through conventional wireline NoC switches. During data routing (Wang and Zhao 2007), data is routed along the ring when the destination core is within two hops on the ring, otherwise the central router takes care of this. Similarly, a hybrid star (see Figure 11.22) is formed by connecting each switch with a centralized wireless base station (WB) through wires and connecting cores with the switch. Thus the subnet is formed. Base stations communicate through wireless channels. In a hybrid tree (see Figure 11.23), a base station is the root and children are the switches, connected with the root by wireline. Cores are connected with each switch. All the base stations in the network are connected through a wireless channel.

11.2.4.3 Irregular topology

To meet some specific application requirements as well as to accommodate different sized cores, some irregular topologies are also used in wireless NoCs. One example topology is shown in Figure 11.24.

11.3 Design of Router

The design of a router directly affects the flow control techniques used during routing. The number of virtual channels required, the buffer organization, the switch design, the pipeline strategy, etc., all depend on the router itself. Therefore, it plays an important role in routing strategy through an NoC. The flow of data through a router is illustrated in Figure 11.26 and a typical router architecture is shown in Figure 11.25.

11.3.1 Channel

A channel is the physical link to carry information throughout the network and it connects one node to another. Let the width of the network channel be W (in bits). Then the bandwidth of a channel is given by

$$\text{BW} = f_{ch} \times W \tag{11.1}$$

where f_{ch} is the channel operating frequency. Moreover, in the absence of contention, increasing W reduces the message latency (L_0). But the channel width cannot be increased unconditionally. If the channel width becomes significantly large then to utilize it more efficiently the virtual channel concept has been introduced.

11.3.2 Virtual Channel

A virtual channel splits a single channel into 2 to 8 channels, virtually providing more paths for the packets to be routed throughout the NoC. Virtual channels do not have physical existence but are logically present. VCs can reduce the network latency at the cost of area, power consumption, and production cost of the NoC implementation. But due to some other advantages offered by virtual channels, viz., offering reduced probability of deadlock, elimination of livelock, optimization of wire utilization, etc., VCs are widely used.

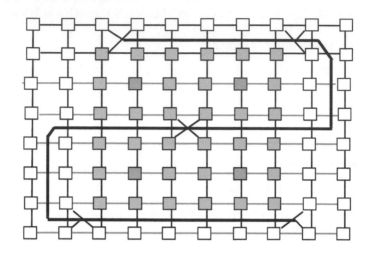

FIGURE 11.20
MORFIC (mesh overlaid with RF interconnect)

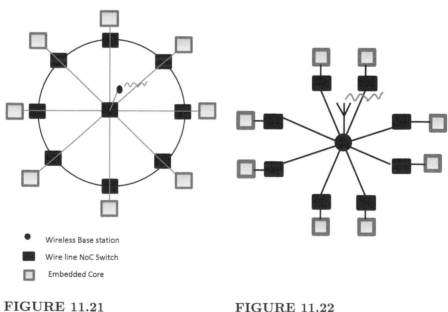

FIGURE 11.21
Hybrid ring

FIGURE 11.22
Hybrid star

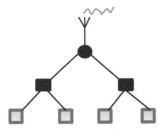

FIGURE 11.23
Hybrid tree

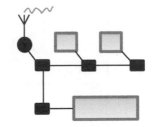

FIGURE 11.24
Hybrid irregular topology

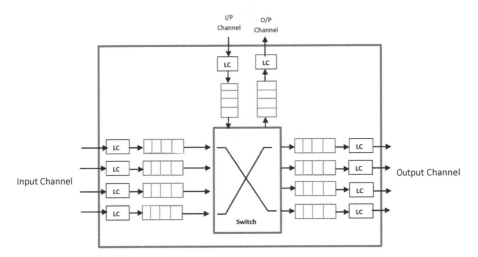

FIGURE 11.25
A typical router architecture

11.3.3 Buffer Organization

Network contention and thereby latency can be drastically reduced by using high capacity buffers in conjunction with VCs in NoCs. But using buffers increases the area as the buffers are area hungry. Thus, the overall use of buffering resources has to be minimized to reduce the implementation overhead in NoCs.

11.4 Switching Techniques

The selection of a particular switching technique to be used inside a network is very important as it affects the performance of routing throughout the NoC.

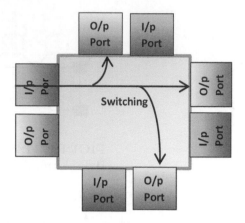

FIGURE 11.26
Router data flow

The routing decisions that are made, the way the switches inside the routers are set/reset, and the method by which the packets are transferred through the switches are dependent on the switching techniques. This also has a direct impact on router architecture and pipeline design. During routing through NoCs, two primary switching techniques are used, viz., circuit switching and packet switching.

11.4.1   Circuit Switching

This approach is better in some ways than the packet switching approach due to its reliability of data delivery, low and predictable latency, etc. Disadvantages include the requirement of control overhead to set up a new connection, and a pretty simple communication protocol unable to handle any complex scenario or support any advanced technique or feature. However, circuit switching does not require buffers, and bandwidth can be improved by increasing the number of wires without significantly compromising the area overhead. This can help to design a less complex and energy-efficient router.

11.4.2   Packet Switching

Packet switching involves the division of the data into small units called *packets*. Unlike circuit switching, no path is established before sending any data. The packets generally include source and destination addresses, order/sequence information, and may possibly include routing information. Each packet of a message is dispatched independently and may take a different path. At the destination, the message is reassembled in the correct order (Ritchie

2008). Packet switching involves buffering of the packets at source and destination, and in intermediate switches.

11.5 Routing Flow Control

A flow control mechanism determines how the network resources, such as channel bandwidth, buffer capacity, virtual channels, etc., should be used during the traversal of packets through the network. Decisions like storing, forwarding, or dropping a packet in a buffer are taken by this mechanism. Therefore it plays an important role toward the improvement of system throughput. Different mechanisms of routing flow control are as follows.

11.5.1 Store-and-Forward

In this mode each packet is treated as a single object; no further division of packets into flits is considered. A packet can be forwarded to the next router only after storing the whole packet in the memory of the current router. The network latency is given by the total time spent in storing and forwarding the entire packet. The buffer memory has to be large enough to store the largest packet in the network.

11.5.2 Virtual Cut-Through

This is a modified version of store-and-forward. Here, forwarding of a packet may begin before the entire packet has been stored into the memory of the current router. Though the required buffer memory size is similar to the store-and forward method, the latency decreases considerably.

11.5.3 Wormhole

In a wormhole flow control process, packets are divided into smaller equal-sized units called *flits*. The header or first flit of a packet contains only routing information and is routed similarly to packets in the virtual cut-through routing process. The remaining flits of the same packet follow the path that was established by the header flit. This process requires less buffer memory and gives a smaller latency but with an increased risk of deadlock.

11.6   Traffic Patterns

Network traffic is an important metric that has to be considered when designing an optimized NoC topology. The performance of a specific topology may vary depending upon the generated traffic pattern, thus making traffic pattern one of the most important parameters during design of application specific routing topologies and routing algorithms. Multiple traffic patterns, viz., uniform, transpose, and hotspot, may be used to check the flexibility of any proposed algorithm. However, traffic patterns may also be classified as either realistic or synthetic. A synthetic traffic pattern refers to an abstract traffic model in an NoC, while a realistic traffic pattern represents traces of the real applications running on NoCs. Different types of traffic patterns are described as follows.

11.6.1   Synthetic Traffic

Uniform Traffic Under Uniform traffic each node sends messages to other nodes with an equal probability and destination nodes are chosen randomly using a uniform distribution. This traffic model is considered as a standard benchmark in network routing studies.

Transpose Traffic Two types of traffic patterns are considered for transpose traffic. With the first transpose traffic pattern, a node (i, j) only sends messages to node $(n - j, n - i)$, where n is the network diameter (e.g., $n \times n$ in the mesh topology). This traffic pattern is very similar to the matrix-transpose, hence the name. In the second traffic pattern, a node (i, j) only sends messages to node (j, i).

Hot Spot Traffic Hot spot nodes are very busy nodes in a network. In the hot spot traffic pattern, each node sends messages to other nodes with an equal probability except for a specific node (the hot spot) that receives messages with a greater probability. The percentage of messages that a hot spot node receives beyond the usual nodes is indicated after the hot spot name (e.g., hot spot 30%).

Complement Traffic In this case each node sends messages only to the ones, complement of its source address. So the destination is given by $(\overline{n_1}\,\overline{n_2}\,\overline{n_3} \ldots \overline{n_{m-2}}\,\overline{n_{m-1}}\,\overline{n_m})$ for an m-bit address.

Bit Reversal Traffic In this type of traffic each node sends packets to the destination node whose address is the bit reversal of the sender's address. The destination is given by $(n_m n_{m-1} n_{m-2} \ldots n_3 n_2 n_1)$, for an m-bit address.

11.6.2 Realistic Traffic

As the performance of an NoC depends on the generated traffic pattern, the most accurate way to assess the characteristics of the NoC would be to invoke the traffic profiles corresponding to all running applications. This realistic traffic is used to analyze the power and delay of NoCs in a realistic situation.

11.7 Routing Algorithms

Routing is the strategy or policy for moving data through the underlying topology, and it determines the path of the data through the network. It affects directly all important network metrics, viz., latency, throughput, power dissipation, and quality of service (QoS). Communication and performance of the entire system are significantly affected by thc routing algorithm. Routing techniques may broadly be categorized into two types: oblivious routing and adaptive routing. A detailed classification is shown in Figure 11.27.

A brief description of different routing policies used in NoCs follows.

11.7.1 Oblivious Routing

In oblivious or deterministic routing the path is determined by the source or destination and the same protocol is followed at all times to take decisions about the path. It is very reliable when traffic is of a known pattern. It has the advantage of simplicity in the router architecture and algorithmic implementation. However, static routing suffers from significant performance degradation in the case of unexpected faults. In addition, it does not consider spatial and temporal variations in injection rates.

11.7.1.1 Dimension Order Routing

XY routing is the preferred technique for the mesh and torus topologies. Each router position is identified with its X-Y coordinates. In this routing strategy, a packet or flit is routed first in the X direction until it reaches the destination column and then in the Y direction until the destination is reached. This is very simple and both deadlock and livelock free. The disadvantages of this technique include large network diameter as well as a very high load on nodes at or near the center of the NoC.

Pseudo adaptive XY routing works in both deterministic or adaptive modes depending on the current congestion state of the network. The algorithm switches from the deterministic state to the adaptive state when the network becomes congested. In adaptive mode, a router assigns priority to each incoming packet when there is more than one packet coming simulta-

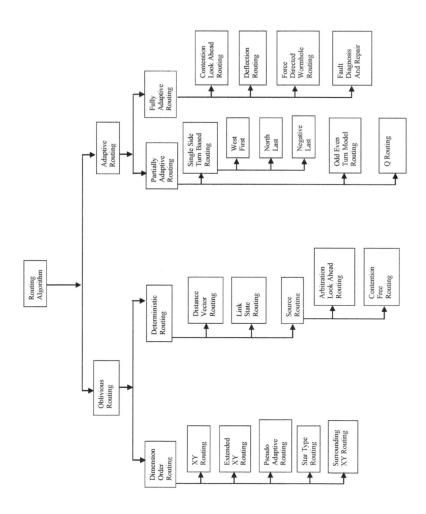

FIGURE 11.27
Different routing policies

neously from different neighboring nodes. Packets are transmitted to the output port based on the assigned priorities or privileges. Pseudo adaptive algorithms distribute the traffic more evenly over the whole network and thus can balance the load over the network, giving better performance than simple XY routing.

Extended XY routing is very similar to simple 2D XY routing and is used in the diametrical 2D mesh (Reshadi et al. n.d.) topology, where some additional diagonal links have been used to reduce the network diameter. Besides simple XY routing, packets may also move on a diagonal path based upon satisfaction of specific criteria to reduce network latency (Reshadi et al. n.d.; Ghosal and Das 2013).

Star type routing is applicable to a star type 2D mesh network (Chen, Peng, and Lai 2010), where a packet may follow different switching policies based upon the position and distance of the destination node relative to the current node. In level-1 routing, a packet may follow either a horizontal or a vertical or a diagonal path to reach the nearest central router of a subnetwork (the size of a subnetwork is 3×3) and in level-2 routing, a packet may follow a simple XY routing to travel from one central node of a subnetwork to the next central node of a subnetwork (Chen, Peng, and Lai 2010).

Surrounding XY routing *S-XY-Routing* (Bobda et al. 2005) is capable of handling routing in an NoC with obstacles created by dynamically placed components. In this routing technique the routers operate in three different modes as follows:

1. Normal XY mode (N-XY),
2. Surround horizontal XY (SH-XY) mode, and
3. Surround vertical XY (SV-XY) mode.

In normal XY mode, the router operates in a similar way to a normal XY router. The router enters the second mode when its left neighbor or its right neighbor is deactivated. In SH-XY, a packet may move through Y-coordinates in the upwards or downwards direction depending on the Y-coordinate of the destination node. The router enters surround vertical XY (SV-XY) mode when its upper neighbor or its lower neighbor is deactivated. In this mode a packet is forwarded through X-coordinates in the leftward or rightward direction. Before sending a packet to any alternate path a stamp-bit is set in the packet to notify the next router that the packet is now operating in either SH-XY or SV-XY mode and to avoid a deadlock situation. This stamp-bit is reset whenever the packet reaches the boundary corner router of any faulty area.

11.7.1.2   Deterministic Routing

In this routing scheme the routing path is fixed. The algorithm performs well in a congestion-free network. Each router has a routing table that includes routes to all other routers in the network. When the network structure changes, every router has to update its routing table. Different deterministic routing policies are as follows.

Distance vector routing The minimum cost route is defined as the route between any two nodes with minimum distance. Each router or node maintains a table of minimum distances to every other node. The routing table also guides the packets to reach their desired node by showing the next node to be visited in the route. Each node shares its routing table information with its immediate neighbors periodically and when there is a change, it updates. Each node does not have to know the whole network structure.

Link state routing This is a modified version of distance vector routing. The mechanism is based upon a distributed map, where each router has a copy of the whole network map and it is regularly updated. The goal of link state routing is for each peer to have an identical picture of the state of the entire network. Link state routing is an efficient but a very complex protocol. The routing tables can grow endlessly with an increase in the number of routers in the network. However, in the NoC context, the complexity of link state routing can be simpler by implementing static routing tables. These tables will never grow based on the fact that no new nodes will be added once the chip is fabricated. In the case when a link fails, only relevant entries need to be removed from the tables. Also, no periodic updates of neighbors are needed as there are no manual interventions to change the topology of the NoCs as in the case of regular data or communications networks (Ali, Welzl, and Hellebrand 2005).

Source routing This routing scheme is widely used because it is a cost effective scheme for network and transport layer design. In this scheme, when a packet is injected into the network, the packet routing path is already determined and the packet is routed through arbitration at each switch hop.

> **Arbitration look ahead scheme:** This scheme offers cost-effective, faster, and deterministic routing. In this routing technique, the routing path of an injected packet is determined prior to sending the packet and the channel is reserved for the packet. This reduces network transport latency considerably (Kim et al. 2005).

> **Contention-free routing:** This offers a guarantee to a certain level of performance in the NoC, using both a routing table and TDM (time division multiplexing) to reserve a connection path for some specific time period for some specific sender-destination pair. The Æthereal

NoC router uses this contention-free routing algorithm (Goossens, Dielissen, and Radulescu 2005).

11.7.2 Adaptive Routing

Adaptive routing algorithms use information about network traffic and channel status to avoid congested or faulty regions of the network. Adaptive routing results in an increase in the complexity of both the router architecture as well as the routing algorithm. However, dynamic routing has the advantage of learning by communicating with the neighbor routers and adapts itself to consider different faults in the network as well as congestion.

Adaptive routing is of two types:

- Partially adaptive routing

- Fully adaptive routing

Partially adaptive and fully adaptive routing algorithms can be distinguished by whether they are minimal or non-minimal. Minimal routing algorithms only use the shortest paths between sender and receiver. Non-minimal routing algorithms do not necessarily use the shortest paths.

11.7.2.1 Partially Adaptive Routing

Partially adaptive routing algorithms use some of the shortest paths between the sender and the receiver, but not all packets are allowed to use the shortest paths. The DyAD router (Dynamic Adaptive Router) (Hu and Marculescu 2004) dynamically switches from deterministic to adaptive routing when congestion is detected. Deterministic routing achieves low packet latency under low packet injection rates. The neighboring nodes send an indication to use adaptive routing when their buffers are filled above a preset threshold.

Single side turn based routing Turn model routing is based on prohibiting certain turns during packet routing to prevent deadlock and livelock. Different turn models proposed are as follows.

> **West first** Prevents any turn toward the west direction, as shown in Figure 11.28 with a dashed line in both the clockwise and anti-clockwise directions.

> **North last** In this routing a north turn is allowed as the last step. So after a turn toward the north direction no further turn is allowed (as shown in Figure 11.29).

> **Negative first** Any turn from a positive direction to a negative direction (both along the X-axis and the Y-axis) is not allowed. So any negative turn should take place first before any positive turn (as shown in Figure 11.30).

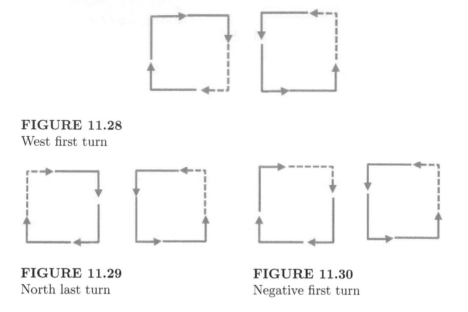

FIGURE 11.28
West first turn

FIGURE 11.29
North last turn

FIGURE 11.30
Negative first turn

Odd even turn model routing This is a partially adaptive, deadlock-free routing algorithm (Chiu 2000) for meshes. This model restricts the locations where some types of turns can be taken. The even adaptive property provided by the odd-even turn model is important under non-uniform traffic.

> **Rule 1** Prohibits any turns from east to north and from east to south at tiles located in even columns.

> **Rule 2** Prohibits any turns from north to west and south to west at tiles located in odd columns.

> In a 2D mesh, a column is called an even (or odd) column if the dimension0 coordinate of the column is an even (or odd) number.

Q-routing Q-routing (Boyan and Littman 1994) proves superior to a non-adaptive algorithm as it is able to route efficiently even when critical aspects of the simulation, e.g., network load, etc., are allowed to vary dynamically. The Q-routing algorithm, related to certain distributed packet routing algorithms, learns a routing policy that balances minimizing the number of *hops* a packet takes with the possibility of congestion along popular routes. It does this by experimenting with different routing policies and gathering statistics regarding the decisions to minimize total delivery time. The learning is continual and online, uses only local information, and is robust in the face of irregular and dynamically changing network connection patterns and loads. This Q-routing algorithm can be adopted to fit in DyNoC (Majer et al. 2005).

11.7.2.2 Fully Adaptive Routing

This technique does not provide too many restrictions to the packet movement like partially adaptive routing. The main focus is on finding congestion-free routing paths that may not be minimal.

Contention-look-ahead scheme A contention-look-ahead routing scheme (Partha Pratim Pande et al. 2012) is one where the current routing decision is helped by monitoring adjacent switches, thus avoiding any possible blockage. When the header flit of a packet arrives at a node, the traffic condition of the neighboring nodes can be acquired through the control signal wires. The traffic signal can be either a 1-bit wire, indicating whether the corresponding switch is busy or free, or multiple-bit signal, indicating the buffer level (queue length) of the input waiting queue. Based on this information, the router can choose the route to the next available (or shortest queue) switch. The local routing decision is performed at every switch once the header flit arrives. It is stored to allow the remaining flits to follow the same path until the tail flit releases the switch.

Hot-potato or deflection routing This routing scheme (Partha Pratim Pande et al. 2009) is based on the idea of delivering a packet to an output channel at each cycle. It is assumed that each switch has an equal number of input and output channels. Therefore, input packets can always find at least one output exit. Under this routing scheme, when congestion occurs and the desired channel is not available, the packet, instead of waiting, will move through any alternative available channels to continue moving to the next switch.

Force-directed wormhole routing (FDWR) This routing algorithm (Ganguly et al. 2011) forces a traffic distribution policy across the entire network and follows wormhole routing principles for packet movement. The advantage of wormhole routing, as discussed earlier, is the small buffer space in switches, and also pipelined data transfer. During routing, the first flit of a packet is used as a look-ahead flit to investigate the buffer status of its neighboring switches.

Fault-diagnosis-and-repair (FDAR) routing of packets may consist of multiple phases. NoC implementations based on Extended Generalized Fat Tree (XGFT) follow two phases of routing. During the first phase, packets are routed upwards in the network until they arrive at the nearest common ancestor of the source and destination leaf nodes. In the next phase, they are routed downwards from this switch node. In the first phase it is possible to route packets upwards in an adaptive way along several alternative routing paths. Because there is only one path available for routing the packets downwards from the nearest common ancestor, packets must be routed downwards in a deterministic way according to their destination addresses. A deterministic routing could be used for routing the packets upwards also in the XGFT and around faulty parts of the XGFT, if the locations of the faults are known. This requires that the faults must first be found and located by diagnosing the network. Therefore, fault-tolerant XGFTs must be implemented with both deterministic and adaptive routing, and also need a suitable fault diagnosis method like the FDAR (Ganguly, Pande, and Belzer 2009).

11.8   Problems of Routing in NoC

Routing in NoC suffers from various problems. The main problems that need special attention to design any routing topology and policy are as follows.

11.8.1   Deadlock

This is caused when packets wait for each other in a cycle. Several situations may arise in different routing techniques, viz., wormhole routing, dimension order routing, etc., that may suffer from this problem. Many techniques have been proposed to resolve this issue and are discussed as follows.

Duato's method is based on generating a Channel Dependency Graph (CDG) (Song, Ma, and Song 2008), where a deadlock is detected when there is a cycle in the CDG. This method restricts some combinations of channels so that cycles could be avoided in the CDG. Such a routing algorithm can be implemented using routing tables inside routers in the network. In dimension-ordered routing, constrained turn rules can solve the deadlock problem, but it cannot avoid contention: packets have to wait for the channel to be free when contention occurs.

Another way to solve the deadlock problem is to use virtual channels. In this approach, one physical channel is split into several virtual channels. Virtual channels can solve the deadlock problem while achieving high performance. But this scheme requires a larger buffer space for the waiting queue of each virtual channel.

11.8.2   Livelock

Livelock is a potential problem in many adaptive routing schemes. It occurs when a message proceeds through the network indefinitely, never arriving at its destination. It is possible only if message routing is adaptive and is non-minimal. Hot potato or deflection routing is the one technique where this problem is predominant.

11.8.3 Starvation

Using different priorities can cause a situation where some packets with lower priorities never reach their destinations. This occurs when the packets with higher priorities reserve the resources needed by the lower priority packets for all of the time. Starvation can be avoided by using a fair routing algorithm or by reserving some bandwidth for low-priority packets.

11.9 Emerging Techniques in NoC Routing

The conventional metallic interconnect is becoming the bottleneck of NoC performance with limited bandwidth, long delays, and high power consumption. Therefore, to mitigate this problem, new emerging technologies are being proposed by researchers. Some of the proposed techniques are discussed below.

11.9.1 Routing in Optical NoC

Silicon-based optical interconnects and routers are being used in optical NoCs for their compatibility with existing CMOS technology. Here bandwidth increases at the cost of latency. Also, wavelength routing introduces a totally new dimension by improving the functionality of the routing devices since it is possible to devise fully contention-free structures. Studies have shown that optical NoCs are a promising candidate to achieve significantly higher bandwidth, lower power, lower interference, and lower delay compared to electrical NoCs. In almost all high performance systems like multicomputer systems, on-board inter-chip interconnect, and the switching fabrics of Internet routers, optical interconnects have shown very promising results. Progress in photonic technologies, especially the development of micro-resonators, makes optical on-chip routers possible and opens a new direction to researchers studying NoCs.

11.9.1.1 Characteristics of Optical Routing

Depending upon the characteristics, routing in optical NoCs may be grouped into the following categories:

Wavelength-based routing In this type of routing, a packet route is solely dependent on the wavelength of its carrier signal. This has motivated researchers to adopt an all-optical solution for data transmission. This can reduce the hardware required for E-O (electro-optic) or O-E (opto-electric) conversion of the signal. This routing technique has been popular in optical local area network (LAN)/wide area network (WAN) technology for this same reason.

TABLE 11.3
The Wavelength Assignment of 4-WRON

W	D_1	D_2	D_3	D_4
S_1	W_2	W_3	W_1	W_4
S_2	W_3	W_4	W_2	W_1
S_3	W_1	W_2	W_4	W_3
S_4	W_4	W_1	W_3	W_2

Oblivious routing In this type of routing the wavelength employed to connect a source-destination pair is invariant for that pair, and does not depend on ongoing transmissions by other nodes, thereby simplifying design and operation.

Passive optical routing In this type of routing a routing pattern is set at design time, allowing for area and power optimization. No time is lost for routing/arbitration decisions.

In optical networks-on-chip the passive optical routing scheme is mainly used.

11.9.1.2   Routing Algorithms in Optical NoCs

Passive Optical routing

Several routing techniques like OXY routing, across-first routing, circular-first routing, etc., fall into this category. OXY is primarily used for mesh and torus topologies (Shacham, Bergman, and Carloni 2008). The Spidergon NoC adopts the across-first routing (Koohi and Hessabi 2009) as its deterministic routing algorithm; circular-first (Koohi, Abdollahi, and Hessabi 2011) was proposed as a novel deterministic routing algorithm for the 2D-HERT architecture.

WRON routing

In WRON (Zhang et al. 2006), each routing path P_i is associated with a tri-tuple $\langle S, D, W \rangle$, where S denotes the address of the source node, D denotes the address of the destination node, and W is the assigned routing wavelength for the data transmission. All the wavelength assignments of a 4-WRON are tabulated in Table 11.3. For instance, to send data from source node S_1 to destination node D_3, only wavelength w_1 can be used. It can also be observed from the table that by using four different wavelengths, S_1 can reach four destinations, and using the same wavelength different sources can reach different destinations without any conflict.

In general, for an N-WRON, given any two of the three parameters, the routing path is uniquely determined and the last parameter can be derived from the two known parameters.

11.9.2 Wireless NoC

Development in the nano-scale fabrication technology of tiny, low-cost antennas, receivers, and transmitters (on chip) has provided us with a new technology known as Radio-on-Chip technology. As a result, an alternative RF/wireless interconnect technology has been introduced for future intra-chip communication. Using this technology, researchers have developed wireless NoCs (WNoCs). WNoCs will overcome the limitations of NoCs by providing higher flexibility, higher bandwidth, reconfigurable integration, and freed-up wiring. Different routing strategies have been reported (Varatkar and Marculescu 2004; Petracca et al. 2009) for routing in WNoCs. Presently this area is being actively researched.

11.10 Conclusion

Networks-on-chip encompasses a broad spectrum of research, ranging from highly abstract software-related issues, across system topology to physical-level implementation. This chapter has mainly focused on topology structures and routing strategies in NoC structures. Throughout this chapter an effort has been made to provide a detailed overview of different techniques being adopted by researchers for topological structures and routing schemes from the early days until today. With the advent of 3D integration technologies, researchers are trying to explore the feasibility of extending the architectures of NoCs into three dimensions. An extended view is also provided on this aspect in this chapter with a brief but relevant discussion on the recent efforts made in the area of three dimensional NoCs.

Another emerging and important trend is the application of optical interconnects for high bandwidth communication between the cores inside a system-on-chip. Due to the trade-off between the speed of development of device technology and interconnect technology, communication efficiency has become a challenging issue inside an NoC. A number of researchers have tried to mitigate this problem with optical interconnects.

Researchers have also been trying to apply wireless technologies for communication inside an NoC. This solution has also come up with several advantages with dynamic remote monitoring as well as optimization capability.

As time passes, communication infrastructure mitigates the growing complexity of interconnects exhibiting superior performance. Technology advancement will allow us to implement photonic and wireless interconnects inside a small chip and route packets through the wireless and photonic channels. Researchers are also considering some hybrid structures that can incorporate the benefits of more than one technology.

12

Efficient Topologies for 3-D Networks-on-Chip

Mohammad Ayoub Khan

Center for Development of Advanced Computing, Noida, India

Abdul Quaiyum Ansari

Department of Electrical Engineering, Jamia Millia Islamia, New Delhi, India

CONTENTS

The Network-on-Chip (NoC) represents a relatively new communication paradigm for increasingly complex on-chip networks. The NoC provides techniques for a generic on-chip interconnection network (IN) realized by routers that connect processing elements (PEs) like ASICs, FPGAs, memories, IP cores, etc. To reduce latency and wire length we need an efficient interconnection architecture. Performance of the network is measured in terms of throughput. The throughput and efficiency of an interconnect depends on the network parameters for a given topology. Therefore, the topology of any communication network has an important role to play for efficient design.

The performance of an interconnection architecture depends on degree and diameter. The cost of an interconnection architecture can be defined as a value of the degree × diameter. Three dimensional (3-D) integrated circuits offer a low interconnect latency and area efficient solution for NoC. The 3-D arrangement offers opportunities for new circuit architectures based on the geometric capacity that provides greater numbers of interconnections among multi-layer active circuits. The 3-D NoC can reduce significant amounts of wire length for local and global interconnects. This chapter investigates 3-D topologies for NoC application. Finally, the chapter considers a new design of efficient topology based on ring structures. We have obtained the degree of proposed topology as $\frac{6 \times (2^n - 1) - 4}{2^n - 1}$, while the diameter of a topology is obtained as $D = 1 + 2 \times (n - 1)$. The degree of the proposed topology is 25% less than the torus along with a drastic reduction in the diameter. The layout of the proposed topology could be easily extended to a 3-D NoC architecture by adding a few extra links.

12.1   Introduction

A Network-on-Chip (NoC) can be defined as a communication paradigm that uses multiple processors on a single chip. The NoC paradigm represents a promising solution for forthcoming complex embedded systems and multimedia applications. The International Technology Roadmap for Semiconductors estimates that NoCs will soon contain billions of transistors running at speeds of many GHz (line rate), operating below 1 V (Khan and Ansari 2011a). A typical NoC application consists of multiple storage components (memory cores) and processing elements, such as general-purpose CPUs, specialized cores, and embedded hardware connected together over a complex communication architecture. The NoC technology has replaced traditional bus architecture with a low-cost point-to-point and packet-based architecture. The NoC solution incorporates a layered network protocol stack analogous to the open system interconnection (OSI) model. The performance of the network is measured in terms of throughput (Abuelrub 2008). The throughput and efficiency of the interconnect depends on network parameters of the topology. We can formally define topology as a physical interconnection of the routing elements (R) by the communication channels (Khan and Ansari 2011c). This may be represented by a network graph $G = (R, C)$, where routing elements are the vertices and channels the edges of the graph. R and C are described in equation (12.1) and (12.2) (Khan and Ansari 2011c).

$$R \in \{r_1, r_2, r_3 \ldots r_n\} \tag{12.1}$$

$$C \in \{c_1, c_2, c_3 \ldots c_n\} \tag{12.2}$$

TABLE 12.1
Classification of NoC Topology (Khan and Ansari 2011c)

Direct	Indirect
Orthogonal (Mesh and Torus, Tree)	Crossbar Switch Fabric
Cube-Connected-Cycles (CCC)	Fully Connected Network
Octagon	Omega
4-Cube	Delta
	Butterfly
	Fat-Tree Topology

12.1.1 Classification of Network Topologies

The topology of a network determines possible and efficient protocol regulations implemented by routing elements. The topology of an NoC affects the scalability, performance, power consumption, area, and complexity of the routing elements (Khan and Ansari 2011c). NoC topologies can be classified broadly in two categories, viz., direct and indirect, as shown in Table 12.1 (Khan and Ansari 2011c). In a direct network, the routing element is directly connected to a limited number of neighboring routing elements using a network interface (NI) (Khan and Ansari 2011c). The NI also connects to IP cores. Each routing element performs routing as well as arbitration. The IP element connected to the routing element injects the messages into the network through an injection channel and removes incoming messages through an ejection channel. The injected messages compete with the messages that pass through the routing element for the use of output channels (Khan and Ansari 2011c).

12.1.2 Topology Properties

Interconnection networks can be static, dynamic, or hybrid in nature, as shown in Figure 12.1. Hybrid networks are those interconnections which have complicated structures such as hierarchical or hyper-graph topologies. In the following section we will present two network families of static and dynamic networks. Figure 12.1 shows the overall classification of interconnection networks. In a static interconnection network, links among different nodes of the system are considered as passive. Thus each node is directly connected to a small subset of nodes by interconnecting links. Each node performs both routing and computations. Here we present some important properties of topologies other than node degree and diameter.

Regularity A network is regular when all the nodes have the same degree.

Symmetry A network is symmetric when it looks the same from each node's
  perspective.

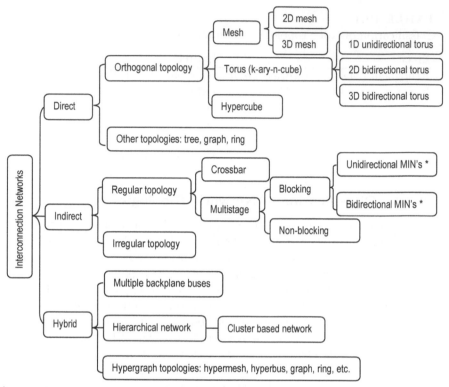

* = multistage interconnection network.

FIGURE 12.1
Classification of interconnection networks

Orthogonal property A network is orthogonal if the nodes and interconnecting links can be arranged in n dimensions such that the link is placed in exactly one dimension. In a weakly orthogonal topology, some nodes may not have any link in some dimensions.

In static networks, the paths for message transmission are selected by a routing algorithm. The switching mechanism determines how inputs are connected to outputs in a node. All the existing switching techniques can also be used in direct networks.

As compared to static networks, in which the interconnection links between the nodes are passive, the linking configuration in a dynamic network is a function of the states in the switching elements (SEs). In layman's terms, the paths between the graph nodes of a dynamic network change with the change in the states of the switching elements. Dynamic networks are built using crossbars (especially of size 2×2). The dynamic network may consist of single stage or multiple intermediate stages for switching. In an indirect

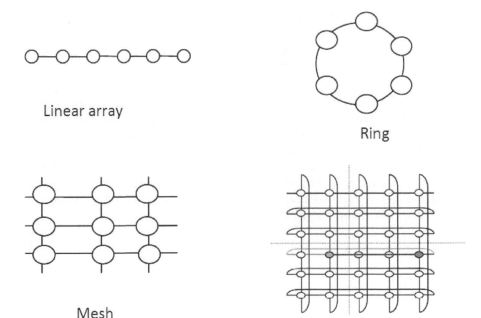

FIGURE 12.2
Basic network topologies

network, the routing elements (switches) are connected to one or many intermediate routing elements. These intermediate nodes are responsible for routing and arbitration. These networks are sometimes referred to as multistage interconnect networks (MIN). A broad classification of NoC topology is shown in Table 12.1 and some basic topologies are illustrated in Figure 12.2.

12.1.3   Performance Evaluation Parameters

Network topologies may be evaluated by the cost (size, degree, diameter, and bisection width) and performance measures (Khan and Ansari 2011c).

Size The size of a network may be defined as the number of vertices in the graph. Here we will define some of the basic terminology of performance evaluation parameters (Khan and Ansari 2011c).

$$S(G) = |R| \tag{12.3}$$

Degree The node degree is defined as the maximum number of physical links emanating from a node. The degree of the network may be defined as follows (Khan and Ansari 2011c):

$$\sum_{k=1}^{n} \text{Degree}(r_i), \forall r_i \in R \qquad (12.4)$$

Diameter The diameter of a network is the maximum inter-node distance, i.e., the maximum distance between any two points in G where distance is the shortest path between (r_i, r_j). This should be relatively small if network latency is to be minimized. The diameter is more important with store-and-forward routing than with wormhole routing (Parhami 1999). Consider a connected graph G, in Figure 12.3, $d(A, E) = 2$ and $Diam(G) = 3$. Therefore, we can define diameter as follows in equation (12.5) (Khan and Ansari 2011c).

$$\text{Diameter}(G) = \max(d(r_i, r_j)) \qquad (12.5)$$

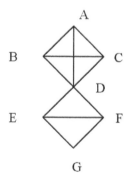

FIGURE 12.3
Diameter in a connected graph

Bisection width The bisection width (BW) of a network is defined as the minimum number of channels or links that must be removed to partition the network into two equal size disconnected networks. This is important when nodes communicate with each other in a random fashion. A small bisection width limits the rate of data transfer between the two halves of the network. This may affect performance of communication-intensive algorithms in SoC.

Number of edges (degree) This represents the number of communication ports (edges) required at each node (processing elements). The degree should be a constant independent of network size if the architecture is to be readily scalable to larger sizes (Parhami 1999). The node degree has a direct effect on the cost of network.

Channel width The number of bits that can be sent simultaneously over a communication channel or link. Sometimes, this is loosely defined as the number of wires in the communication channel or link.

Channel rate This defines the peak rate at which a single wire can deliver bits (Khan and Ansari 2011c).

Channel bandwidth This defines the peak rate at which a communication channel or link can deliver bits (Khan and Ansari 2011c).

Cost of network This defines the total number of communication links (Khan and Ansari 2011c).

12.1.4 Basic 3-D Topologies

Various new topologies have been explored to implement 3-D NoC. Some of these are based on basic topologies like mesh and torus that are extensively used in 2-D designs. In a 3-D mesh architecture multiple 2-D meshes are connected using vertical links, i.e., one more dimension Z is added to the regular X and Y dimensions. Similarly a 3-D torus network is the same as a 3-D mesh network except that it contains wrap-around edges at the terminal nodes of all axes (Kini, Kumar, and Mruthyunjaya 2009; Rahman and Horiguchi 2004). Consequently the degree increases in both the 3-D networks. The mesh architecture is the most regular and simple architecture that is used in the design of NoCs. The implementation of a mesh network is simple to understand and verify. In a mesh architecture every node is connected to four of its neighbors. In an N-dimensional mesh network every node is connected to $2N$ of the neighboring nodes. Thus, the degree of a node in an N-dimensional mesh is $2 \times N$. The number of connections per node remains constant in a mesh network even though the size of the network increases. The performance of a large mesh network degrades due to the increase in diameter. The diameter of a 3-D mesh can be defined as $D = d \times (k-1)$, where d represents dimension and k is the number of nodes in a plane (Khan and Ansari 2011b). A torus network is the same as a mesh network with boundary nodes connected by wrap-around edges. These wrap-around edges significantly reduce the overall diameter of the network and thus improve the throughput and latency. The diameter and network cost of the torus are just half of the mesh topology. The degree of the network is 4, the number of nodes $N = n \times n$, and the network cost is $(4 \times n)$ in an $n \times n$ torus (Ki, Lee, and Oh 2009). The architecture shown in Figure 12.4 has an asymmetric number of nodes in planes. There are many SoC applications where we place unequal numbers of IP cores in the planes. Therefore, every plane has a different number of processing elements, and thus produces a different diameter for each plane. Therefore, we have logically partitioned the torus space into quadrants and selected the nearest wrap-around edge to connect the destination node (Khan and Ansari 2011b).

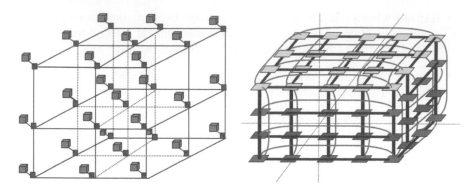

FIGURE 12.4
3-D mesh and torus topologies (Khan and Ansari 2011b)

The diameter of an asymmetric torus can be defined as follows:

$$D = d \times \left(\left\lfloor \frac{n_x}{2} \right\rfloor + \left\lfloor \frac{n_y}{2} \right\rfloor + \left\lfloor \frac{n_z}{2} \right\rfloor \right) \tag{12.6}$$

where n_x, n_y, n_z is the number of nodes in plane x, y, and z, respectively.

12.1.5   Power Consumption Issues in 3-D Topologies

Power dissipation is an important issue in 3-D circuits. The interconnection network has substantial amounts of power dissipation due to interconnects and buffers. For example, the MIT Raw on-chip network consumes 36% of the total chip power and 20% of the total power of the Alpha 21364 microprocessor due to the interconnection network (Soteriou and Peh 2004). Therefore, there is a need for power-aware interconnections and efficient topologies. The existing traditional approaches for power saving are not sufficient to address the needs of power issues in current SoC designs. Power consumption is also affected by IC technology. 3-D IC technology is expected to have lower power consumption than 2-D circuits due to shorter global interconnects. The topology of the NoC also affects the power consumption of the network in many ways. Power consumption can be reduced by using a topology that has minimum network cost. Pavlidis and Friedman (2007) have shown the effect on power for three different types of topologies. When the authors have used 2-D IC technology for a 3-D NoC they found that power consumption is decreased in this topology by reducing the number of hops for packet switching. 3-D topology can reduce power even in small networks where the number of IPs is very small. However, the power savings are greater in larger networks. In a second approach, the authors have experimented with 3-D IC technology for 2-D NoC, where horizontal bus length has been made shorter by implementing the IPs in more than one physical plane. The greater number of physical planes integrated in a 3-D IC technology provides optimum value for power regardless of the network size and operating frequency. In their third approach the authors have used 3-D IC technology for 3-D topology, where they observed the greatest savings in power in addition to the minimum delay.

12.2   Related Work

Researchers have considered butterfly fat tree (BFT), generic fat tree based interconnection networks for NoC applications (Greenberg and Guan 1997; Grecu et al. 2004; Guerrier and Greiner 2000). Feero and Pande (2009) have experimented with a 64-IP SoC with BFT topology that contains 28 switches. Each switch in a BFT network consists of six ports, one to each of four child nodes and two to parent nodes, with the exception of the switches at the topmost layer. When the authors mapped to a 2-D structure, the longest interswitch wire length for a BFT based NoC is $l_2DIC/3$, where l_2DIC is the die length on one side (Grecu et al. 2004; P. P. Pande et al. 2005). P. P. Pande et al. (2005) have found that if the NoC is spread over a 20 mm × 20 mm die, then the longest interswitch wire is 10 mm. On the other hand, when the authors mapped the same BFT network onto a four-layer 3D SoC, wire routing became simpler, and the longest inter-switch wire length was reduced by at least a factor of two (Feero and Pande 2009). In this work, the load on the router varies. At the bottom layer we have more IPs connected to the router while there are fewer routers at the top layer. Therefore, the numbers of input/output ports and the power dissipation also vary. The diameter of the FAT tree is large with varying node degree. The FAT tree based topology may not be optimum for many NoC applications. Also, a significant amount of research has been conducted with respect to off-chip networks (Pinkston and Duato 2006; Dally and Towles 2004; Duato, Yalamanchili, and Ni 2003). The basic concepts of off-chip networks can be applied to NoCs. The majority of NoC topologies gravitate toward either ring or mesh. The IBM Cell processor, the first product with an NoC, is built on ring topology. Ring topology is largely being used for design simplicity, ordering properties, and low power consumption. The IBM Cell architecture (Hofstee 2005; Gschwind, D'Amora, and Eichenberger 2006) is a joint effort between IBM, Sony, and Toshiba to design a power-efficient family of chips targeting game systems. The IBM Cell has been designed on a 90 nm, 221 mm^2 chip that can run at frequencies above 4 GHz. This consists of one IBM 64-Four ring that is used to boost the bandwidth which in turn alleviates the latency problem in the network. The Intel Larrabee is also based on two-ring topology (Seiler et al. 2009). Balfour and Dally (2006) have presented a comparison of various on-chip network topologies including mesh, concentrated mesh, torus, and fat tree. The MIT

FIGURE 12.5
Binary tree (Khan and Ansari 2011c)

Raw chip has also used a multiple mesh structure on a Tilera TILE64 chip (Wentzlaff et al. 2007).

12.3   Binary Search Tree Based Ring Topology

A binary tree is an ordered rooted structure where every node has at most two nodes designated as left or right child node. The maximum depth (height) of a binary tree of n nodes is $(n-1)$ (every non-leaf node has exactly one child) (Cormen et al. 2010). The minimum depth of a binary tree of n nodes is $(n > 0)$, $\lceil \log_2 n \rceil$ (every non-leaf node has exactly two children, that is, the tree is balanced). In what follows we consider a few examples of binary trees with different permutations. A binary search tree (BST) is a tree that satisfies the following criteria. The left node is always less than the root and the right node is always greater than the root value (Cormen et al. 2010). A BST algorithm is applied to find out any element in the tree. The BST, by nature, allows us to apply the divide-and-conquer technique easily.

$$\forall y \text{ in left subtree of } x \text{ then } [y] \leq [x] \tag{12.7}$$

$$\forall y \text{ in right subtree of } x \text{ then } [y] \geq [x] \tag{12.8}$$

In this work the authors have constructed a modified structure of a binary tree that reduces the network diameter and degree. The basic module of the proposed tree is shown in Figure 12.5 (Khan and Ansari 2011c). The structure contains only three nodes. Every node in the basic module is capable of communicating with other nodes directly without any hop. Thus, the diameter of the basic module is 1 only. In Figure 12.6, we have also shown a ring interconnection with $l = 1$, $l = 2$, and $l = 3$. The level is formed by extending terminal nodes. In Figure 12.7, the authors have shown a ring interconnection that has a maximum of 21 nodes at level $l = 3$ (Khan and Ansari 2011c). In the next section, we present the derivation for the total number of nodes, average degree, and diameter of the network.

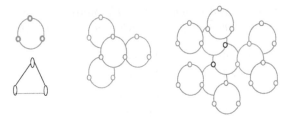

FIGURE 12.6
Proposed topology with different levels ($l = 1, 2$, and 3) (Khan and Ansari 2011c)

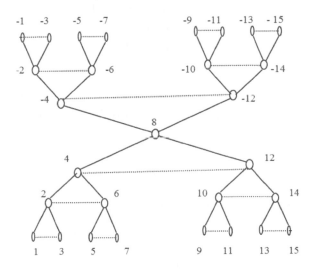

FIGURE 12.7
Ring based tree topology (Khan and Ansari 2011c)

12.3.1 Number of Nodes (N) at l^{th} Level

Theorem 12.1 *For a ring based tree having level l and N nodes, then:*

1. *The total number of nodes N having level l is $3(2^l - 1)$*

2. *The total number of terminal nodes T at level l is $3(2^{l-1})$*

Proof 12.1 *The number of nodes at any level l can be derived using induction as follows:*
Base case: *At level $l = 1$, the number of nodes is $N = 3$. This is clearly true for all as the base module has three nodes.*
Induction Hypothesis: *If we move to the next level $l = 2$, then the next*

level has 3 old nodes and 6 new nodes. Therefore, $N = 3 + 6$. Similarly, we can derive for other levels as follows (Khan and Ansari 2011c):

$$l = 3 \quad , \quad N = 9 + 12$$
$$l = 4 \quad , \quad N = 21 + 24$$
$$l = 5 \quad , \quad N = 45 + 48$$
$$l = 6 \quad , \quad N = 93 + 96$$
$$l = 7 \quad , \quad N = 189 + 192$$

$$\vdots \quad \vdots$$

$$l = n \quad , \quad N = (3(2^{n-1}) - 3) + (3(2^{n-1}))$$
$$, \quad N = 3(2.2n - 1 - 1)$$
$$, \quad N = 3(2n - 1 + 1 - 1)$$
$$, \quad N = 3(2^n - 1)$$
$$l = n \quad , \quad N = 3(2^n - 1)$$

Therefore, the number of nodes at level l can be written as follows:

$$\boxed{N = 3(2^n - 1)} \tag{12.9}$$

Proof 12.2 *The number of terminal nodes at any level can be derived by induction.*
Base case: *At level $l = 1$, the number of terminal nodes is $T = 3$. Therefore, at level $l = 1$, $T = 3 \times (2^l - 1)$. This is clearly true for all as the base module has three terminal nodes (Khan and Ansari 2011c).*

$$l = 1 \quad , \quad T = 3$$
$$l = 2 \quad , \quad T = 6 = 3(2)$$
$$l = 3 \quad , \quad T = 12 = 3(4) = 3(2^2)$$
$$l = 4 \quad , \quad T = 24 = 3(8) = 3(2^3)$$

$$\vdots \quad \vdots$$

$$l = n \quad , \quad T = 3(2^{n-1})$$

Therefore, the number of terminal nodes at level l can be written as follows:

$$\boxed{T = 3(2^{n-1})} \tag{12.10}$$

12.3.2   Average Degree (d) of the Network at l^{th} Level

The degree of a node represents the number of communication ports (edges) required at each node (processing elements). The node degree has a direct effect on the cost of each node, with the effect being more significant for parallel ports containing several wires.

Proof 12.3 *In the topology, every internal node has a degree of 4, while every terminal node has a degree of 2. Therefore, the degree varies between 2 and 4. The total number of terminal nodes at the lth level is $3(2^{n-1})$. The total number of internal nodes at the lth level is $(3 \times 2^{n-1} - 3)$. Therefore, the degree of internal and terminal nodes can be calculated as follows (Khan and Ansari 2011c):*
The degree of internal nodes in the network is

$$4 \times (3(2^{n-1}) - 3$$

The degree of terminal nodes in the network is

$$2 \times (3(2^{n-1}))$$

Total degree of the network would be

$$4 \times (3(2^{n-1}) - 3) + (2 \times (3(2^{n-1})) = 18 \times (2^{n-1}) - 12$$

Therefore, the average degree can be defined as:

$$\frac{Total\ Degree\ of\ Network}{Number\ of\ Nodes} = \frac{18 \times (2^{n-1}) - 12}{N}$$
$$= \frac{18 \times (2^{n-1}) - 12}{3(2^n - 1)}$$
$$= \frac{6 \times (2^{n-1}) - 4}{2^n - 1}$$

The average degree of the network at the lth level is as follows:

$$\boxed{D = \frac{6 \times (2^n - 1) - 4}{2^n - 1}} \qquad (12.11)$$

12.3.3 Diameter (D) of Level l Network

The diameter of a network is the maximum inter-node distance, i.e., the maximum distance between any two points in the topology where distance is the shortest path between (r_i, r_j). We have derived D using induction as follows (Khan and Ansari 2011c):

Base case: At level $l = 1$, every node is connected to all the nodes at unit distance. Therefore, at level $l = 1$, $D = 1 + 2 \times (n - 1)$, where $l = n$, is clearly true for all as the diameter of the base module is 1.

Proof 12.4 *At the next level, let $l = 2$, diameter D could be written as a summation of the diameter of the base module and distance of the left and right networks from the base module. Therefore, at level $l = 2$, the distance of the left and right networks from the base module is 2 only.*
Hence, $D = 1$(base module) $+ 1$(left network) $+ 1$(right network). Similarly, we can verify for the remaining values of l.

$$
\begin{aligned}
l &= 3, &\quad D &= 1 + 2 + 2 = 5 \\
l &= 4, &\quad D &= 1 + 3 + 3 = 7 \\
l &= 5, &\quad D &= 1 + 4 + 4 = 9 \\
l &= 6, &\quad D &= 1 + 5 + 5 = 11 \\
&\;\vdots & &\quad\vdots \\
l &= n, &\quad D &= 1 + (n-1) + (n-1) \\
& & D &= 1 + (2n - 2) \\
& & D &= 1 + 2 \times (n-1)
\end{aligned}
$$

Therefore, the diameter of the network can be written as follows:

$$\boxed{D = 1 + 2 \times (n-1)} \tag{12.12}$$

12.4   Layout and Implementation

Ring topology is simple but it has poor performance when compared to higher-dimensional networks like the mesh, torus, tree, etc. The latency, throughput, energy, and reliability of higher-dimensional networks are good. A ring has a node degree of two while a mesh or torus has a node degree of four, where node degree in an NoC refers to the number of links (physical ports) in and out of a node. The mesh and torus require more links at routers. The topologies, featured in Figure 12.4, are three-dimensional topologies that map readily to a multiple metal layer. The torus has to be physically arranged in a folded form to equalize wire lengths instead of employing long wrap-around links between edge nodes. 3-D torus topology has a lower hop count (which leads to lower delay and energy) compared to a mesh. On the other hand, tree topology has the advantage of lower diameter. The topology shown in Figure 12.7 at level $l = 3.5$ has 29 nodes. Sometimes, traffic across the subnetwork moves through the root node only. The remaining 28 nodes in the network are divided into 4 subnetworks, as shown in Figure 12.8. Each subnetwork has 7 nodes with a local root node. To minimize the wiring length, a mesh and torus structure has been adopted as shown. The placement shown of the proposed topology offers simple routing regulations like XY or XYZ for a three dimensional arrangement.

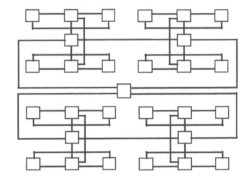

FIGURE 12.8
Layout of the proposed topology (Khan and Ansari 2011c)

TABLE 12.2
Analysis of Network Parameters for Base Module (Khan and Ansari 2011c)

l	N^i	T^i	Average Degree (d)	D
1	3	3	2	1
2	9	6	2.6	3
3	21	12	2.8	5
4	45	24	2.9	7
5	93	48	2.9	9
6	189	96	2.9	11
$\vdots$	$\vdots$	$\vdots$	$\vdots$	$\vdots$
20	3,145,725	1,572,864	2.9	39

12.5 Discussion and Analysis

In Figure 12.7, we have shown a tree with $l = 3.5$ that contains a total of 29 nodes. The equivalent VLSI layout is shown in Figure 12.8. We have presented an exhaustive analysis of the network parameters for the proposed topology, as shown in Table 12.2. Based on mathematical analysis we present a graphical analysis among level, number of nodes, degree, and diameter, as shown in Figures 12.9 and 12.10. The proposed tree topology can have a large number of nodes if sufficient levels are chosen. If we choose $l = 1$, then we have a total of 3 nodes while the topology can support 3,145,725 nodes for $l = 20$. The extended 3-D mesh tree topology as shown in Figure 12.11 will have little increased diameter as $3 + 2 \times (1 + 2 \times (n - 1))$. Here, $2 \times (1 + 2 \times (n - 1))$ is the diameter for the source and destination subnetworks, while 3 is the diameter of 3-D mesh.

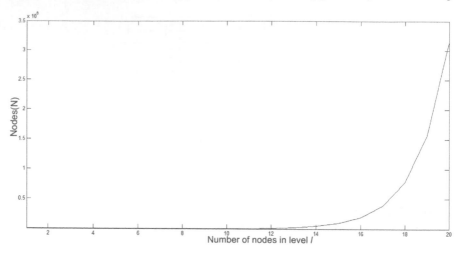

FIGURE 12.9
Number of nodes in level l (Khan and Ansari 2011c)

12.6   Conclusions

The topology determines possible and efficient protocol strategies implemented by routing elements. The topology of NoC affects the scalability, performance, power consumption, area, and complexity of the routing elements. In this chapter, we have constructed a modified structure of a ring based binary tree that reduces the network diameter and degree drastically. We have found that the degree of the proposed topology is 25% less than the torus along with a drastic reduction in the diameter of the proposed topology. For an SoC of node 3,145,725, the diameter of the proposed tree is 39. The diameter of a torus topology for the same number of nodes is approximately 1800, that is, too large for an NoC application. We also found that the degree of the presented topology varies between 2 and 3 while the torus has a fixed degree regardless of the number of nodes in the topology. This chapter has demonstrated that both mesh and tree based NoCs are capable of achieving better performance when instantiated in a 3-D IC environment compared to more traditional 2-D implementations. However, the proposed tree based topology shows significant performance gains in terms of network diameter, degree, and number of nodes. Tree based NoCs achieve significant gain in energy dissipation and area overhead without any change in throughput and latency. The Network-on-Chip (NoC) paradigm continues to attract significant research attention in both academia and industry. With the advent of 3-D IC technology, the achievable performance benefits from NoC methodology will be more pronounced, as this chapter has shown. Consequently this will also facilitate adoption of the NoC paradigm as a mainstream design solution for larger multicore systems.

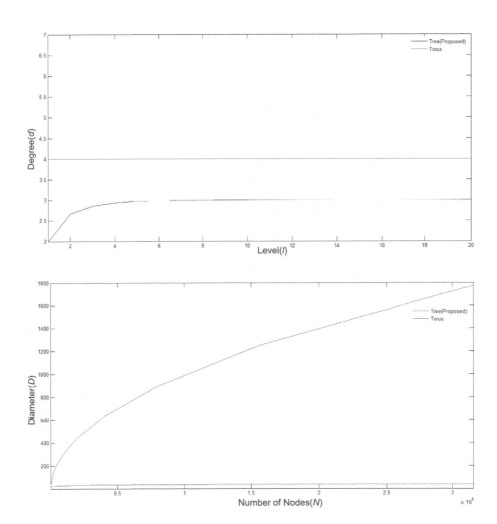

FIGURE 12.10
Degree and diameter analysis of the proposed topology (Khan and Ansari 2011c)

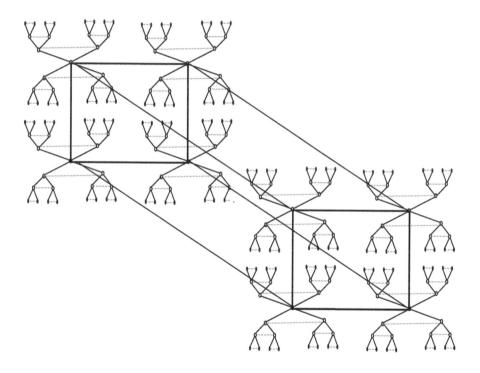

FIGURE 12.11
3-D tree mesh

12.7 Glossary

Topology: Topology defines logical structure and interconnection between nodes in a network.

Multicore: The multicore processor is a single computing component with two or more independent processors (called 'cores') on a single chip. Multiple instructions can be executed in multicore at the same time, increasing overall speed for programs amenable to parallel computing.

Application Specific Integrated Circuits (ASICs): An ASIC is an integrated circuit (IC) that is customized for a particular use, rather than intended for general-purpose use. For example, a chip designed solely to run a bluetooth transceiver is an ASIC.

Field Programmable Gate Arrays (FPGAs): The FPGA's function is defined by a user's program (VHDL/Verilog/Netlist) rather than by the manufacturer of the device. A typical integrated circuit performs a particular function defined at the time of manufacture. In contrast, the FPGA's function is defined by a program written by someone other than the device manufacturer.

System on Chip (SoC): The SoC is a new paradigm for design of a VLSI system. The SoC is an integrated circuit (IC) that integrates all components of a computer or other electronic system into a single chip. The SoC may contain digital, analog, mixed signal, and often radio frequency functions all on a single chip substrate.

Binary Search Tree (BST): The BST is an ordered placement of nodes of binary tree. In a BST, the left node is always less than the root, while the right node is always greater than the root.

Intellectual Property Cores (IP): An IP core is a reusable logic cell, or chip layout design that is the intellectual property of one party.

Open System Interconnection (OSI): The OSI mode was invented by the International Organization for Standardization (ISO). It is a prescription for characterizing and standardizing the functions of a communications system in terms of abstraction layers.

13

Network-on-Chip Performance Evaluation Using an Analytical Method

Sahar Foroutan, Abbas Sheibanyrad, and Frédéric Pétrot

TIMA Laboratory, Grenoble, France

CONTENTS

13.1 Introduction

Semiconductor technology improvements enable designers to integrate complex systems composed of tens and even hundreds of modules (blocks or components) into a single chip, called a System-on-Chip (SoC). An on-chip module can be an Intellectual Property core (IP such as a digital signal processor (DSP), a general purpose processor, or a specific custom hardware module), an embedded memory, a peripheral controller (such as a USB controller), or even a complex sub-system composed itself of IPs, memory blocks, and a local interconnect. SoC applications span from dedicated real-time systems to massively parallel general-purpose ones. To address this latter, Multi-Processor Systems-on-Chip (MP-SoCs) have been developed. Such huge Systems-on-

Chip require high-bandwidth interconnecting structures for communication, which have become a challenge in System-on-Chip design.

The history of on-chip communication began with the proposal of various bus-based interconnecting templates. From a single shared bus to complex hierarchical buses, these templates suffer from the strong drawback of poor scalability with system size. Bus bandwidth is shared between all connected modules and therefore it can serve only a limited number of connected blocks, whereas according to the International Technology Roadmap on Semiconductors (*The International Technology Roadmap for Semiconductors* 2008), by the end of 2022, SoCs will exploit more than 1300 processing engines! This huge growth in the number of embedded components coupled with on-chip communication requirements such as high bandwidth and low communication latency have led to a new scalable interconnecting platform: Network-on-Chip (NoC). By providing scalable performance and a higher degree of communication parallelism (compared to previous interconnects) NoCs have emerged as suitable interconnect platforms for current and future SoC communication requirements.

The NoC paradigm has opened a new and active area in semiconductor research both in academia and industry, and led to the definition of many NoC architectures and design strategies. The design process of an NoC comprises several phases like specification, architectural design, performance evaluation, implementation, and validation. NoC performance evaluation is necessary to estimate the network performance metrics (e.g., network latency, network throughput, average buffer utilization, etc.) for the purpose of design space exploration (in the sense of both architectural choices and application mapping) in order to meet the requirements of the entire system.

NoC performance evaluation is traditionally done by simulation at various levels of abstraction. Even though a higher level of abstraction provides shorter simulation time by hiding more details, simulation-based methods do not scale with the increasing network size and communication parallelism because of prohibitive simulation runtime. For example, in one of our experiments, the simulation of a 10×10 two-dimensional mesh NoC, at cycle-accurate level with the injection of 1000 packets per node and per offered load, did not terminate even after 48 hours (Foroutan et al. 2010). Besides, simulation-based methods are not exhaustive; i.e., we can never cover all possible communication paths. Finally, simulation results are achieved relatively late in the design flow because making an appropriate simulation platform is very time-consuming and necessitates the implementation and validation of all NoC components like IPs, router, and network interface in a hardware modeling or description language like SystemC, Verilog, or VHDL. In order to deal with these issues and to cope with time-to-market pressures, analytical methods are increasingly being used for performance estimation of Networks-on-Chip. As abstract and elegant mathematical-based techniques, analytical methods are proving to be powerful tools to provide an early estimate of desired performance metrics of NoCs. Therefore analytical methods can be a good alternative for time-consuming simulation-based methods.

The analytical performance evaluation method presented in this chapter addresses the network layer of NoC-based systems and permits an architectural exploration of this layer for a given application whose tasks have been mapped onto the NoC resources. Additionally, for a given network architecture, the method permits examination of the performance of different mappings of the application onto the NoC. It enables a comparison between different NoC architectures by measuring performance metrics, such as packet latency, saturation threshold, and average buffer utilization. The proposed method is based on the computation of probabilities and contention delays between packets competing for shared resources. It provides a comprehensive delay analysis of the network layer. In each router of the NoC, the mutual impact of competing packets is obtained by implementing a new iterative computation technique. The dependency graph between the latencies of a sequence of adjacent routers is considered and implemented through a recursive algorithm. Our methodology is generic in the sense that it supports arbitrary network topologies and deadlock-free routing algorithms with arbitrary packet and buffer lengths. Regarding applications, we address traffic with known spatial distribution on the NoC. Experimental results provide a precise analysis of various delay components of a router, a specified path, or the entire network. They show how the method enables an analysis of the network layer. Analytical results are compared with the results obtained from a corresponding SystemC cycle-accurate and bit-accurate (CABA) simulation platform. This comparison shows that the analytical results are in line with cycle-accurate and bit-accurate simulations.

13.2 Network-on-Chip Concepts

The introduction of Network-on-Chip as a new interconnecting solution for SoCs has led to the definition of various NoC architectures, implementation strategies, and network performance evaluation methods. The SPIN micro-network presented by Guerrier et al. (Guerrier and Greiner 2000) was the first published NoC architecture. Since then a large number of Networks-on-Chip have been proposed. Some examples are Dally's NoC (Dally and Towles 2001), AEthereal presented by Goossens, Dielissen, and Radulescu (2005), xPipes by Dall'Osso et al. (2003), CLICHÉ by S. Kumar et al. (2002), Nostrum by Millberg et al. (2004), aSoC by Liang, Swaminathan, and Tessier (2000), ANoC by Beigne et al. (2005), QNoC by Bolotin et al. (2004), Octagon by Karim, Nguyen, and Dey (2002), Spidergon by Coppola et al. (2004), Nexus by Lines (2004), SoCBUS by Wiklund and Liu (2003), Chain by Bainbridge and Furber (2002), QoS by Feliciian and Furber (2004), SoCIN by

Zeferino and Susin (2003), HERMES by Moraes et al. (2004), BFT by P. P. Pande et al. (2003), BONE by Lee et al. (2003), Proteo by Saastamoinen, Siguenza-Tortosa, and Nurmi (2002), MANGO by Bjerregaard and Sparso (2005), DSPIN by Sheibanyrad, Greiner, and Miro-Panades (2008), and AS-PIN by Sheibanyrad, Panades, and Greiner (2007), etc.

System performance is fundamentally affected by the network architecture, which provides the communication infrastructure for the resources. Relating to the network architecture, a crucial problem is the trade-off between generality and performance. Generality provides reusability of hardware, operating systems, and development practices, while performance is achieved by using application-specific structures (S. Kumar et al. 2002). Although it is argued that an interconnect should be specific to an SoC application because the communication requirements depend on the composition of IPs, which is application specific (Goossens et al. 2005), according to the fact that time-to-market is very tight and design costs are very high, today the tendency is toward general-purpose NoC architectures that could be used as a pre-fabricated part of any MPSoC design.

13.2.1   Architectural Parameters

Different NoC architectures can be identified by considering five key parameters: topology, routing algorithm, switching strategy, (low-level) flow control, and arbitration mechanism.

13.2.1.1   Topology

The network *topology* is the study of the arrangement and connectivity of routers. Crossbar, Binary Tree, Fat-Tree, Butterfly Fat-Tree, k-ary n-cube, Ring, 2D-Ring (Torus), Chordal Ring, and 2D-Array (Mesh) are the most usual topologies for on-chip interconnects. As stated in Culler, Singh, and Gupta (1999) and Jantsch (2003), different topologies can be compared formally by analyzing some critical parameters. The total number of routers needed for a given number of cores, the degree of each router, the diameter of the network, the total number of channels, and the number of bisection channels, explained below, are the most important parameters of the network topology.

Router degree is the number of router I/O ports. A lower router degree means reduced design complexity and consequently lower silicon area and power consumption.

Network diameter is the maximum shortest path of the network measured in hops (i.e., routers). Lower network diameter means the possibility of lower (average) communication latency.

Number of channels determines the maximum simultaneous communications that can occur in the network. A higher number of channels means higher aggregate throughput and better saturation threshold, but results in greater use of silicon area.

Number of bisection channels (or bisection width) is the minimum number of channels which, if removed, will divide the network into two equal parts. A higher number of bisection channels means a higher possibility of concurrent global communications and a higher global saturation threshold.

A useful analysis of the most utilized interconnect topologies regarding the above parameters is presented in Sheibanyrad (2008). According to Sheibanyrad (2008) and Ogras and Marculescu (2006) a regular two-dimensional mesh (grid like) is the most conventional and the most utilized topology for general purpose NoCs (for example DSPIN (Sheibanyrad, Panades, and Greiner 2007)), while the best choice for an application-specific NoC is usually a customized irregular topology derived by mixing different forms in a hierarchical, hybrid, or asymmetric fashion. For example, in Ogras and Marculescu (2006) and Ogras (2007) the authors address partial topology customization through long-range link insertion and present a methodology for inserting such links in the standard 2D-mesh topologies, in an application-specific manner, for the purpose of NoC performance optimization. Ogras et al. believe that 'the selection of the interconnect topology has a dramatic impact on the overall performance, area, and power consumption. Hence, constraining the network architecture only to a small set of topologies produces sub-optimal operating points' (Ogras and Marculescu (2006) and Ogras (2007)).

Nevertheless, although customized topologies may improve the overall communication latency by providing fast paths between distant nodes, they require a specific routing algorithm which can be difficult to implement. According to Ogras and Marculescu (2006), better logical connectivity due to customized topologies comes at the expense of long wire problems like crosstalk, timing closure, and wire backend routing.

13.2.1.2 Routing Algorithm

Routing is the process of selecting paths in the network between a source and a destination. Regular topologies allow an algorithmic routing, as opposed to static or statistical routings, which often need to use a Look-Up Table (LUT). A routing algorithm can be implemented either by source routing or distributed routing. The first solution results in potentially smaller routers but usually suffers from the lack of scalability caused by routing information overhead which must be added to each packet header and may also degrade performance. Routing can be either deterministic or adaptive. An adaptive routing can route packets around blocked nodes and channels. This ability is essential for fault tolerance but comes at the cost of a more complex design that possibly works more slowly. Moreover, an adaptive routing could not guarantee the in-order-delivery property in the network layer. In this case this property is usually guaranteed by implementing an end-to-end flow control mechanism. Routing algorithms must also avoid two major problems of *live-lock* and *dead-lock*.

13.2.1.3 Switching Strategy

A routing algorithm defines a path between source and destination nodes, while *switching strategy* addresses the transport of data and determines how the data of a message traverses this route. Basically two switching strategies exist: *circuit switching* and *packet switching*. In circuit switching a route from a source to a destination is established prior to message transport and stays reserved until the message is completely transferred. In packet switching the message is broken into a sequence of packets. Packets are independently routed through the network, and they are assembled into the original message at the destination network interface. In-order-delivery is an essential property that should be provided by packet switching. Usually packet switching is used to provide best effort service, as opposed to guaranteed service. Best effort service does not give any performance guaranties in terms of latency and effective throughput. The majority of NoC architectures are based on packet switching due to the fact that it is a cost-effective solution and often allows better utilization of network resources, since buffers and channels are only occupied while a packet is traversing them (Sheibanyrad 2008). However, even though the saturation threshold in Networks-on-Chip is much higher than in previous interconnects, any interconnect will saturate if a large number of cores generate traffic and the average offered load exceeds its saturation threshold. As the network latency in best effort networks becomes unpredictable, the quality of some delay-sensitive applications which need a guaranteed latency will be reduced. For example, a data stream from a camera to an MPEG encoder in a high quality video application requires high throughput with low, stable, and predictable delay. If the time interval between frames exceeds a certain limit, the quality of service of the application could not be guaranteed. NoCs providing circuit switching (i.e., connection-oriented) service (e.g., TDM: Time Division Multiplexing, VC: Virtual Channels, etc.) give an opportunity for this kind of real-time application. The ability of a network to provide guaranteed service to specific connections is often denoted as QoS or Quality-of-Service.

13.2.1.4 Flow Control

Flow control strategies can be defined in different layers of a network-based system. At the network level between routers, the flow control (sometimes called low-level or link-level flow control) is defined as the mechanism that determines how and when each flit of a packet moves from a router to another. A *flit* (flow control unit) is the minimum unit of information that can be transferred across a router at one time. Flow control mainly influences communication performance. There are three different flow control strategies:

store-and-forward, *virtual cut-through*, and *wormhole*. Regarding the communication latency of the network, with a store-and-forward strategy, latency is the product of D and L/W, which means

$$T_{SF} \propto (D \times L/W)$$

where D is the number of hops between the source and the destination, L is the message length, and W is the channel width. With wormhole routing, latency is reduced to the sum of D and L/W (Dally 1990):

$$T_{wh} \propto (D + L/W)$$

The storage capacity between neighbor routers can be implemented as input buffering or output buffering. The use of Virtual Channels (VCs) provides an alternative way of organizing the storage capacity between routers.

13.2.1.5   Arbitration Mechanism

An *arbitration strategy* (i.e., output scheduling) determines the priority order of candidate packets which compete for the same output port. The arbitration mechanism increases the router latency and thus it should do the job as fast as possible. *Static priority* and *random priority* are the simplest techniques with the least delay, but they can cause indefinite latencies for some incoming packets. Starvation prevention is the main property that must be considered in an arbitration mechanism. *Oldest-first* (first-come first-served) is an efficient starvation-free arbitration that selects packets according to the time they have been waiting. This solution tends to have the same average latency as random assignment, but significantly reduces the variance in latencies (Dally 1992). Another simple effective starvation free arbitration is *round-robin*, in which packets take the highest priority in sequence, turn-by-turn. The round-robin policy serves all inputs equally in a circular order. Since round-robin scheduling is easy to implement, it is the most widely used arbitration mechanism in NoCs using best effort traffic.

13.2.2   Layered Concept

In the NoC literature the term *methodology* covers a vast number of design aspects ranging from issues in low-level physical implementation to high-level abstract modeling. Problems acquire different meanings as we move from one design level to another (in the OSI layered concept, as explained later). For example, system software issues do not involve physical problems such as clock skew or long wire delays. Similarly, at the physical layer level, the designer may have no idea about handling massive high-level parallelism.

As a consequence, the Open Systems Interconnection (OSI) model of layered network communication can be adapted for NoC usage (Jayadevappa, Shankar, and Mahgoub 2004). The aim is to shield each level of the system from issues of other levels and thereby to allow communication between independently developed layers. Note that awareness of lower levels can be beneficial as it can lead to higher performance. In an NoC-based system the layers are more closely dependent than in a conventional data communications network and often have physically related characteristics. The design of a system using Network-on-Chip architecture could be considered in four different layers as follows:

Application (OSI: Session, Presentation, Application) This layer provides application specific functions so that the system can use these abstract communication functions without any concern for network details. All system software issues are related to the application level of a Network-on-Chip.

Network Interface (OSI: Transport) The main function of the network interface layer of an NoC is to deal with the decomposition of messages into packets at the source and their assembly at the destination. Furthermore this layer ensures independence regarding network implementation and translates the communication protocol of upper levels to the network compliant protocol.

Network (OSI: Data Link, Network) The network layer provides a topological view of the communications. The main function of this layer is to determine how messages are routed from a source to a destination. This can be customized by selecting appropriate architectural parameters from the network layer design space such as the routing algorithm, the switching strategy, and the link-level flow control, which all have significant impact on the performance of the network. Performance metrics such as packet latency (latency of a packet from the moment it enters the network), network throughput, and link capacity are related to this layer.

Physical (OSI: Physical) This layer is concerned with the physical details of transmitting data on wires. It defines signal timings and all related problems like long wire delays, clock skew, and synchronization failure, which are all aggravated as technology scales down. Globally-Asynchronous Locally-Synchronous (GALS) (Chapiro 1985) design issues are also related to this layer of the network. Implementing the system as a synchronous circuit or an asynchronous one is a design decision that is made in this layer. Power consumption and silicon area are two of the most important performance cost-related metrics of this layer.

Figure 13.1 shows these layers in an NoC-based architecture. Typically each subsystem contains one or several processors, one or several physical memories, optional dedicated IP cores (Hardware Coprocessors, I/O Controllers ...), etc., which may communicate together via a local interconnect. As shown, all these components are considered in the application layer of the system. The subsystems are connected to the network by a Network Interface Controller (NIC), which is the only access. The NIC is involved in the issues of the

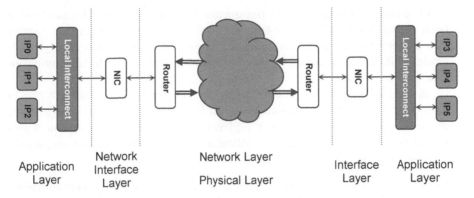

FIGURE 13.1
Operational layered concept of an NoC-based SoC architecture

interface layer. Thanks to this component, even if the architecture is physically clusterized,[1] all processors in all clusters may share the same flat address space and any processor in the system can address any target or peripheral IP core.

A network consists of a number of switches and so the network layer problems include the design of the network switches and their connectivity. Precisely, the network level of design defines the concept of switches and determines the geometric topology of the system. The switching module of an NoC is often called a *router* and its hardware must be implemented according to the physical issues. Thus, a router may belong to both network and physical layers. Even though the term NoC essentially means a complete network-based on-chip system composed of all of the above layers, the term is commonly used to signify particularly the network layer.

13.2.3 Levels of Abstraction

During the design of an SoC, the system is typically modeled at several abstraction levels. Once the model satisfies the constraints of a more abstract level, then it is refined toward a more detailed one. A lower level of abstraction gives more accurate results but causes more complexity in modeling. Raising the level of abstraction is on the basis of hiding unnecessary details of an implementation by summarizing the important parameters into a more abstract model. Therefore a higher level of abstraction can result in more details being ignored and consequently a reduction in the accuracy of the modeling. But above all it provides faster results by enhancing critical parameters like simulation or analysis speed, flexibility, time to develop, code length, and the ease of evaluation. Jayadevappa, Shankar, and Mahgoub (2004) compared the

---

[1] *Clusterized*: when the system is physically separated into different parts called *clusters*.

effect of abstraction on all these parameters for a peripheral device modeled at both system level and register-transfer level (RTL), and showed that moving to a higher level of abstraction does not necessarily affect the accuracy of the model while it certainly has a positive impact in terms of reducing simulation time and code length.

A level of abstraction is determined by the type of objects that can be manipulated and the operations that can be performed and modeled. Ranging from functional to transistor representations, each level introduces new model details. In SoC design the following levels of abstractions are usually considered (Coppola et al. 2003):

Functional Level: No notion of resource sharing or delays, executed instantaneously at each clock edge (sometimes called a zero-delay model).

Transaction Level (TL): Structural models with atomic transactions.

Cycle Accurate Bit Accurate (CABA) also called BCA (Bus Cycle Accurate): transactions mapped to a clock cycle, timed properties may be accurately modeled.

Register Transfer Level (RTL): A time domain, bit-accurate data type, pin-accurate interface.

Gate Level: Similar to RTL models with additional information, e.g., layout configuration.

Transistor Level: Electronic simulation below gate level in which the behavior of individual transistors, and electrical effects such as capacitance, are modeled.

We refer henceforth to the transaction and CABA levels together as *system level* modeling. Note that a level of abstraction may be confused with the modeling language. For example, SystemC is commonly referred to as system level modeling. It is true that often the modeling language and the abstraction level of the model are tightly related, but it should be remembered that they are two separate and independent concepts.

Modeling abstraction level is a concept somehow dependent on the method used for verification or for performance evaluation. The abstraction levels cited above are according to the terminology usually used in a traditional design flow and for models built for simulation-based methods. Indeed, analytical methods need their own proper abstract and mathematical models. Such mathematical models could be equivalent to some of the above abstraction levels and replace the corresponding simulation for the purpose of performance evaluation. For example, a queuing model could be defined for message arrival/consumption rates or for flit arrival/consumption rates, somehow equivalent to transaction or cycle accurate abstraction levels, respectively.

Recall that the NoC layers are the operational layers of an NoC-based system and are independent from the modeling levels. In fact, the abstraction

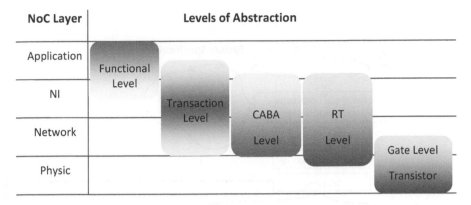

FIGURE 13.2
The relation between NoC layers and levels of abstraction from a performance evaluation viewpoint

level of a model expresses its granularity, i.e., how much the model and its communications are detailed, while the layering concept means in each layer, we are concerned with different issues of the NoC. For each layer of an NoC, designers take care of all details of that layer. Hence, the issues of a particular layer may be expressed at different levels of abstraction. For example, a flow control protocol of a Network-Interface layer can be modeled at the functional level, transaction level, or cycle-accurate bit-accurate level, regarding the purpose of modeling. In fact the choice of abstraction level depends on the purpose of modeling. For example if the functionality of a protocol is targeted, functional modeling can be efficient, while this level of modeling is too abstract to explore performance characteristics of NoCs.

From a performance evaluation perspective, there is an implicit relation between these two concepts (Figure 13.2). For example, functional level modeling may be used to explore performance issues of the application layer but it is not suitable for lower layers of an NoC. The TLM and CABA levels (together system level) have been widely used to verify the performance of both Network-Interface and Network layers. Although register level modeling of the entire system is mandatory as one approaches physical implementation, it is not very appropriate for the performance evaluation of the entire system since simulation at this level is very slow and the number of traces[2] is huge. Gate and transistor level models, which are obtained almost automatically from RTL models, help to explore the performance metrics of the physical layer of the system, such as timing, power consumption, and silicon area.

---

[2]A *trace* refers to information about the time-dependent inter-core communications that is captured during a simulation run.

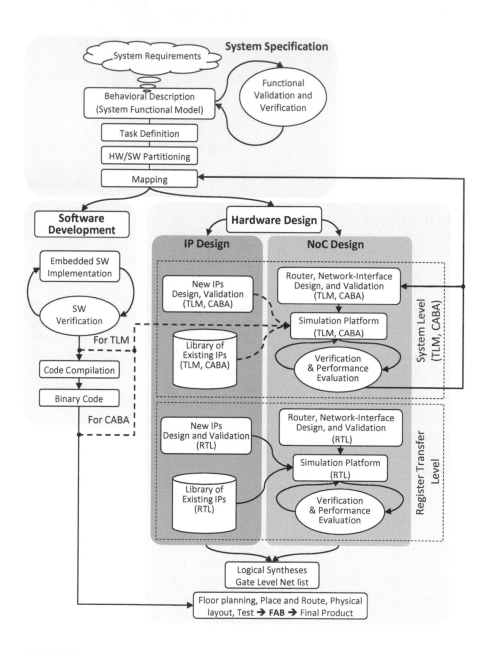

FIGURE 13.3
A generic design flow for an NoC-based system

13.2.4   Design Flow

The design of an NoC-based system is a complex process involving algorithms and techniques for developing both embedded software and the hardware architecture of the system. In order to do this, a conventional design flow performs several tasks; each has an essential role in the design and is done either sequentially or in parallel with others. For example, two of the principal tasks in each design flow are *functional verification* and *performance evaluation* of the entire system and all of its components. Functional verification consists of verifying whether the designed model conforms to the system specification and satisfies its functional properties. For example, freedom from deadlock of a routing algorithm or a flow control protocol is a typical property which is verified during the design of an NoC-based system. Performance evaluation is the crucial and heavy task of evaluating the impact of a vast number of design-space tradeoffs (containing both application and architectural issues) on the performance and the cost of the system. Since the main motivation for the use of NoC in on-chip systems is to achieve a higher communication performance (compared to bus-based interconnects), it is very important to rapidly and precisely estimate the performance of the system. This task requires fast, efficient, and reliable methods and tools.

As shown in Figure 13.3, the SoC design starts by capturing the system specifications at the highest level of abstraction and proceeds toward an efficient system integration and silicon implementation. Usually an SoC design flow contains at least these general steps: system specification, task definition, hardware/software partitioning and mapping, software development and hardware design, and physical implementation. As the design proceeds toward physical implementation, it becomes more and more expensive and time-consuming to make changes. Therefore, although a later performance evaluation gives results closer to reality, it is very important to analyze the performance of a system as soon as possible in the flow. On the other hand, as discussed before, very high-level functional models are not adequate for exploring architectural performance metrics because they do not deal with architectural details of the system. Therefore it is very important to determine which step of the design flow is more suitable for performance evaluation of the entire system.

13.2.4.1   System Specification and Performance Requirements

A conventional design process starts with an informal definition of the major objectives, environment, and requirements of the system. The steps of this part of the design (i.e., before hardware design) essentially aim to determine the performance requirements of the system rather than to evaluate its performance metrics.

Behavioral Description

At first, a behavioral description of the system allows designers to rapidly verify the whole system functionality and feasibility. Specification languages can be used to provide a formal specification of the system and are more adequate for this purpose than programming languages which are used to implement the system. Having such a formal specification of the system, it is possible to use formal verification techniques (for example, model checking) to demonstrate that the candidate system is correct with respect to the specification. In this step there is no difference between software and hardware parts and the entire system application is considered as a whole. Therefore, although this step is very appropriate for functional verification purposes, the performance evaluation of the system does not make sense in this step since the system specification essentially aims to describe *what* the system should do, not *how* the system does it.

Task Definition and Making an Application Graph

The next step is to make an *application graph* (task graph) by partitioning the entire application into separate communicating tasks. In this step there is no difference between software and hardware tasks and each task (node of the graph) can later be replaced by a software program or a hardware processing block. Nodes are connected by simple point to point links (edges of the graph, also called channels). Models of Computation (MoCs) (Soininen and Hensala 2003) are formal models used to describe the application graph. Synchronous Data Flow Graphs (SDFG) (Lee and Messerschmitt 1987b) and Communication Architecture Graphs (CAG or core graph) (Benini and De Micheli 2006) are two examples of MoCs used particularly for application graph modeling. The key design challenge in this step is to expose application graph issues such as task-level parallelism, and to capture formally concurrent communication in MoCs. It could help to define the performance requirements of the application graph, such as the memory requirement between different tasks, the data rate transmitted between different tasks, the bandwidth needed between them, or the maximum (worst-case) task-to-task latency that the application can tolerate. For example, in Geilen, Basten, and Stuijk (2005), the authors aim to minimize the memory requirements of SDFGs using a model checking method. Therefore, NoC performance evaluation is not meaningful at this stage since the communicating infrastructure is still invisible.

Hardware-Software Partitioning

In this step designers assign to each task of the application graph either a specific hardware block or a software program executed on a processing unit. Then several hardware and software components may be grouped to construct *local subsystems*. Usually this step does not yet deal with the architectural choices of global communicating infrastructure, so performance evaluation is no more adequate than in the previous steps. Nevertheless, the construction of local subsystems gives insights into the type of interconnect platform which is needed (bus, crossbar, or NoC) and its size (i.e., the number of nodes). Henceforth these are local subsystems that communicate together, and not tasks of the application graph. So it is possible to determine the communication requirements between local subsystems.

Mapping or Topological Placement of Subsystems on the Network

This step mainly aims to determine a primitive mapping of local subsystems on the nodes of the network. The difference from the previous step is that the topological placement of the subsystems is now determined. To do so the size and the topology of the network are needed. Clearly the subsystems which may communicate more frequently should be mapped on the nodes which are relatively closer. From the network layer point of view, *local* communication within a single subsystem (i.e., *intra-cluster* communication) is hidden, whereas the *global* communication between subsystems (i.e., *inter-cluster* communication) is visible. The global communication determines the distribution of total traffic over the network, i.e., it determines the amount of traffic each cluster of the network sends to other clusters. Note that we define the *traffic distribution* for clusters of the network and not for the local IP blocks of a subsystem. Therefore, when the mapping changes, the traffic distribution changes too (see Figure 13.4). The initial mapping and consequently the initial traffic distribution, determined in this step, are not fixed, and according to feedback from performance evaluation they may be changed to satisfy performance requirements.

Performance Requirements, Performance Evaluation, and Performance Verification

In fact all the steps mentioned above aim to determine the performance requirements of the system. These requirements could be expressed as minimum, maximum, or average values. For example, the minimum throughput required between each pair of local subsystems or each pair of application tasks that a real-time application needs to be executed can be determined. Timing constraints of the application can be defined, for instance, the maximum message latency under which the application could properly operate. Briefly, the main requirements and conditions for executing the application must be explicitly defined, without involving network architectural issues. As demonstrated in Figure 13.4, once the network architecture and topological mapping are selected, a suitable performance evaluation method enables designers to measure or estimate the performance metrics of the NoC. Then a performance verification technique checks whether these measured performance metrics satisfy the performance requirements or not. In this sense performance evaluation and performance verification are two separate tasks. As an example of performance verification, Goossens et al. (2005) propose an independent (from

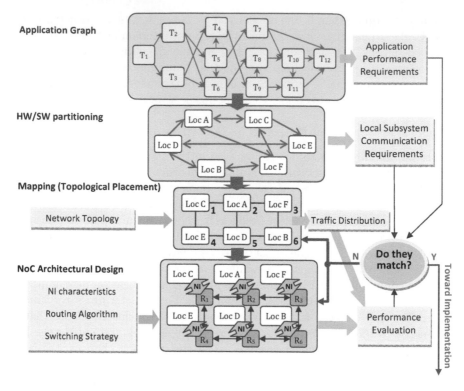

FIGURE 13.4
Performance requirements versus performance analysis

performance evaluation) performance verification tool that verifies analytically that the NoC instance (hardware) and its configuration (software) together meet the application performance requirements.

13.2.4.2 NoC Design and Performance Evaluation

Once the hardware/software partition of tasks is determined, two parallel design flows start for the software development and hardware design. Software development consists of programming and compiling for the target processing elements. SoC hardware design is composed of two (usually) parallel flows of IP design and NoC design in both register-transfer (RT) and CABA levels of abstractions.

An NoC (in the sense of a network layer) is characterized by its architectural parameters and its performance metrics for a given mapped application. A vast number of architectural parameters, such as topology, routing algorithm, switching strategy, and flow control mechanism, as well as the repartition of the traffic, influence the NoC performance metrics such as latency and throughput. The NoC design space is the set of all possible combinations of the above parameters. The NoC design is the complex process of finding the best solution from the design space in a way that either the NoC performance satisfies the application requirements (in the case of application specific NoCs) or the NoC provides a generally satisfactory performance (in the case of general purpose NoCs). In both cases, the performance evaluation of an NoC (from a network layer point of view) is meaningful once its architectural parameters are determined and the traffic distribution is given. Irrespective of application specific or general purpose NoC design, the role of a performance evaluation method remains the same, namely, to determine the NoC performance metrics in order to find the best solution from the architectural design space (mostly in the case of application specific NoCs) and/or the best mapping of the application (mostly in the case of general purpose NoCs where the architecture is rather fixed). So if the obtained performance results are not satisfying, either the mapping or the architectural parameters can be changed to converge toward the desired performance. This *optimization loop* continues until the desired performance is achieved. Since the task of NoC performance evaluation is traditionally done by simulation, before arguing why analytical methods may be more suitable for this task, let us discuss the specifics of different levels of simulation.

Simulation: System Level versus RTL

NoC hardware design begins with system level modeling, simulation, and sending feedback to earlier steps. Once designers achieve the desired performance, the NoC architectural parameters become fixed and the NoC is remodeled in RTL, which can be automatically synthesized to a gate level description. SystemC and SystemVerilog are the two most popular languages used for system level modeling.

System level modeling was a great shift in SoC design; simulation at this level permits capture of the entire system performance metrics such as communication latency and aggregated bandwidth in a high level of abstraction and before RTL description. In fact this step is the most convenient design step for the performance evaluation of the entire system. It is not possible during previous design steps when NoC architectural parameters are not yet determined; on the other hand, RTL simulation is not adequate for performance evaluation of an entire large system because this level involves too much detail and the number of traces is huge, simulation time is much longer than at system level, and sometimes simulation is impossible because of excessive run times. Nevertheless, RTL modeling is still mandatory as a design approaches physical implementation since system level synthesis is not yet sufficiently mature to allow it to be omitted.

Finally physical implementation contains all the processes of chip logic synthesis and verification for building the chip layout for fabrication (i.e., back-end processes). Some important cost-related metrics are obtained during physical implementation such as electrical delays, power consumption, and silicon area. These metrics are related to the physical layer of an NoC-based system.

Traffic Generators

In order to speed up the NoC performance evaluation and keep it independent from IP design it is common to design and use traffic generators instead of real IPs or subsystems. According to a study on NoC performance evaluation presented by Salminen, Kulmala, and Hamalainen (2007), real applications are only used in one third of performance analysis. Real application (application-driven) traffic requires modeling and validation of all the network clients (IPs) and then mapping them onto the communication platform. This is based on full system simulation and communication traces (Lu 2007), which may be too complex and time-consuming in the early phases of development.

Instead, traffic generators can replace the subsystems mapped onto the clusters of the network and mimic the behavior of real applications according to a *synthetic* traffic model and without performing actual computation. Synthetic traffic captures the prominent aspects of the application-driven workload and can be easily manipulated. In order to obtain fast performance results it is also usual to omit the dependency between different subsystems when using traffic generators. The concurrent development of IPs and NoC due to the separation of NoC design (communication) from IP design (computation) speeds up NoC performance evaluation but leaves the development of the communication platform without sufficient traffic knowledge. Therefore network performance should be evaluated with different types of traffic.

A traffic pattern can be parameterized by its spatial distribution, its temporal distribution, and by changing the offered load per source node. Temporal distribution determines the timing behavior of packet generation (in a more general term, data generation) per terminal. It could have, for example, a Poisson, self-similar, or bursty nature (a non-constant data rate is called bursty traffic; for example, Thid, Sander, and Jantsch (2006) presented an efficient way to generate bursty traffic loads artificially). As demonstrated in the next section, and to the best of our knowledge, until now all proposed analytical performance techniques in the literature are based on a Poisson assumption. Spatial distribution of the traffic is the pattern according to which the destination is selected. In the case of real applications, it is usually represented by a matrix, called a *traffic distribution matrix*. Well known spatial distributions for synthetic traffic patterns are uniform and localized destination distributions. Finally, *offered load* means the rate of traffic injected to the network by each source node. To avoid all ambiguity both temporal and spatial distribution and the definition of offered load (inclusion of header or payload only) must be explicitly given in every performance study (Salminen, Kulmala, and Hamalainen 2007). Whereas in almost all performance analysis methodology, the temporal distribution of traffic follows the Poisson process, the most commonly used spatial distributions are uniform and localized. Also, a few performance analyses are based on transposed, bit reversal, or hot-spot traffic.

Uniform Traffic Pattern

Uniform traffic pattern is the most common workload used for network performance evaluation and particularly it is an adequate workload to determine the load-latency curves and the saturation point of the networks (Salminen, Kulmala, and Hamalainen 2007). According to this pattern each source node sends data to all other destination nodes with the same probability. Specifically, as we will see in the following section, most of the proposed analytical methods are based on (and are limited to) this traffic pattern, especially because of its symmetric nature (Pande, Grecu, Jones, et al. 2005).

Localized Traffic Pattern

The assumption of spatially uniform traffic is not very realistic in an SoC environment since usually subsystems are mapped onto the network such that the subsystems which communicate more frequently are situated closer together. Thus the total communication exhibits a highly *localized* behavior and so the NoC performance must also be evaluated under the localized distribution. Different types of localized traffic have been proposed. For example, in Ogras and Marculescu (2007), each node communicates only with the nodes that are located within a forwarding radius. Furthermore if the distance between the source and destination nodes is given by $\mathrm{dist}(s, d)$, then the forwarding probability $p_f(s, d)$ is proportional to the inverse of the distance $(p_f(s, d) \propto 1/\mathrm{dist}(s, d))$. Weldezion et al. (2009) use another type of localized traffic in which all nodes communicate with all other nodes, but the forwarding probability decreases as the distance increases.

Hot Spot Traffic

A *hot spot* arises when a number of nodes direct the largest fraction of their generated messages to a few destinations. For instance, global synchronization is a typical situation that can produce hot spots, since each node in the system sends a synchronization message to a distinct node (Ould-Khaoua and Sarbazi-Azad 2001). Another example of hot-spot traffic can be found in shared memory systems. In some cache coherency protocols, a message is sent to all nodes having a dirty copy of the cache line to perform write-invalidation. Those nodes should then send an acknowledgment back to the host node to maintain memory consistency correctly, which results in a hot-spot traffic distribution (Loucif, Ould-Khaoua, and Min 2005). Ould-Khaoua and Sarbazi-Azad (2001) and Loucif, Ould-Khaoua, and Min (2005) present examples of analytical methods addressing hot-spot traffic.

Bit-Reversal Permutation

Bit-reversal permutation is generally defined for k-ary n-cubes. When the traffic pattern is generated according to bit-reversal permutation, a message generated in the source node $x = x_1 x_2 \ldots x_n$ is destined for the destination node $B(x) = x_n x_{n-1} \ldots x_2 x_1$, which is the reversed order of the bit pattern x. Sarbazi-Azad, Ould-Khaoua, and Mackenzie (2001) generate a mixed workload for the network with both bit-reversal and uniform traffic. When a message is generated at a source node it has a probability β of being a bit-reversal message and probability $(1 - \beta)$ of being uniform.

13.2.4.3   Analytical Methods in Design Flow

In order to reduce the total time of the optimization loop and to rapidly verify several mapping and/or several architectural parameters from design space, an adequate analytical method can be used instead of simulation for performance evaluation purposes.

Making an adequate simulation platform, even at a high level of abstraction, is time consuming in the sense that it needs a huge amount of development. In fact it necessitates the design, modeling, and validation of all NoC components such as router and network interfaces and also IPs or traffic generators. Furthermore, the accuracy of simulation results depends tightly on simulation run-time; for example, to obtain the average latency, the simulation must be repeated for transmission of thousands of packets (or messages). Some other phenomena appear only with long enough execution; for example, the NoC saturation threshold cannot be detected during the first packet transmissions, i.e., in a warmup period (Salminen, Kulmala, and Hamalainen 2008). Another major drawback of simulation-based approaches is that they are not exhaustive in the sense that it is impossible to cover all communication traces.

One of the advantages of analytical techniques is that they provide approximate performance metrics with a minimum amount of effort in development and adaptation. The accuracy of analytical results depends on the mathematical methods used for traffic and network modeling. Analytical results can be imprecise due to the abstractions used but it gives an initial and quick estimation (Lu 2007), which is very useful to rapidly explore different architectural and mapping combinations. In brief, the popularity of analytical methods in performance evaluation is due to a good balance between the rapidity in obtaining and relative accuracy of results on one hand, and the speed and efficiency of method development and adaptation on the other hand.

As presented above, NoC design consists of finding the best solution from the design space such that the NoC performance satisfies application requirements. But the design space is too large to allow a complete exploration (Salminen, Kulmala, and Hamalainen 2007). Hence, in order to explore more and more architectural solutions, analytical methods are needed to obtain fast and reliable performance results. By using such methods, time consuming sim-

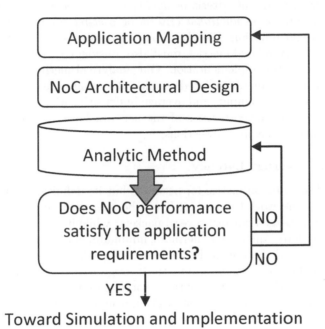

FIGURE 13.5
Optimization loop: architectural exploration and mapping exploration

ulations can be postponed to the later stages of the design, when the design space is efficiently reduced.

13.2.5　Parameters Addressed in NoC Performance Evaluation

We call *mapping exploration* the search for the best mapping for a given architectural solution so that the NoC provides the desired performance. *Architectural exploration* means here to investigate the impact of different architectural parameters of the network layer (such as topology, routing algorithm, packet and buffer size) on the quality-related performance metrics, and so to choose the best combination. In Figure 13.5 the inner loop depicts architectural exploration and the outer one depicts mapping optimization.

To compare and contrast different NoC architectures, a standard set of performance metrics can be used. For example, it is always desirable that the NoC exhibits high throughput, low latency, energy efficiency, and low area (Pande, Grecu, Jones, et al. 2005). Lu (2007) separates NoC performance metrics into *functional* criteria or *quality-related* metrics such as latency and throughput and *non-functional* criteria or *cost-related* metrics such as area and power consumption. Clearly a complete design space exploration, in addition to quality-related metrics, must also consider cost-related metrics which are obtained after implementation. Our analytical method presented in this chapter addresses quality-related metrics. It computes average latency, maximum accepted throughput, and average buffer utilization, which are the most important performance metrics of the network layer. These metrics are explained in the following subsections.

13.2.5.1   Average Latency

Latency is a common metric for networks but its definition is ambiguous and varies from one work to another. It can be measured per data word, header, packet, or message transfer (several packets) while including or excluding the time in the source queue. Furthermore, minimum, maximum, or average values may be reported (Salminen, Kulmala, and Hamalainen 2007).

Our method aims to determine the average *packet latency* of a given NoC architecture and traffic distribution. Packet latency is the time spent by a packet to traverse the NoC from a given source to a given destination. It is an essential metric in NoC-based system performance particularly because it can help to estimate the application runtime and is a key issue for comparing different NoC architectures. Path latency, communication latency, end-to-end latency, and source-to-destination latency are other terms sometimes used for this purpose.

When the packet latency is expressed as a function of offered load it could demonstrate the maximum acceptable throughput (saturation threshold) of a network and allows designers to check which architecture can accept a higher amount of offered load. To do so, one should inject different amounts of offered load to the network according to a particular traffic distribution and examine at which level of load the network saturates. Saturation is detected when the latency tends to infinity, if the latency includes the waiting time in source queues. If the latency does not include the waiting time in source queues, the latency remains constant after saturation as the NoC does not accept more load. In fact, after saturation only a fraction of the offered load can be absorbed by the network so the non-accepted load begins to accumulate in source queues. Thus the waiting time in source queues tends to infinity, while the latency after arriving to the network remains constant. Synthetic traffic patterns like uniform and localized distributions are used for this purpose since they can be configured for different levels of offered load.

The method presented here is also able to determine the latency of packets traversing each router of the NoC. The router latency is decomposed into different delay components and each delay component can be analyzed separately. The mean latency of the entire NoC can be obtained by analyzing the latencies of different classes of path.

13.2.5.2 Saturation Threshold or Maximum Accepted Throughput

In the network layer of an NoC, throughput signifies the maximum amount of offered load sustainable by the network (i.e., the load that the network is capable of physically handling) (Pande, Grecu, Jones, et al. 2005). In this sense, the term *accepted load* is sometime used for *throughput*. Ideally throughput should increase linearly with the offered load, but due to the limitation of network resources, throughput will saturate at a certain amount of offered load. This point is called the *saturation threshold* or the *maximum accepted throughput* of the network. In fact, below the saturation threshold the offered load and the throughput (accepted load) are equal.

The offered load directly affects the average latency. As the offered load approaches the saturation threshold, there will be more contention in the network and latency will increase and become unpredictable. Consequently, the desirable point of operation for an NoC-based system should be well below its saturation threshold (Pande, Grecu, Jones, et al. 2005).

Scalability in NoCs means that the value of the saturation threshold seems to be roughly independent of the number of communication units. The main motivation supporting the NoC paradigm is the fact that classical interconnects such as shared busses have a low saturation threshold and do not scale with the number of connected units. They can serve a limited number of units, and beyond that the average communication latency strongly increases (Sheibanyrad 2008).

Therefore the saturation threshold of a network is one of the most important performance issues of the network layer and should be carefully estimated. The exact value of the saturation threshold depends on the offered load, average packet length, traffic distribution, and buffering capacity. These arguments should be taken into account as parameters of a generic performance analysis in order to evaluate the value of the saturation threshold.

13.2.5.3 Mean Buffer Utilization

In NoC design a non-negligible silicon area overhead arises due to the presence of buffering space between routers (more precisely the buffering space is physically located in the input or output ports of a router, or both). On the other hand, their existence usually has a positive impact on communication performance (latency and throughput). Therefore, the analysis and measurement of required storage capacity and finding a suitable balance between silicon consumption and a higher performance (i.e., a cost-performance tradeoff) is one of the important NoC design decisions.

The approximation of average buffer utilization under a given traffic pattern, gives helpful insights into optimal link capacity allocation. It can help for allocating buffer space over the NoC because the average buffer utilization displays the aggregation of traffic over different parts of the NoC.

13.3   The State-of-the-Art in NoC Performance Evaluation

Valuable studies about NoC performance evaluation exist which provide basic useful guidelines for methods and techniques of evaluation. Some of the studies discuss how NoCs are and should be compared, which aspects of performance have been analyzed, and which need more research (Coppola et al. 2004; Wiklund and Liu 2003; Bainbridge and Furber 2002; Feliciian and Furber 2004; Loucif, Ould-Khaoua, and Min 2005; Sarbazi-Azad, Ould-Khaoua, and Mackenzie 2001; Salminen, Kulmala, and Hamalainen 2008; Dally and Towles 2004; Salminen et al. 2007; Kermani and Kleinrock 1979; Agarwal 1991). These studies identify various trade-offs between performance and design parameters and sum up many pitfalls that must be considered when evaluating the performance of complex Systems-on-Chip. According to these studies the basic methods for performance comparison are analysis and simulation. In simulation-based methods different traffic scenarios are often modeled with traffic generators instead of using real applications. Networks, especially NoCs, are often compared and benchmarked via their latency versus load behavior (Dally and Towles 2004). Salminen, Kulmala, and Hamalainen (2008) study the impact of various simulation and NoC setups in common load-latency curves that are used for performance evaluation. According to them, in 18% of about 140 NoC publications, load-latency curves were analyzed. Both load (offered load) and latency are ambiguous terms which need to be clearly defined in each study. Salminen, Kulmala, and Hamalainen (2008) suggest that to deal with this ambiguity three basic choices in load-latency measurement must be defined:

1. Whether latency is measured for the header or for the whole packet.

2. Whether latency includes or excludes the time waiting at source queues.

3. Whether offered load includes the header or just the payload.

According to Salminen, Kulmala, and Hamalainen (2008) the most suitable choices for latency are average t_{avg} and percentile (or box-plot) $t(j)$, for example, $t(90\%)$ or $t(99\%)$. Latency percentiles are reported when dealing with real-time constraints as the average latency is an inadequate metric in those cases. The units of latency and offered load must be explicitly defined.

13.3.1   Simulation-Based Methods

NoC performance evaluation is traditionally based on *simulation*. In a simulation-based approach designers build network and traffic models (either real IPs or traffic generators) and connect them to construct a simulation platform. Then the network operation is simulated by loading the traffic into the network. Designers generally use simulation to analyze system operation and performance in different abstraction levels prior to physical design. Pande, Grecu, Jones, et al. (2005) explore design tradeoffs that characterize the NoC architecture and affect its performance. The authors compare the performance of a variety of NoC architectures such as SPIN, CLICHÉ, OCTAGON, etc., with regard to latency, throughput, energy dissipation, and silicon area. They use a wormhole router simulator to compare the NoC topologies in terms of throughput and latency. Their simulator handles variable message length and can generate traffic according to both uniform and localized traffic patterns (as spatial distribution) and using both Poisson and self-similar traffic injection distributions (as temporal distribution). They assume that messages arriving at the destination are immediately consumed at the rate of one flit per time step; i.e., no blocking is encountered at the destinations (in other words destinations behave like sinks). They also study the impact of virtual channels on throughput and average latency. Average latency increases with the number of virtual channels while a higher throughput is provided. They show that a system with four virtual channels strikes an appropriate balance between high throughput, low latency, and conservation of silicon area. Salminen et al. (2007) present a simulation-based comparison of generic, synthesizable single bus, hierarchical bus, and 2-dimensional mesh on-chip networks while addressing a cut-through two-dimensional mesh NoC.

In general, the NoC design space is too big to explore by simulation (Ogras 2007). For instance, there are $n!$ different ways to map a given application to a network with n nodes. Ogras (2007) compares an analytical approach with a simulation-based one to find the best mapping of a given application on a 4×4 2D mesh network. According to the experiment presented in Ogras and Marculescu (2007) it takes about 22 hours to find the best mapping through simulation, whereas the analytical approach proposed by Ogras and Marculescu (2007) completes the analysis in about 7 seconds, which is about 4 orders of magnitude faster. However, analytical performance methods are usually validated against simulation methods.

13.3.2   Analytical Methods

The earliest work (and the majority of it) about performance analysis methods comes from the parallel computing and macro-network research communities, while NoC performance evaluation is mainly based on simulation techniques. In fact the emergence of the NoC paradigm was more or less in parallel with the emergence of high-level simulation techniques such as, for example, SystemC modeling (enabling both CABA and TLM[3]modeling). Thus the earliest NoC performance evaluations were mostly based on these high-level simulation techniques. Nowadays, while increasing system size results in limitations

---

[3]TLM, for Transaction-Level Modeling, is a high-level approach to modeling digital systems where details of communication among modules are separated from the details of the implementation of functional units or of the communication architecture.

for the use of simulation-based techniques even at high levels of abstraction, analytical performance methods have attracted the attention of NoC designers.

In the following we review most of the existing analytical performance methods, in three different groups. The first group contains the earliest and the most cited performance analysis methods. The second group gathers a set of similar works which are more or less based on the same approach (derived from the first group) and the same set of assumptions. Finally, in the third group we review more recent works focusing on NoC performance analysis (while the first two groups are mostly about off-chip networks). The methods of this last group are based on the approaches of the first and second groups, but adjusted with on-chip requirements.

Some of the most important characteristics of the works studied are extracted and summarized in Table 13.1 on page 387.

13.3.2.1   Early Methods

The papers of this group are those which are known and cited as the first analytical approaches basically proposed for latency and delay analysis in networks with wormhole routing. The analytical models prior to this group address other routing strategies, for example, Kermani and Kleinrock (1979) and Agarwal (1991), which address virtual cut-through routing. The majority of these models have been proposed for off-chip networks. They mostly apply to communication which predominantly consists of long messages, whereas in multi-processor networks the main bulk of communications is due to short cache-line requests (Hennessy and Patterson 2003).

Dally (1990) was the first to propose an analytical performance model for wormhole routing (Draper and Ghosh 1994). Benini and De Micheli (2002) presented an analytical performance approach for analyzing VLSI communication networks. This approach, which is based on numerical methods, is certainly one of the most cited references in the domain of network performance analysis. The main objective of the paper was to show that low-dimensional k-ary n-cube networks outperform high-dimensional networks with the same bisection width. Therefore Dally (1992) investigated network throughput and latency under wormhole routing and a deterministic algorithm (the e-cube algorithm, which is similar to the x-first algorithm but for networks with higher dimensions). Concerning the approximation of latency and throughput, this approach is strictly limited to k-ary n-cube network topology with the assumption that both n and k are even integers. Based on the symmetry due to this kind of topology and also due to the uniform traffic assumption, the paper proposes a simplified model for average network latency which cannot be extended to other topologies or traffic patterns. In addition, it assumes no buffering space in the network, whereas in fact buffering space has a great impact on network latency and a suitable performance method must be able to model buffers of different lengths. Furthermore, it estimates the average waiting time for a channel under the assumption that a message waits upon the service completion of, at most, one other message before acquiring the channel. For example, it cannot cover the case in which a message is blocked by two other messages. Since this approximation holds only for low-dimension networks when the utilization approaches zero, this model becomes inaccurate for large networks and high traffic rates (Draper and Ghosh 1994). This work analyzes the latency of the network layer, i.e., from the moment messages enter the network; thus message latency does not include the waiting time in the source queue.

Hu and Kleinrock (1997) develop a queuing model for wormhole routing with finite-size network buffers and infinite-size source buffers. This model estimates both output link contention delay (equivalent to port acquisition delay in our model to be presented in the next section) and buffer queuing delay (equivalent to link transfer delay in our model). The finite-size buffer is approximated with an $M/G/1/K^4$ queue. The model does not hold for virtual channels. It addresses arbitrary packet size distribution with a Poisson arrival process. They address networks with a system of back-pressure control which means when a switch (router) is in blocking mode (occupied by a packet), it informs up-stream switches to stop transmission. Considering the timing and area limitation, the implementation of such complex routers for on-chip purposes is not so promising. The main drawback of this model is that it is designed for huge networks (like LANs) with large buffer size (hundreds of bytes) which could hold more than one packet. Thus the model is not adequate for a network with short buffer size that can only hold a portion of a packet (which is the case in NoCs). The authors use iterative methods to solve the interdependency between different variables of the model (link occupation time and blocking time).

Guan, Tsai, and Blough (1993) present a queuing-theory analysis to evaluate the routing and switch (router) delays in wormhole multicomputer interconnection networks. The approach can be applied to arbitrary network topologies and deterministic routing algorithms. Any deterministic traffic pattern can be modeled with the assumption that message generation at each node is a Poisson process independent from other nodes. It presents a comprehensive delay analysis (in the sense of the components of the latency) of the network. The main drawback of this approach is that it is restricted to channels with only one flit buffer, and thus not suitable for studying the impact of the buffer length on the network performance. In fact, for on-chip networks the design choices are much more limited compared to off-chip networks. One

---

[4] A queuing model A/S/N represents a single queue system, where A is the arrival process of requests to the queue, S represents the service process of requests in the server, and N represents the number of servers. For example, the $M/M/1$ model is the most commonly used, in which there is only one server, and arrival and service processes are Poisson (i.e., request inter-arrival and service time both follow exponential distributions). In a $G/G/1$ model, request inter-arrivals are independent and identically distributed with a general distribution.

of the most important architectural choices which can be changed to improve the network performance is the buffer length. Thus an NoC performance evaluation method must certainly be able to model different buffer lengths. The methods which put limitations on buffer (and also packet) length are indeed not suitable for NoC performance evaluation. Furthermore, the following assumption taken by Guan, Tsai, and Blough (1993) results in a non-negligible inaccuracy in router and path latency. In fact, for computing the contention probability and delay, the router model presented by Guan, Tsai, and Blough (1993) (which is modeled with a Markov chain) does not distinguish which flit of a competing message is occupying the output port, but it is always considered that the competing message has just arrived. So the time to transmit the competing message out of the channel (equivalent to contention delay in our model) is always equal to the time of the transfer of an entire message – from the header to the last flit. An accurate router delay model must also consider the case of contention with a portion of a message. With this hypothesis their delay model provides a sort of worst-case approximation, while the objective is to obtain the average latency. Additionally, as Guan, Tsai, and Blough (1993) mention themselves, the complexity of the model is rather high, barely manageable when applied to high-dimensional networks.

Draper and Ghosh (1994) present a model, based on queuing theory, for performance evaluation of wormhole routing in unidirectional k-ary n-cube networks with virtual channels and a deterministic routing algorithm (e-cube or left-right). In a network queuing model, each channel is regarded as a server with its service time depending on blocking in subsequent channels. In this approach each channel is modeled as the server of an $M/G/1$ queuing system. The model applies to communication with long messages and is limited to a uniform traffic pattern. Based on the uniform communication assumption (uniform traffic and symmetric topology), they derive equations of the model which are not extendable to asymmetric topologies or traffic patterns. For example, their method is based on the hypothesis that the statistical characteristic of a channel in a given dimension is identical to those of any other channel in the same dimension. For instance, the average utilization of all channels of the ith dimension are statistically the same, irrespective of the position of the router to which the channel is connected. Because of these assumptions, their method is applicable only to symmetric communication.

13.3.2.2 k-ary n-Cubes

Almost all methods of this group address the average message latency in different types of k-ary n-cube multicomputer networks, except two of them, which are partly adapted for on-chip purposes, as will be discussed. All these methods are based on a complete symmetry in the network and so limited to symmetric topologies (k-ary n-cube) and (mostly to) uniform traffic patterns. For example, the method proposed by Kiasari et al. (2008) works for a torus topology but is not adapted for a 2D mesh topology.

Assumptions

All these methods are limited to only one-flit-deep virtual channels. Using such methods, again, the impact of buffer depth on network performance cannot be investigated. The message length is assumed to be longer than the total buffer length over the path, so the head of a message reaches its destination before its tail leaves the source node. This assumption can be considered as a special kind of circuit switching in which the message establishes its path itself. Some other common assumptions and characteristics of this group are listed below.

- Nodes generate and inject messages independently of each other, following a Poisson process.

- Messages are transmitted to the local nodes as soon as they arrive at their destination.

- Message length is fixed and equal to M flits. A flit is transmitted through a physical channel in one cycle.

- The source queue at the injection channel has infinite capacity.

- An arbitrary number of virtual channels may be used per physical channel.

Approach

In all these methods, at first the mean message latency is obtained. The mean message latency is composed of the mean waiting time seen by a message in the source queue plus the mean network latency, which is the time to traverse the network. Then, to capture the effect of virtual channel multiplexing (multiple virtual channels share a physical channel bandwidth in a time multiplexed manner), the mean message latency is scaled by the average degree of virtual channel multiplexing. The approaches are based on queuing theory and combination theory. The queuing theory is used to determine the mean waiting time in a queue (for both source and network queues). The combination theory is used to compute the blocking probability at a channel. There are several inter-dependencies between the different variables of these models. For example the average service time of a channel is a function of the mean waiting time to acquire a virtual channel and vice versa. They use iterative techniques to solve the equations and determine the inter-dependent variables. Although all of these approaches are similar, some of them have some particularities which are explained in the following.

Sarbazi-Azad, Khonsari, and Ould-Khaoua (2002); Khonsari, Ould-Khaoua, and Ferguson (2003); and Najaf-Abadi and Sarbazi-Azad (2004): these three papers use the same analytical approach for networks with an arbitrary number of virtual channels. Their approach is an average case analysis which is limited to k-ary n-cube topologies. Sarbazi-Azad, Khonsari, and Ould-Khaoua (2002) address a deterministic routing algorithm, in which for organizing virtual channels, a restricted version of Duato's methodology (Duato, Yalamanchili, and Ni 2003) adapted for the case of deterministic routing is used. Khonsari, Ould-Khaoua, and Ferguson (2003) address exactly the adaptive routing algorithm of Duato. Both approaches are restricted to a uniform traffic pattern. Using an M/G/1 queuing model, they obtain the waiting time to acquire a virtual channel and also the waiting time in the source queue. By assuming a complete symmetry and considering the average value for all parameters, these methods obtain directly the average latency of the whole network (i.e., over all communication in the system). For example, they consider that all messages traverse on average d hops to cross the network, or they compute a unique average rate of message arrival for all channels of the network, or they assume that a message experiences the same mean waiting time and mean service time across all channels, regardless of the position of channels in the network. These assumptions simplify the computation but cannot provide the average latency due to a specific path of the network. However, an adequate performance model should obtain the average latency due to any desired path (end-to-end communication). Another work with a similar approach is presented by Najaf-Abadi and Sarbazi-Azad (2004) which targets torus multicomputer networks with a deterministic routing algorithm.

Kiasari et al. (2008) on-chip networks: this paper is based on exactly the same approach as previous ones, but adapted for on-chip networks. Its main contribution is that the proposed model is also used to estimate the power consumption of routers. However, this method, like the previous ones, is completely restricted to torus network topology, uniform traffic pattern, and an XY (x-first) routing algorithm. Nevertheless, comparing it to the work presented by Sarbazi-Azad, Khonsari, and Ould-Khaoua (2002) and Khonsari, Ould-Khaoua, and Ferguson (2003), the authors of Kiasari et al. (2008) propose a more sophisticated delay model, in the sense that the average message latency is decomposed to different delay components and then each delay component is separately analyzed. They use a backward latency computation order (i.e., from the destination to the source). To estimate power consumption of the entire network, this approach considers constant static power P_s (when the router is empty) and dynamic power dissipated P_d (when there is a message transmission) per router. Static power consumption of the network is $5k^2VP_s$ where k is the torus dimension, 5 is the number of input channels per router, and V the number of virtual channels per physical channel. Assuming that every message is of M flits and traverses on average d hops and the message arrival rate at routers is λ_g, they estimate the dynamic power consumption as $\lambda_g M(d+1)P_d$.

Ould-Khaoua and Sarbazi-Azad (2001); Sarbazi-Azad, Ould-Khaoua, and Mackenzie (2001); Loucif, Ould-Khaoua, and Min (2005); Loucif and Ould-Khaoua (2004): one of the distinctive features of these papers is the use of non-uniform traffic patterns. Ould-Khaoua et al. (Ould-Khaoua and Sarbazi-Azad 2001) address hypercube topology networks with virtual channels and Duato's adaptive routing in the presence of hot spot traffic. They assume that there is a single hot spot in the network. The traffic model proposed by Pfister and Norton (1985) is used to generate hot spot traffic. In this model, each generated message has a finite probability α of being directed to the hot spot node and probability $(1 - \alpha)$ of being directed to other network nodes. The presence of hot spot traffic causes an asymmetry in the network, i.e., service time varies from one channel to another due to the non-uniformity of traffic rates on the channels. For this reason Ould-Khaoua and Sarbazi-Azad (2001) compute the average service time and mean message latency separately for hot spot messages and regular messages and then take the average of them. However, as the authors state themselves, to simplify the development of the method, they propose a rough approximation of the variance of the service time at a given channel, by ignoring the interdependencies between service times at successive channels. This simplification causes inconsistency in results when the network is carrying heavy traffic. Loucif, Ould-Khaoua, and Min (2005) and Loucif and Ould-Khaoua (2004) suggest an analytical model, very similar to the one proposed by Ould-Khaoua and Sarbazi-Azad (2001). The only difference is that they address deterministic routing while Ould-Khaoua and Sarbazi-Azad (2001) address adaptive routing. Another similar paper is by Sarbazi-Azad, Ould-Khaoua, and Mackenzie (2001) in which the authors use the same approach but in the presence of bit-reversal permutation traffic (instead of hot spot) in hypercube networks with adaptive routing.

13.3.2.3   NoC Performance Analysis

The performance methods gathered in this group are basically derived from the previous ones. However, they are more or less adapted for being used in NoC design.

Kiasari et al. (2008) derive an analytical performance model for on-chip networks. However, its overall objective is to map IPs on a given network architecture in a way that the average communications delay is minimized. To this end, the authors propose a performance-aware mapping algorithm called PERMAP. They present an average-message-latency evaluation method which can be used for any arbitrary network topology with wormhole routing under an arbitrary traffic pattern and an arbitrary deterministic routing algorithm. A distinctive feature of the method is the use of a G/G/1 queue to model the channels of the network (against the common M/G/1 queue model). According to the authors, this means that both packet injection rate and packet arrival rate to a channel are distributed with arbitrary distribution (i.e., non-Poissonial injection and arrival rate). The model is, however, limited to one-flit router input channels and fixed priority arbitration mechanism. This paper is one of the few analytical studies which address also a real application as a case study, but the results presented are not concrete in the sense of packet latency comparison. In fact, the experimental result presented is reduced to a comparison between simulation and analytic runtimes from a mapping exploration point of view, without any demonstration to show the accuracy of the latency model.

Elmiligi et al. (2007) present an analytical model based on queuing theory for a 2D mesh input-queue (input buffer) router. To analyze the entire system (a router) they model each of the five input queues of the router separately with an M/M/1/B queue. Then they use an iterative approach to satisfy all boundary conditions. They address an input queue with finite length, but with the assumption of packet loss, which means when the input queue is full a new arriving packet is discarded. This assumption is rarely true in NoCs. In fact, their approach does not provide a delay analysis of routers but it mainly analyzes the probability of packet loss for different rates of packet arrival.

Guz et al. (2007, 2006): The main objective of these two papers is to propose a link capacity allocation algorithm in application-specific wormhole NoCs. The papers present first an analytical delay model to approximate the end-to-end transfer delay under application-specific flows. Then they apply the delay model to a capacity allocation algorithm which assigns link capacities such that packet delay requirements for each flow are satisfied. This approach is generic in terms of network topology, deterministic routing algorithm, the model of traffic, and the number of virtual channels. Therefore, this delay model is one of the few which could cover different characteristics (of both architecture and application) of NoCs. Furthermore, it can approximate the delay of any desired end-to-end flow (path). However, the model suffers from some serious drawbacks. The delay model is based on the assumption that packets are larger than the total buffer length along with their path, which is not the usual case even in application-specific NoCs. Besides, this assumption means that the header of a packet reaches the destination before its tail leaves the source. The model addresses low to medium loads and does not attempt to achieve high accuracy under very high traffic, whereas a delay model must support high loads because even a specific application may generate a high load when it is mapped onto the network. In addition, the model is limited to one-flit-deep channels. The main drawback of the model, however, is that it ignores link acquisition time by assuming that there are as many virtual channels as the number of flows sharing the same physical link (so every head flit of any flow can acquire a VC instantaneously on every link it traverses). Thus by using queuing models (M/D/1 queue) they just approximate the link transmission time by accounting for the time attributed to other virtual channels that interleave over the same physical link.

Lu, Jantsch, and Sander (2005): although this work does not address directly the performance analysis of the network layer, it is worth citing it here because it presents an application of network layer analysis for estimating the performance metrics of upper layers. In fact, it makes a rough estimation of network latency and uses this estimate to examine the feasibility of messages. It addresses the feasibility analysis of both real-time (RT) and non-real-time (NT) messages in a wormhole-routed network-on-chip. It examines the pass ratio of both RT and NT messages. Pass ratio is the percentage of the messages that pass the feasibility test. A message is feasible if its own timing constraint is satisfied. Timing constraint is expressed as a deterministic bound for RT and a probabilistic bound (depicted by average value) for NT messages. Therefore for each message an end-to-end delay constraint D is given and the message is feasible if the average network latency (in the case of NT) or the worst-case network latency (in the case of RT) is equal or inferior to its delay constraint D. Network latency (either worst-case or average) is composed of a non-contentional delay, which is the base latency, and a contentional delay, which is the blocking time (average or worst-case for NT or RT, respectively) due to contention. In mentioning that analytically estimating the contention delay is a difficult task, the authors just propose a very rough estimation for contention delay, by considering infinite buffers. In fact, their paper investigates the impact of bandwidth sharing between RT and NT messages on the pass ratio of each type of message, by making a very rough estimate of network latency.

Ogras and Marculescu (2007) present a generalized router model, based on queuing theory, which addresses wormhole NoCs. The router model supports arbitrary message size and buffer length, arbitrary network topology, and any deterministic routing algorithm. It covers deterministic traffic (i.e., driver application with its mapping on the network). The router model allows us to analyze each router of the network and to compute the average number of packets at each buffer in the network as a function of the traffic arrival process. It provides three performance metrics: average buffer utilization, average packet latency per flow (i.e., per path), and network throughput. Their approach is based on the generalization of the traditional delay model of single queues to the case of multiple queues per router by determining the average number of packets at each input queue. According to the authors, this generalization is the main contribution of their performance model. Since their approach is based on separated router delay models, it provides not only aggregate performance metrics such as average latency and throughput (of the entire network), but also feedback about the network behavior at a fine level of granularity, such as average utilization of all buffers and average latency per router and per flow. Such a method can be used in the optimization loops of NoC design for the purpose of application mapping, network architecture synthesis, and buffer space allocation.

By providing a case study Ogras and Marculescu (2007) illustrate the use of their router model as a tool for router design. Indeed, thanks to providing separately the delay analysis of each router, this method can be used by designers to evaluate possible trade-offs offered by different router design choices (e.g., buffer size and channel width) that are nowadays mostly pre-determined in an ad hoc manner. Note that previous performance methods cannot provide this ability since they address directly the entire network without providing a delay analysis at the router level.

13.3.2.4   The Synthesis of Reviewed Analytical Methods

Some of the most important characteristics and properties of the papers reviewed above are summarized in Table 13.1. We can extract the following conclusions about the reviewed analytical performance methods:

- As has been seen, all models are developed under the assumption of having an infinite buffer at source nodes and sink behavior at destination nodes. Messages are generated based on a Poisson process in source nodes, independent of each other.

- Most of them are limited to some fixed parameters. They mostly address different kinds of k-ary n-cube topologies and the uniform traffic pattern and put some restricting limits on buffer or message length. The most generalized performance model is the one presented by Ogras and Marculescu (2007).

- All methods presented compute the average value of desired metrics. Thus other kinds of measurement such as maximum or minimum (for example, worst-case or best-case latency) are not targeted by existing analytical methods.

- Almost all analytical methods address average latency (according to their own definitions). The only work which in addition to latency proposes a not so accurate approach to approximate power consumption is Kiasari et al. (2008), while Dally (1990) and Ogras and Marculescu (2007) determine the throughput of the network in addition to its latency. Furthermore, the latter compute the average buffer utilization for every router buffer.

- Some authors use their latency model for some special purposes. For example, Kiasari et al. (2008) and Ogras and Marculescu (2007) use their latency model for the purpose of mapping exploration, while Guz et al. (2007, 2006) use their model in a capacity allocation algorithm.

- All performance models of the second group and also most of the models of the first and the third groups compute directly the latency of the entire network, at once. In other words, they compute the average number of hops that a message traverses (on average) through the network to reach its destination. Then they approximate the average latency of this average path which is in fact the average latency of the network. In fact, an adequate NoC performance evaluation method must provide the per-path (or per-flow) latency. To the best of our knowledge, the only methods which could compute per-path average latency are presented by Ogras and Marculescu (2007) and Guz et al. (2007, 2006).

TABLE 13.1
Characteristics of Analytical Methods

	1	2	3	4	5	6	7	8	9	10	11	12	13	14	15
Network type	Off-chip							NoC	Off-chip				NoC		
Topology[a]	K	A	D	K				T	H	T	H	A	2D-mesh	A	A
Routing Algorithm[b]	E	D		E	D	a		X-Y	Duato	D (X-Y)	a	D		D	D
Traffic Pattern[c]	U	—	A			U				Hot-Spot	BR	A		A	A
Temporal Distribution	Poisson											generic	—		Poisson
Infinite Source Buffer	No	Yes	—			Yes								Yes	
Network Buffer Depth	—	finite							1-flit				finite	1-flit	A
Lossless						Yes						No	No	Yes	Yes
VC	No					Yes						No		Yes	No
Technique[d]	n	Q	Q		L				QC				Q		
Metrics or Property[e]	LT			L	L			LP		L		LM	Lm	LC	LTBM
Sink Destination							Yes	Yes						Yes	
Latency									network						per-path

1 Dally (1990)
2 Hu and Kleinrock (1997)
3 Guan, Tsai, and Blough (1993)
4 Draper and Ghosh (1994)
5 Sarbazi-Azad, Khonsari, and Ould-Khaoua (2002)
6 Khonsari, Ould-Khaoua, and Ferguson (2003)
7 Najaf-Abadi and Sarbazi-Azad (2004)
8 Kiasari et al. (2008)
9 Ould-Khaoua and Sarbazi-Azad (2001)
10 Loucif and Ould-Khaoua (2004); Loucif, Ould-Khaoua, and Min (2005)
11 Sarbazi-Azad, Ould-Khaoua, and Mackenzie (2001)
12 Elmiligi et al. (2007)
13 Guz et al. (2007, 2006)
14 Ogras and Marculescu (2007)
15 Kiasari et al. (2008)

[a]K = k-ary; t = torus; h = hypercube; A = arbitrary
[b]D = deterministic; E = e-cube; a = adaptive; BR = bit-reversal
[c]U = uniform; A = arbitrary; BR = bit-reversal
[d]n = numeric; Q = queueing theory; C = combinational theory
[e]L = latency; T = throughput; m = miss ratio; C = capacity allocation; P = power consumption; B = average buffer utilization

13.4   An Analytical Performance Evaluation Method

Our method addresses the network layer of an NoC-based system. Consequently, in our model, applications (local subsystems) and network-interface layers are abstractly represented by *cores* that generate packets (i.e., source cores) according to a given *traffic distribution*. Cores are also consumers of packets (i.e., destination cores). The network is constituted from a set of routers arranged according to a particular topology and connected by links with limited buffering capacity. The method is generic in terms of network topology, deterministic routing algorithm, traffic distribution, buffer length, and packet length. Further assumptions are listed below, and Table 13.2 lists the symbols used.

1. We assume that the target application, mapped on the network, enables us to characterize the spatial distribution of the traffic in a matrix, called a *traffic distribution matrix* $T = [t_{sd}]$. The temporal distribution of the traffic follows the Poisson process. Cores generate traffic independently of each other following the average data rate of t flits/cycle determined by T (element t_{sd} of T indicates the average data rate source s sends to the destination d). This assumption is quite common in performance analyses of networks (especially in almost all of the queuing-theory-based approaches such as Ogras and Marculescu (2007), Dally (1990), Draper and Ghosh (1994), *SoClib* (*Open Platform for Virtual Prototyping of Multi-Processor Systems-on-Chip*), etc.).

2. The method addresses deadlock-free deterministic routing algorithms. It does not involve implementation choices such as source routing or distributed routing. The only thing that is important is the deterministic path between each pair of source and destination.

3. The data arrival process to a given router input port is approximated by an independent Poisson law.

4. The fragments of different packets cannot be interleaved. When the header flit of a packet arrives at a port, its body flits arrive immediately after it in a pipeline mode.

5. As we address best-effort NoC, to control contention we assume a no-priority starvation-free arbitration for each router output port (such as round-robin). The arbitration is symmetric in terms of the probability of selection of simultaneous packets on the input (so in the long run it gives an equal chance to all demanding inputs for the acquisition of the shared output). Also we assume that arbitration is necessarily performed for each packet. This means that two consecutive packets from the same input cannot be routed with a single allocation when there is another demanding input; they require two arbitrations.

6. Lossless data flow: a minimum link level flow control is assumed that means when a buffer is full it does not accept further data by signaling its *upstream* (i.e., previous) buffer (by emitting a *full* signal, for example). So the data is not lost but stays blocked until the *downstream* (i.e., following) buffer has at least one flit space to accept the incoming data.

7. The time unit is equal to the time duration needed for a flit transmission which could be expressed in cycles in synchronous-logic architectures and as a fraction of a second in asynchronous-logic architectures.

TABLE 13.2
Parameters of the Analytical Performance Evaluation Method Presented in Section 13.4

Symbol	Meaning
l_{i2o}^r	Average latency traversing router r from input port i to output port o
O_r	Set of output ports of r
I_r	Set of input ports of r
λ_{i2o}^r	Forwarding rate, i.e., the data rate passing through input i to output o of router r
F_i^r	Forwarding set, i.e., the set of all output ports r that a packet coming from i can be forwarded to
f_{i2o}^r	Forwarding probability indicating with which probability a packet coming from i is forwarded to output port o of router r
$R_{s \to d}(i, o)$	The routing indicator
P_i	Tagged packet coming from input i
P_j	Disrupting packet coming from input j
τ	Router service time (constant delay)
$\widehat{\tau}$	Buffer constant delay
ρ_{i2o}^r	Port acquisition delay experienced by P_i
β_{i2o}^r	Buffer transfer delay experienced by P_i
γ_{j2o}^r	Presence time of P_j, i.e., the total time duration that the output port o is blocked by it
π_{j2o}^r	Contention probability between P_i and P_j
δ_{j2o}^r	Contention delay caused by P_j to release output o
$\widehat{\delta}_{j2o}^r$	Buffer occupancy caused by P_j
$\widehat{\pi}_{j2o}^r$	Buffer occupancy rate caused by P_j
$\nu_o^{r'}$	Blocking delay at the next router
φ_o^r	Back pressure delay caused by the retro-propagated flits of output buffer o
θ_o^r	Average buffer utilization

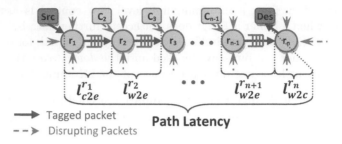

FIGURE 13.6

At each router of the path, disrupting packets appear probabilistically in front of the tagged packet

13.4.1   The Method

The method aims to analyze the latency of any desired path. Other performance metrics are obtained either during the latency analysis procedure (such as buffer utilization and router delay components) or are results of it (such as saturation threshold).

Path latency is defined as the average latency of a *tagged packet* crossing a user-specific path, obtained from the sum of the average latencies of the path routers. For example, in Figure 13.6

$$\text{Average path latency} = l_{c2e}^{r_1} + l_{w2e}^{r_2} + \cdots + l_{w2e}^{r_{n+1}} + l_{w2c}^{r_n}$$

where l_{i2o}^r is the average latency traversing router r from its input port i to output o (for 2D meshes, the I/O ports are identified by their directions, e.g., c for core, e for east, etc.). Due to resource sharing in best-effort NoCs, the tagged packet may have contention with packets coming from other flows. These packets are called *disrupting packets*. We distinguish two kinds of contention, called *direct* and *indirect contention*:

- Direct contention happens in *one* router and causes a *cyclic dependency* in the computation of latencies of different flows incoming to the router. This is physically explained by the reciprocal impact that different flows produce on the average latency of each other when they are in direct contention. The cyclic dependency is resolved by an *iterative technique*.

- Indirect contention happens in a *sequence of routers*. In a best effort wormhole network a chain of packets with different destinations may stay blocked one after the other over a sequence of routers. It means that the latency of each router is a function of the latency of its following routers (downstream routers in the sequence). To deal with this *acyclic dependency*, we build the *dependency tree* and then recursively compute router latencies from the leaves of the tree backward to its root.

13.4.1.1   Dependency Tree

The dependency tree computes the latency of the source router represented by the root of the tree. A node of the tree, illustrated by the couple $(r, i2o)$, represents the latency l_{i2o}^r. The edges going out of a node determine its dependencies. Thus the children of each node represent the latencies of the following router to which the node is dependent (i.e., the latencies between the corresponding input port and all possible outputs of the downstream router toward which the disrupting packets can be forwarded). The children of node $(r, i2o)$ are determined from the topology and routing algorithm. We determine the set of all output ports of the router r' that a packet coming from input port i' can be forwarded to, called the *Forwarding set* $(F_{i'}^{r'})$:

$$F_{i'}^{r'} = \{o' \in O_{r'}, \Gamma_{r'}(i', o') \neq 0\}$$
$$\Gamma_r(i, o) = \sum_s \sum_d R_{s \rightarrow d}(i, o) \tag{13.1}$$

where for a given router r, O_r represents the set of its output ports, $\Gamma_r(i, o)$ the number of flows crossing through input port i and output port o of r, and $R_{s \rightarrow d}(i, o)$ the routing algorithm, which is equal to 1 if the data flow from source s to destination d is routed through input port i and output port o of r, and equal to 0 otherwise (s and d belong to the set of all routers of the NoC).

$$R_{s \rightarrow d}(i, o) = \begin{cases} 1 & \text{if flow } s \text{ to } d \text{ passes through ports } i \text{ and } o \\ 0 & \text{otherwise} \end{cases} \tag{13.2}$$

Since we address deadlock-free (and livelock-free) routing algorithms, the dependency tree does not contain any cycles and they eventually terminate to cores, simply because cores are not followed by any router. So using a Depth First Search and by retaining the reverse order of the dependency, we traverse each node of the tree and compute its corresponding latency and use it for computing the mean latency of its (backward) upstream node. Figure 13.7 shows two examples. In Figure 13.7(a), the first disrupting packet coming from north of $r_{3,3}$ may be routed (according to its destination) toward the core or south output ports (according to the x-first routing algorithm packets are routed first on the x axis then on the y axis, and so there is no connection from vertical ports to horizontal ones). Each possible direction may yet be blocked by other disrupting packets. So, $l_{3,4}^r$ (represented by the root) depends on both $l_{3,3}^r$ and $l_{3,3}^r$ (respectively represented by nodes $((3,3), n2c)$ and $((3,3), n2s)$ as the children of the root). Whereas $l_{3,3}^r$ can be independently computed, $l_{3,3}^r$ is, in the same way, dependent on $l_{3,2}^r$ and $l_{3,2}^r$. As shown, the dependency tree extends until $l_{3,1}^r$. So node $((3,1), n2c)$ is the end of the dependency because regarding the x-first routing, the core output port is the only output choice for the packets arriving from the north input of $r_{3,1}$.

In Figure 13.7(b) the dependency tree is larger because the sequence of

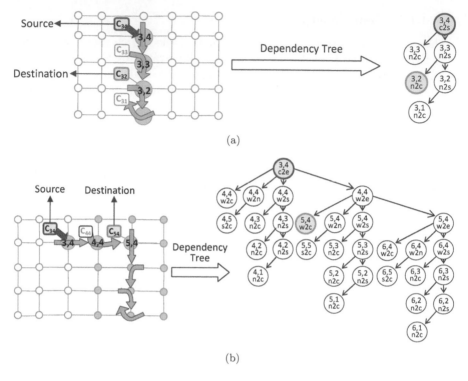

(a)

(b)

FIGURE 13.7

Dependency trees corresponding to the latency (a) 'core to south' and (b) 'core to east' of $r_{3,4}$ in a 6×5 2D mesh NoC with the x-first routing algorithm

blocked packets may spread not only toward the east direction but also toward the south and north directions (parts of the NoC which are shown with filled circles in Figure 13.7(b)). The tagged packet (coming from $C_{3,4}$ toward the east direction) may be blocked by disrupting packets that can be routed toward the core, north, south, or finally west directions of $r_{4,4}$. So the chain of blocked packets may be spread over all those directions (represented by four branches of the tree). For example, the tagged packet may be blocked by a sequence of packets as demonstrated in Figure 13.7(b) by wide arrows. Therefore the root has four children that demonstrate these dependencies.

13.4.1.2   Router Delay Model

Throughout this section we assume that P_i represents the tagged packet coming from input port i of router r addressing output o of that router and P_j represents a disrupting packet coming from any input port j of r (other than i) addressing the same output port in competition with P_i. Router latency l^r_{i2o} is the average latency of the header of P_i to cross r and the buffer located on

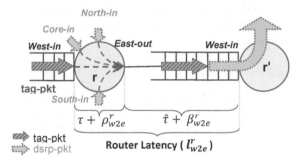

FIGURE 13.8
Router delay model related to a 2D mesh NoC

the output link o (i.e., the buffer between r and its downstream router r'). In our abstract router model, the physical implementation of buffers in outputs or inputs of routers is not visible. What is important is the total buffer space, located between two neighbor routers r and r'. Therefore, depending on the situation, we may use the terms *output* or *input* buffer.

As shown in Figure 13.8, the router latency consists of two major delays. The first is related to crossing the router, i.e., from the moment the header of P_i is present at the input port until the moment its output is allocated and thus the header can be written to the output buffer. This delay is composed of *router service time* (τ) and *port acquisition delay* (ρ_{i2o}^r) as defined below. The second is related to transmission of the header of P_i through the output buffer and thus arriving at the input port of r' and is the sum of two delay components, namely, *buffer constant delay* ($\hat{\tau}$) and *buffer transfer delay* (β_{i2o}^r) as defined in the following. As a result, the router latency can be expressed as the sum of the four aforementioned delay components: $l_{i2o}^r = \tau + \rho_{i2o}^r + \hat{\tau} + \beta_{i2o}^r$.

- τ and $\hat{\tau}$ are constant delays representing, respectively, router service time as the pure delay a router takes to route a packet even when there is no contention (this depends on the router architecture) and *buffer constant delay* as the delay required for a flit to be transferred through a buffer when it is empty (this depends on the buffer architecture).

- ρ_{i2o}^r: *port acquisition delay* is the average delay P_i experiences at the input port of r while waiting for other disrupting packets (demanding simultaneously the same output port) to pass through the router. In other words, it is the router waiting time.

- β_{i2o}^r: *buffer transfer delay* is the delay caused by the flits previously accumulated in the output buffer at the moment the header of P_i is written in the buffer. In other words, it is the time that the header remains in the buffer until it arrives at the input port of r.

Assuming that τ and $\hat{\tau}$ are given to the method as input, the challenging part of the router model is the computation of ρ^r_{i2o} and β^r_{i2o} described in the two following subsections.

ρ^r_{i2o}: Port Acquisition Delay

ρ^r_{i2o} is a function of *contention probability* (π^r_{j2o}) between the P_i and any P_j and also *contention delay* (δ^r_{j2o}), which is the delay that P_j takes until it releases shared output port o (i.e., until the tail of P_j leaves the router):

$$\rho^r_{i2o} = \sum_{j \in I_r} \pi^r_{j2o} \times \delta^r_{j2o} \qquad (13.3)$$

where I_r is the set of input ports of r and π^r_{j2o} and δ^r_{j2o} are defined below.

δ^p_{j2o} : Contention Delay

δ^p_{j2o} is defined as the average delay P_j produces in front of P_i, before releasing shared output port o. P_i may arrive at any moment of the *presence time* of P_j (γ^r_{j2o}), that is, the total time during which output port o is blocked by P_j. γ^r_{j2o} is composed of the following delay components:

$$\gamma^r_{j2o} = L^r_{j2o} + \rho^r_{j2o} + \varphi^r_o \qquad (13.4)$$

L^r_{j2o} is the average length of P_j in flits. Since the unit of time is equivalent to the transmission time of one flit, every flit of P_j (scheduled before P_i) takes at least one unit of time to be transferred. Thus P_i is at least blocked by the flits of P_j.

ρ^r_{j2o} is the port acquisition delay met by P_j due to contentions with other packets (including packets coming from i) which are reciprocally considered as disrupting packets for P_j. This delay must be considered because P_i may arrive at r just when the header of P_j is waiting at the router input port to be served.

φ^r_o is called back pressure delay related to output buffer o. It represents the back pressure impact which arises when contention in the network is very high. In this case data flits accumulated in the output buffer o may exceed the buffer size. Since we assume lossless link level flow control, these extra flits retro-propagate in the preceding input buffers. Therefore the waiting packet (here P_j) has to wait until the retro-propagated flits are transferred. On page 399 we will see how φ^r_o is computed.

However, P_i does not always meet all the presence time[5] of a pre-scheduled P_j. Instead, according to its arrival time, it meets the *average* delay taken

---

[5]The *presence time* of a packet is the time duration that the packet is present at the head of the input port buffer.

by P_j, called contention delay (δ_{j2o}^r). To compute δ_{j2o}^r we separate two cases according to the arrival time of P_i:

$$\delta_{j2o}^r = \left(\frac{L_{j2o}^r + \varphi_o^r}{\gamma_{j2o}^r} \cdot \frac{L_{j2o}^r + \varphi_o^r + 1}{2} \right) + \left(\frac{\rho_{j2o}^r}{\gamma_{j2o}^r} \cdot (L_{j2o}^r + \varphi_o^r) \right) \quad (13.5)$$

Case A: With a probability of $\left(L_{j2o}^r + \varphi_o^r \right)/\gamma_{j2o}^r$, P_i falls on the body of P_j (L_{j2o}^r) or on the remaining flits of previous transmissions because of back pressure impact (φ_o^r). In this case blocking flits ($L_{j2o}^r + \varphi_o^r$) are in moving mode[6] so when P_i arrives it meets a stream of flits crossing the router in a pipeline fashion. Thus the delay experienced by P_i is equivalent to the average number of flits left to be transferred, which is equal to the mean value of $L_{j2o}^r + \varphi_o^r$. In (13.5), the first parenthesized term on the right represents this case.

Case B: With a probability of $\rho_{j2o}^r/\gamma_{j2o}^r$, P_i arrives while the header of P_j is waiting for output allocation. In this case P_i is exactly stalled by the total flits of $L_{j2o}^r + \varphi_o^r$. In (13.5), the second parenthesized term on the right represents this case.

As can be deduced from (13.3) and (13.5), a cyclic dependency arises here between the delay components related to flows coming from different inputs of r (e.g., ρ_{i2o}^r in (13.3) is a function of δ_{j2o}^r, which, according to (13.5), is a function of ρ_{j2o}^r, and the latter is similarly a function of δ_{i2o}^r and thus a function of ρ_{i2o}^r. This is because just as P_j is considered as a disrupting packet for P_i, a packet arriving from i is considered as a disrupting packet for P_j. This dependency exists between all inputs of r which compete to obtain output port o and is solved by an iterative fixed-point technique.

π_{j2o}^p: Contention Probability

In general the probability of contention between P_i and P_j for obtaining output o is represented by $\pi_{\{i,j\}2o}^p$:

$$\pi_{\{i,j\}2o}^r = \alpha_{i2o}^r \cdot \alpha_{j2o}^r \quad (13.6)$$

where α_{i2o}^r and α_{j2o}^r represent the probabilities of the presence of P_i and P_j on input ports i and j both toward o. Since in our model we assume that the tagged packet (i.e., P_i) is already present, α_{i2o}^r is always equal to 1 and thus:

$$\pi_{\{i,j\}2o}^r = \alpha_{j2o}^r \quad \text{(At the presence of the } P_i\text{)}$$

For reasons of consistency with other variables of the model, we denote $\pi_{\{i,j\}2o}^r$ simply with π_{j2o}^r as used in (13.3). Note that $\pi_{i2o}^r = 0$ (i.e., $\pi_{\{i,i\}2o}^r = 0$) since i cannot be in contention with itself (packets arrive at the router in a FIFO mode). At quasi-zero load levels when the contention in the network is very

---

[6] *Moving mode* is when the packet is not stalled and its flits are shifting (or moving) forward.

low, a good approximation for π_{j2o}^r would be the *forwarding rate j* to o (λ_{j2o}^r), that is, the data rate passing through input j to output o. But in practice when the load increases, all delays imposed to the header of a disrupting packet P_j such as ρ_{j2o}^r and φ_o result in an increase of the presence time of P_j and consequently an increase in the contention probability:

$$\pi_{j2o}^r = \lambda_{j2o}^r \frac{\gamma_{j2o}^r}{L_{j2o}^r} \tag{13.7}$$

where $\gamma_{j2o}^r / L_{j2o}^r$ gives the average presence time per flit for P_j, which is multiplied by λ_{j2o}^r to give the probability of the presence of data on port j addressing port o.

The forwarding rate λ_{j2o}^r is obtained by the following equation:

$$\lambda_{i2o}^r = \sum_s \sum_d R_{s \to d}(i, o) \cdot t_{sd} \tag{13.8}$$

where t_{sd} is the $(s, d)^{\text{th}}$ element of traffic distribution matrix T and $R_{s \to d}(i, o)$ is the routing indicator as defined in (13.2).

β_{i2o}^r: Buffer Transfer Delay

β_{i2o}^r is the time duration counted from the moment the header of P_i is written into the output buffer until the moment it arrives at the input port of the next router (r'). In a wormhole flow control strategy, as soon as the last flit of any P_j is transferred through router r, the header of P_i could be written to the output buffer (assuming that the output buffer has enough space to accept at least one flit of P_i). Then the header of P_i shifts after the accumulated flits of previous transmissions. This pipeline shifting takes an amount of time (i.e., β_{i2o}^r) which is equivalent to the average number of flits accumulated in the buffer just when the header of P_i is written to the buffer:

$$\beta_{i2o}^r = \sum_{j \in I_r} \pi_{j2o}^r \cdot \widehat{\delta}_{j2o}^r + \sum_{j \in I_r} \widehat{\pi}_{j2o}^r \cdot \frac{\widehat{\delta}_{j2o}^r}{2} \tag{13.9}$$

where $\widehat{\delta}_{j2o}^r$ is the *buffer occupancy* and $\widehat{\pi}_{j2o}^r$ the *buffer occupancy rate* related to any previous P_j, both explained later, and π_{j2o}^r is the contention probability, (13.7). In the computation of β_{i2o}^r two cases are considered (corresponding to $\sum_{j \in I_r} \pi_{j2o}^r \cdot \widehat{\delta}_{j2o}^r$ and $\sum_{j \in I_r} \widehat{\pi}_{j2o}^r \cdot \frac{\widehat{\delta}_{j2o}^r}{2}$ in (13.9).

Case 1: $\sum_{j \in I_r} \pi_{j2o}^r \cdot \widehat{\delta}_{j2o}^r$ in (13.9) covers the cases in which P_i is in direct contention with any P_j. So when the output is allocated to P_i, P_j has already traversed the router and some of its flits are still in the buffer. P_i is shifted behind the tail of P_j and thus until arrival at the input port of r' it experiences a delay caused by the buffer space occupied by P_j (i.e., $\widehat{\delta}_{j2o}^r$). This case is demonstrated in Figure 13.9 and happens with the probability of contention between P_i and P_j (i.e., π_{j2o}^r).

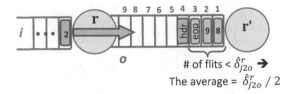

# of flits $< \widehat{\delta}_{j2o}^{r}$ →
The average $= \widehat{\delta}_{j2o}^{r} / 2$

hdr = header
eop = end of packet

FIGURE 13.9
The average number of accumulated flits in the output buffer at the arrival of P_i when there is no header contention

Case 2: $\sum_{j \in I_r} \widehat{\pi}_{j2o}^{r} \cdot \frac{\widehat{\delta}_{j2o}^{r}}{2}$ in (13.9) covers the cases in which P_i arrives after a transmission already accomplished and so traverses the router without any contention. When the header of P_i is written into the output buffer, the buffer may be empty or there may be some flits from previous transmissions. This happens with probability equal to the rate of buffer occupancy by packets coming from any j ($\widehat{\pi}_{j2o}^{r}$). In this case P_i meets the mean buffer space occupied by P_j (i.e., $\widehat{\delta}_{j2o}^{r}/2$). This case is demonstrated in Figure 13.10.

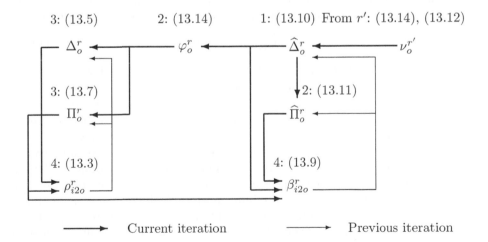

FIGURE 13.10
The order of delay component computation in one iteration

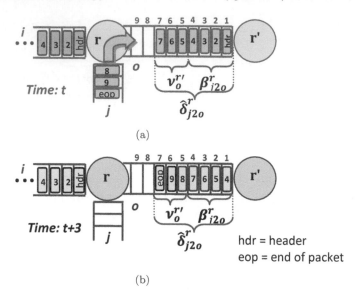

FIGURE 13.11
Buffer occupancy caused by P_j at time instant (a) t and (b) $t+3$ when P_j is transferred and P_i can be written into the buffer

$\widehat{\delta}^r_{j2o}$: *Buffer Occupancy*

Buffer occupancy is the buffer space occupied after the transmission of one packet from any input port j and is obtained by the following equation:

$$\widehat{\delta}^r_{j2o} = \beta^r_{j2o} + \nu^{r'}_o \tag{13.10}$$

where β^r_{j2o} is the buffer transfer delay experienced by P_j and $\nu^{r'}_o$ is the delay due to the blocking of the header of P_j at r', called *next node blocking delay* (explained in the following). This is because, similarly to P_i, when P_j is written into the output buffer, its header meets the average number of flits already accumulated in the buffer (i.e., β^r_{j2o}). Thus its header shifts toward r' after the accumulated flits and experiences a delay equal to β^r_{j2o}. Simultaneously, its body flits are written and shifted behind it. When the header arrives at the input port of r', there are exactly a number of body flits equal to β^r_{j2o} accumulated after it. Then at the input port of r', the header of P_j is stalled for a time duration equal to ν^r_o before routing. Meanwhile more body flits are accumulated behind it. Therefore, as shown in Figure 13.11, at this instant the buffer space occupied by P_j is equal to β^r_{j2o} plus ν^r_o. Then the header of P_j is routed toward the proper output and its body flits follow it. Finally, when the last flit of P_j traverses r (Figure 13.11(b)), P_i is written into the buffer and meets $\widehat{\delta}^r_{j2o}$ number of flits in the buffer.

Note that a cyclic dependency appears between the delay components of buffer transfer delay (β, obtained from (13.9), and buffer occupancy ($\widehat{\delta}$, obtained from (13.10)) related to different competing inputs.

$\widehat{\pi}^r_{j2o}$: Buffer Occupancy Rate

Buffer occupancy rate represents the probability of having some flits remaining from a previous transmission when P_i crosses the router without any contention. Contrary to π^r_{j2o}, $\widehat{\pi}^r_{j2o}$ is not zero for input port i because P_i may conflict with the residual flits left from a previous packet from its own input port. $\widehat{\pi}^r_{j2o}$ is proportional to λ_{jo} and obtained as

$$\widehat{\pi}^r_{j2o} = \lambda^r_{j2o}\frac{\widehat{\delta}^r_{j2o}}{L^r_{j2o}} \tag{13.11}$$

where the fraction $\widehat{\delta}^r_{j2o}/L^r_{j2o}$ determines the occupied buffer space per flit.

$\nu^{r'}_o$: Next Node Blocking Delay

$\nu^{r'}_o$ is equal to the average service time needed for P_j to be routed through r' and thus depends on the delay components of r'. As P_j can be routed to different output ports of r', we have:

$$\nu^{r'}_o = \sum_{o' \in F^{r'}_{i'}} f^{r'}_{i'2o'} \cdot \left(\tau + \rho^{r'}_{i'2o'} + \varphi^{r'}_{o'}\right) \tag{13.12}$$

where, assuming that P_j arrives at input port i' of r', $F_{i'}$ is the forwarding set of i' (13.1) and $f'_{i'2o'}$ is the *forwarding probability* indicating the probability with which P_j is forwarded to output port o' of r':

$$f^{r'}_{i'2o'} = \frac{\lambda^{r'}_{i'2o'}}{\sum_{(o \in F'_i)} \lambda_{i'2or'}} \tag{13.13}$$

The sum of the three delay components in (13.12) is equivalent to the waiting time of P_j to acquire its output port at r'. τ and ρ have been explained earlier; in the following we describe how the back pressure delay is computed.

φ^r_o: Back Pressure Delay

φ^r_o is equivalent to the number of flits retro-propagated in the preceding input buffers and thus the waiting packet is blocked until these flits can be transferred. This arises under high traffic load when the network is beginning to saturate. In our computation, the *average buffer utilization* (θ^r_o) can be theoretically larger than the buffer size (B^r_o). This imaginary extra buffer space is interpreted as the back pressure delay:

$$\varphi^r_o = \begin{cases} 0 & \theta^r_o \leq B^r_o \\ \theta^r_o - B^r_o & \text{otherwise} \end{cases} \tag{13.14}$$

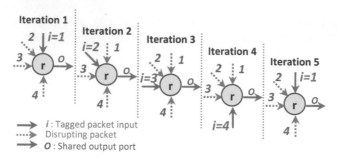

FIGURE 13.12
Iterative computation for inputs $\{1, 2, 3, 4\}$ of router r

where $\theta_o^r = \sum_{i \in I_r} \lambda_{i2o}^r \times \left(1 + \widehat{\delta}_{i2o}^r\right)$.

In practice φ_o^r indicates the aggregate time durations during which the output port is allocated to the demanding packet but the output buffer is full and thus the packet flits cannot be written into the buffer.

13.4.1.3 Cyclic Dependency: Iterative Technique

The cyclic dependencies that exist between the delay components of a router r are demonstrated in Figure 13.12, in which the order of computation of delay components is shown by numbers beside the related equation. Assuming that all the downstream nodes in the dependency tree have been traversed and thus all of the required delay components of the following router (r') are available, we can use a fixed-point iteration technique to compute delay components of r.

In Figure 13.10 $\Pi_o^r = \left(\pi_{j2o}^r\right)_{1 \times k}$, $\Delta_o^r = \left(\delta_{j2o}^r\right)_{1 \times k}$, $\widehat{\Pi}_o^r = \left(\widehat{\pi}_{j2o}^r\right)_{1 \times k}$, $\widehat{\Delta}_o^r = \left(\widehat{\delta}_{j2o}^r\right)_{1 \times k}$ are row vectors representing, respectively, contention probabilities, contention delays, buffer occupancy rates, and buffer occupancies for all input ports of r addressing output o, $j \in I_r$, $(I_r) = k$. Therefore (13.3) and (13.9) become

$$\rho_{i2o}^r = \Pi_o^r \cdot \Delta_o^T \quad \text{and} \quad \beta_{i2o}^r = \Pi_o^r \cdot \widehat{\Delta}_o^T + \widehat{\Pi}_o^r \cdot \frac{\widehat{\Delta}_o^T}{2} \qquad (13.15)$$

As shown in Figure 13.12, in each iteration one of the inputs is marked i as the tagged packet (P_i) input and the others are considered as the disrupting packet inputs. Thus port acquisition and buffer transfer delays are computed for i (i.e., ρ_{i2o}^r and β_{i2o}^r) which are used in the next iteration to update the aforementioned row vectors. Then the position of i changes in the next iteration. For example, in Figure 13.12 input 1 is the tagged input in iteration 1 (i.e., $i = 1$) while it is considered as a disrupting packet input in iteration 2, 3, and 4, until iteration 5, when it is again marked i. Before the beginning of the iterations, $\nu_o^{r'}$ is computed from the delay components of the downstream router. In iteration 1 we build the row vectors (according to the order shown in Figure 13.10) where β_{i2o}^r and ρ_{j2o}^r are initialized to 0 for all $j \in I_r$. Then ρ_{12o}^r and β_{12o}^r are obtained. In the second iteration, $i = 2$ and input 1 is a disrupting input so we first obtain $\widehat{\delta}_{12o}^r$ and update $\widehat{\Delta}_o^r$ with new $\widehat{\delta}_{12o}^r$ while other elements of $\widehat{\Delta}_o^r$ do not change. In the same way the first element of the other row vectors are updated and finally ρ_{22o}^r and β_{22o}^r are computed. In this way we fill the row vectors with proper values obtained from iterations. The more we progress in iterations, the more the result becomes accurate. As soon as the difference between the results of two successive iterations for the same input i is less than a given constant ϵ (determined by the desired accuracy) the computation stops for that input. Since the equations of the router in the dependency cycle of Figure 13.12 make a monotonically increasing sequence when the network gets saturated, in this situation the iterative computation converges to infinity. The iterative approach enables us to determine, for a given router, the latencies between all inputs and the specified output port o. For example, in Figure 13.12 the iterative computation gives ρ_{j2o}^r and β_{j2o}^r for all $j \in I_r$.

13.4.2   Validation of the Method

Analytical methods are usually compared and validated against simulation methods. To this end we present the results obtained by our method against the results obtained by a SystemC CABA simulator. We use the open source DSPIN-NoC simulator, available in the SoCLib library (*SoClib (Open Platform for Virtual Prototyping of Multi-Processor Systems-on-Chip)*), as our simulation platform.

DSPIN (Distributed Scalable Packet-Switching Integrated Network) (Sheibanyrad, Greiner, and Miro-Panades 2008) is a general purpose NoC designed to support large-scale clusterized shared memory MP-SoCs. It is a best-effort NoC with 2D mesh topology and wormhole packet switching. It uses the x-first routing algorithm that routes packets first on x then on y. The DSPIN routers have input buffers and use round-robin arbitration. Each cluster of the simulation platform is composed of a DSPIN router, a traffic generator (as the source core), and a traffic consumer (as the destination core) and is identified by its (x, y) coordinates. The traffic generator can produce uniform and localized traffic. Traffic consumers receive packets and compute the packet latency according to the timing information written in the packet.

Uniform distribution is the simplest traffic pattern that can mimic the homogenous communication in general purpose NoCs. According to this pattern, each traffic generator sends packets randomly (with the same probability) to all external traffic consumers. The offered load is equivalent for all packet generators. The *offered load* is the aggregate data rate that each traffic generator sends to all its destinations (equivalent to the sum of the raw vector components of that traffic generator in traffic distribution matrix T). However, from a design perspective, it is more efficient to place frequently communicating

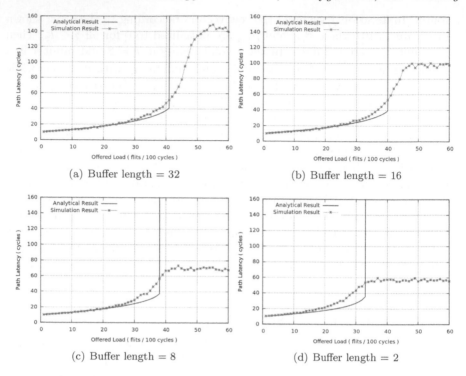

(a) Buffer length = 32

(b) Buffer length = 16

(c) Buffer length = 8

(d) Buffer length = 2

FIGURE 13.13

Latency/load curves for the path $r_{2,4} \to r_{4,2}$ with buffer lengths in flits as indicated and uniform traffic (path latency excludes the source queue waiting time)

resources close to each other. This leads to the use of the localized traffic pattern, which is closer to real application behavior. In the localized traffic model, the destination is assigned to a new generated packet according to the localized probability taken from Weldezion et al. (2009), which assigns the closer destinations to packets with a higher probability.

13.4.2.1 Latency/Load Curves

Figures 13.13 and 13.14 depict latency/load curves (i.e., average path latency as a function of offered load) provided by our analytical model plotted against DSPIN simulation results under, respectively, uniform traffic and localized traffic. The results presented belong to path $r_{2,4} \to r_{4,2}$ in a 5×5 DSPIN NoC. As mentioned, DSPIN is a general purpose NoC suitable for homogenous shared-memory MPSoC architectures. In such a system the majority of packets communicating over the NoC are cache lines and so the packet length is set to 16 flits, which is considered as the usual cache line size. The four plots within Figures 13.13 and 13.14 correspond, respectively, to buffer lengths of

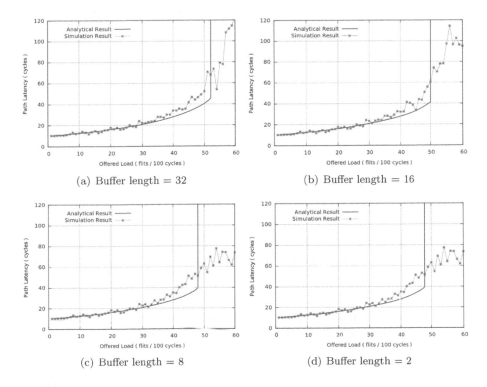

FIGURE 13.14
Latency/load curves for the path $r_{2,4} \rightarrow r_{4,2}$ with buffer lengths in flits as indicated and localized traffic

32, 16, 8, and 2 flits. We assume that all routers have the same buffer length. However, it is possible to assign different buffer lengths to different routers, since the analytical method is based on the decentralized router delay model. The curves are obtained with steps of 1% offered load. This is why the analytical results seem to be discrete at the saturation point. When we obtain the same results for steps of, for example, 0.01%, the curves increase more smoothly, as illustrated in Figure 13.15 (a zoomed view of the curves of Figure 13.14).

These latency/load curves provide two performance metrics:

1. the path latency for any desired offered load

2. the saturation point, explained in the following

Path latency

In both simulation and the analytical method the path latency is defined as the mean latency that a packet needs to traverse the path excluding the source queue waiting time (i.e., from the arrival time of the packet header

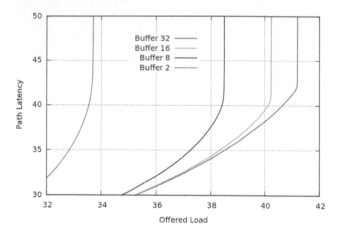

FIGURE 13.15
Analytical method for different buffer lengths and 0.01% offered load steps

onto the network until the header exits from the network). In simulations the average path latency for each offered load is computed after the reception of 1000 packets at the destination router for uniform traffic and 100 packets for localized traffic. This is because in simulations with localized traffic, far away destinations are addressed less frequently than closer ones, so simulation runtime is much longer. For example, for 100 received packets by a six-hop distant destination (such as path $r_{2,4} \rightarrow r_{4,2}$), the simulations of Figure 13.14 took about 48 hours, which confirms again the necessity of using an analytical method. This is why the localized simulation results in Figure 13.14 are not as smooth as those of the uniform results in Figure 13.13.

Throughput and Saturation Threshold

Throughput (also called accepted load) is the traffic load accepted by the network. Like the offered load, the unit of measure for throughput is flits/cycle. Ideally throughput increases linearly with the offered load, but due to the limitation of network resources, it will saturate at a certain fraction of the offered load (Pande, Grecu, Jones, et al. 2005). This point indicates the maximum sustainable throughput (or accepted load) of the network, called here the *saturation threshold*, and is one of the important metrics in NoC performance comparison (Pande, Grecu, Jones, et al. 2005; Salminen, Kulmala, and Hamalainen 2007). Since the offered load directly affects the probability of contention between different flows and thus the average latency, to determine the saturation threshold, we increase the offered load and observe the latency variations. As shown in Figures 13.13 and 13.14, the saturation point in the analytical results is equal to the offered load at which the latency tends to infinity (which means that the iterative computation converges to infinity).

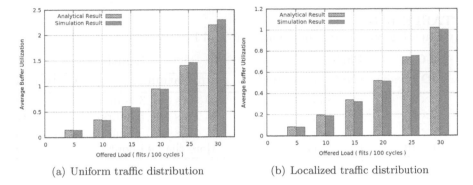

(a) Uniform traffic distribution   (b) Localized traffic distribution

FIGURE 13.16

The average utilization of buffer $r_{3,4} \rightarrow r_{4,4}$ under two traffic distributions

The saturation point determined by simulation is equal to the offered load after which the latency stays constant (because the path latency excludes the source queue waiting time). When the network is saturated, the maximum throughput is achieved and thus any new packet arriving onto a specific path takes a constant latency to traverse it, whatever the offered load is (however, it differs from one path to another). Meanwhile, the difference between offered load and accepted load is accumulated in source queues (and so the source queues and source queue waiting times start growing infinitely). This is why during simulations for load-latency measurements large (infinite, in theory) buffers must be placed between traffic sources and the network in order to avoid self-throttling (Salminen, Kulmala, and Hamalainen 2007). This prevents the propagation of back pressure effect from the network to the traffic sources and thus allows them to continue generating data according to the desired offered load without being affected by network throughput limitations.

13.4.2.2 Average Buffer Utilization

Figure 13.16 shows the average utilization of buffer $r_{3,4} \rightarrow r_{4,4}$ for uniform (a) and localized (b) traffic. It demonstrates that the results obtained by the analytical method match the simulation results. It is observed that the localized traffic loads the buffer half as much as the uniform traffic. The average buffer utilization gives helpful insights into optimal link capacity allocation and the distribution of the traffic over network buffers.

13.4.2.3 Runtime Comparison

Since the main motivation for the use of analytical methods is their relatively lower runtime, in Table 13.3 we compare the runtime of the simulation and analytical methods for obtaining the average latency of the diameter path (between the bottom-right and the upper-left nodes) in the DSPIN NoC, with the uniform traffic and for different dimensions. For the 10×10 DSPIN

TABLE 13.3
Comparing Simulation and Analytical Tool Runtimes

NoC	Path	Runtime	
(2D mesh)	(NoC diameter)	Simulation	Analytical
4×4	$r_{4,1} \to r_{1,4}$	$9\,332$ s ($\approx$ 3 hours)	1 s
5×5	$r_{5,1} \to r_{1,5}$	$16{,}210$ s ($\approx$ 5 hours)	2 s
6×6	$r_{6,1} \to r_{1,6}$	$34{,}415$ s ($\approx$ 10 hours)	6 s
10×10	$r_{10,1} \to r_{1,10}$	More than 48 hours	46 s

the simulation did not terminate even after 48 hours (for the reception of 1000 packets per destination and offered loads between 1 and 50) whereas the desired results are obtained within 46 seconds by using the analytical method. This experiment confirms the fact that the use of analytical methods will be indispensable in NoC design in the near future. This experiment is in fact optimistic about simulation runtime, because, for example, with localized traffic or in real applications, distant destinations are addressed much less frequently than in the case of uniform traffic and thus the simulation runtime for obtaining the average latency of long paths within an acceptable accuracy will be much longer.

13.5   Conclusion

It has been accepted in the semiconductor industry that the future of on-chip communication will rely on NoC technology. One of the important tasks in the design of an NoC is performance evaluation. This task consists of estimating the performance metrics in order to verify whether the NoC satisfies the cost and performance requirements of the entire system. If not, depending on critical design factors, either a new set of architectural parameters or a new mapping of applications, or even a mix of both solutions, may be tried to improve the performance of the system. NoC performance evaluation is traditionally done by time-consuming simulation-based methods. With the rapid development of CMOS technology and thus the integration of thousands of communicating IPs in a single chip, and also tighter and tighter time to market constraints, simulation-based methods are no longer efficient enough to perform this task. A promising alternative is the use of analytical methods which help to accelerate the task of performance evaluation and subsequently enable rapid architectural and mapping explorations in the earliest stages of the design. Thus, the use of simulation-based methods can be pushed to later steps of the design when the design space is already reduced to some practical architectural and mapping choices, in order to obtain more precise results.

Considering the increasing need for rapid performance estimation in on-chip system design, this chapter focused on the role of analytical methods and approaches in performance evaluation and thus design flow of NoC-based embedded systems. The chapter presented a short introduction about the history of NoCs, the most important architecture and communication parameters that make NoCs different from one another, and also the design process of NoC-based embedded systems. In this chapter we also reviewed related work in the field of network performance evaluation (for both off-chip and on-chip networks) with particular attention to analytical performance evaluation methods. The chapter also presented an iteration-based analytical method for performance evaluation of the network layer of NoCs. This method can be used for both architectural and mapping exploration purposes in the NoC design flow. In general, analytical methods enable designers to study the impact of a set of parameters such as topology, routing algorithm, buffer length, packet length, and traffic distribution on network performance metrics such as packet latency, saturation threshold (maximum accepted load), buffer filling etc. This has been presented in a case study in which, by using the proposed analytical method, we have analyzed the performance of a 2D mesh NoC with the x-first routing algorithm. The comparison of result accuracy and analysis time between the simulation-based and analytical methods showed that using a fast analytical technique, such as the one presented in this chapter, one can speed up the task of performance evaluation and design space exploration by orders of magnitude.

Bibliography

Abuelrub, Emad. 2008. "A Comparative Study on the Topological Properties of Hyper-Mesh Interconnection Network." In *World Congress on Engineering,* edited by S. I. Ao, L. Gelman, D. W. L. Hukins, A. Hunter, and A. M. Korsunsky, 1:616–621. Lecture Notes in Engineering and Computer Science. World Congress on Engineering 2008, Imperial College London, UK, July 2-4. International Association of Engineers.

Adiga, N. R., G. Almasi, G. S. Almasi, Y. Aridor, R. Barik, D. Beece, R. Bellofatto, et al. 2002. "An Overview of the BlueGene/L Supercomputer." In *Supercomputing '02: Proceedings of the 2002 ACM/IEEE Conference on Supercomputing,* 1–22. Baltimore, Maryland: IEEE Computer Society Press.

Advanced Micro Devices, Inc. 2007. *AMD64 Architecture Programmer's Manual, Volume 2: System Programming.* Advanced Micro Devices, Inc.

———. 2009. *Multi-Core Processors from AMD.* Advanced Micro Devices, Inc. http://multicore.amd.com/.

Agarwal, A. 1991. "Limits on Interconnection Network Performance." *IEEE Transactions on Parallel and Distributed Systems* 2, no. 4 (October): 398–412.

Akesson, B., S. Stuijk, A. Molnos, M. Koedam, R. Stefan, A. Nelson, A. Nedad, and K. Goossens. 2012. "Virtual Platforms for Mixed Time-Criticality Applications: The CoMPSoC Architecture and SDF3 Design Flow." In *Proceedings of Workshop on Quo Vadis, Virtual Platforms: Challenges and Solutions for Today and Tomorrow.* Oldenburg, Germany. http://qvvp12.offis.de/.

Albonesi, David H. 1999. "Selective Cache Ways: on-Demand Cache Resource Allocation." In *Proceedings 32nd Annual International Symposium on Microarchitecture (MICRO-32),* 248–259. IEEE TC MARCH; ACM SIG-MICRO, November.

Alfalou, A., M. Elbouz, M. Jridi, and A. Loussert. 2009. "A New Simultaneous Compression and Encryption Method for Images Suitable to Recognize Form by Optical Correlation." *Proceedings of SPIE* 7486. doi:10.1117/12.830180.

Ali, M., M. Welzl, and S. Hellebrand. 2005. "A Dynamic Routing Mechanism for Network on Chip." In *Proceedings of the 23rd NORCHIP Conference, 2005.* 70–73. November 21–22. doi:10.1109/NORCHP.2005.1596991.

Amdahl, Gene M. 1967. "Validity of the Single Processor Approach to Achieving Large Scale Computing Capabilities." In *Proceedings of the April 18–20, 1967, Spring Joint Computer Conference,* 483–485. ACM.

Anand, Christopher K., and Wolfram Kahl. 2008. "Synthesising and Verifying Multi-Core Parallelism in Categories of Nested Code Graphs." In *Process Algebra for Parallel and Distributed Processing,* edited by Michael Alexander and William Gardner. Boca Raton/London: Chapman & Hall/CRC.

Anis, E., and N. Nicolici. 2007. "On Using Lossless Compression of Debug Data in Embedded Logic Analysis." In *Proceedings of IEEE International Test Conference,* 1–10. IEEE. doi:10.1109/TEST.2007.4437613.

Annavaram, Murali, Ed Grochowski, and John Shen. 2005. "Mitigating Amdahl's Law through EPI Throttling." *SIGARCH Computer Architecture News* (New York, NY, USA) 33 (2): 298–309. doi:10.1145/1080695.1069995.

Araujo, C., M. Gomes, E. Barros, S. Rigo, R. Azevedo, and G. Araujo. 2005. "Platform Designer: An Approach for Modeling Multiprocessor Platforms Based on SystemC." *Design Automation for Embedded Systems* 10 (4): 253–283.

Arcas, Oriol, Philipp Kirchhofer, Nehir Sonmez, Martin Schindewolf, Wolfgang Karl, Osman S. Unsal, and Adrian Cristal. 2012. "A Low-Overhead Profiling and Visualization Framework for Hybrid Transactional Memory." In *Proceedings of 20th Annual IEEE International Symposium on Field-Programmable Custom Computing Machines (FCCM 2012),* 1–8. Toronto, Canada, May.

ArchC — The Architecture Description Language. http://archc.sourceforge.net.

Arden, W., M. Brillouët, P. Cogez, M. Graef, B. Huizing, and R. Mahnkopf. 2010. "More-than-Moore." *White Paper: International Technology Roadmap for Semiconductors, ITRS.*

ARM Ltd. 2010a. *CoreSight for Cortex-A Series Processors,* March. http://www.arm.com/products/system-ip/debug-trace/coresight-for-cortex-a.php.

———. 2010b. *RealView Development Suite Documentation.* http://infocenter.arm.com/help/topic/com.arm.doc.subset.swdev.rvds/.

August, D., J. Chang, S. Girbal., D. Gracia-Perez., G. Mouchard, D. Penry, O. Temam, and N. Vachharajani. 2007. "UNISIM: An Open Simulation Environment and Library for Complex Architecture Design and Collaborative Development." *Computer Architecture Letters* 6 (2): 45–48. doi:10.1109/L-CA.2007.12.

Austin, Todd, Eric Larson, and Dan Ernst. 2002. "SimpleScalar: An Infrastructure for Computer System Modeling." *Computer* 35 (2): 59–67. doi:10.1109/2.982917.

Awasthi, M., K. Sudan, R. Balasubramonian, and J. Carter. 2009. "Dynamic Hardware-Assisted Software-Controlled Page Placement to Manage Capacity Allocation and Sharing Within Large Caches." In *Proceedings of the International Symposium on High Performance Computer Architecture, 2009 (HPCA'09)*, 250–261. IEEE.

Azevedo, R., S. Rigo, M. Bartholomeu, G. Araujo, C. Araujo, and E. Barros. 2005. "The ArchC Architecture Description Language and Tools." *International Journal of Parallel Programming* 33 (5): 453–484.

Bach, Moshe (Maury), Mark Charney, Robert Cohn, Elena Demikhovsky, Tevi Devor, Kim Hazelwood, Aamer Jaleel, et al. 2010. "Analyzing Parallel Programs with Pin." *Computer* 43 (3): 34–41. doi:10.1109/MC.2010.60.

Bainbridge, J., and S. B. Furber. 2002. "Chain: A Delay-Insensitive Chip Area Interconnect." *IEEE Micro* 22 (5): 16–23.

Balfour, J., and W. J. Dally. 2006. "Design Tradeoffs for Tiled CMP on-Chip Networks." In *Proceedings of the 20th Annual International Conference on Supercomputing*, 187–198. ACM.

Banerjee, K., S. Im, and N. Srivastava. 2006. "Can Carbon Nanotubes Extend the Lifetime of on-Chip Electrical Interconnections?" In *Proceedings of the 1st International Conference on Nano-Networks and Workshops, 2006 (NanoNet'06)*, 1–9. IEEE.

Barroso, Luis Andre, and Michel Dubois. 1991. "Cache Coherence on a Slotted Ring." In *Proceedings of the International Conference on Parallel Processing*, 1:230–237.

Bartzas, Alexandros, Lazaros Papadopoulos, and Dimitrios Soudris. 2009. "A System-Level Design Methodology for Application-Specific Networks-on-Chip." *Journal of Embedded Computing* 3 (3): 167–177.

Bartzas, Alexandros, N. Skalis, K. Siozios, and Dimitrios Soudris. 2007. "Exploration of Alternative Topologies for Application-Specific 3D Networks-on-Chip." In *Proceedings of WASP.* http://proteas.microlab.ntua.gr/ksiop/pdf/wasp2007.pdf.

Bechara, C., A. Berhault, N. Ventroux, S. Chevobbe, Y. Lhuillier, R. David, and D. Etiemble. 2011. "A Small Footprint Interleaved Multithreaded Processor for Embedded Systems." In *Proceedings of IEEE International Conference on Electronics, Circuits, and Systems (ICECS)*. Beirut, Lebanon, December.

Bechara, C., N. Ventroux, and D. Etiemble. 2010. "Towards a Parameterizable Cycle-Accurate ISS in ArchC." In *Proceedings of ACS/IEEE International Conference on Computer Systems and Applications (AICCSA)*, 1–7. Hammamet, Tunisia, May.

———. 2011. "A TLM-based Multithreaded Instruction Set Simulator for MPSoC Simulation Environment." In *Proceedings of International Workshop on Rapid Simulation and Performance Evaluation: Methods and Tools (RAPIDO)*. Crete, Greece, January.

Beigne, E., F. Clermidy, P. Vivet, A. Clouard, and M. Renaudin. 2005. "An Asynchronous NOC Architecture Providing Low Latency Service and Its Multi-Level Design Framework." In *Proceedings of the 11th IEEE International Symposium on Asynchronous Circuits and Systems*, 54–63.

Beltrame, G., C. Bolchini, L. Fossati, A. Miele, and D. Sciuto. 2008. "ReSP: A Non-Intrusive Transaction-Level Reflective MPSoC Simulation Platform for Design Space Exploration." In *Proceedings of Asia and South Pacific Design Automation Conference (ASPDAC)*, 673–678. Seoul, Korea, January. doi:10.1109/ASPDAC.2008.4484036.

Benini, L., D. Bertozzi, A. Bogliolo, F. Menichelli, and M. Olivieri. 2005. "MPARM: Exploring the Multi-Processor SoC Design Space with SystemC." *Journal on VLSI Signal Processing Systems* 41 (2): 169–182.

Benini, L., and G. De Micheli. 2002. "Networks on Chips: A New SoC Paradigm." *Computer* 35, no. 1 (January): 70–78.

———. 2006. *Networks on Chips: Technology and Tools*. San Francisco, CA: Morgan Kaufmann.

Bennett, Jon C. R., and Hui Zhang. 1996. "WF^2Q: Worst-case Fair Weighted Fair Queueing." In *Proceedings IEEE INFOCOM'96. Fifteenth Annual Joint Conference of the IEEE Computer Societies. Networking the Next Generation*. 1:120–128. IEEE, March.

Bentley, J., D. Sleator, R. Tarjan, and V. Wei. 1986. "A Locally Adaptive Data Compression Scheme." *Communications of the ACM* 29, no. 4 (April): 320–330. doi:10.1145/5684.5688.

Berg, E., H. Zeffer, and E. Hagersten. 2006. "A Statistical Multiprocessor Cache Model." In *Proceedings of IEEE International Symposium on Performance Analysis of Systems and Software (ISPASS '06)*, 89–99.

Bertogna, M., M. Cirinei, and G. Lipari. 2008. "Schedulability Analysis of Global Scheduling Algorithms on Multiprocessor Platforms." *IEEE Transactions on Parallel and Distributed Systems* 20, no. 4 (April): 553–566.

Bhattacharyya, S. S., P. K. Murthy, and E. A. Lee. 1996. *Software Synthesis from Dataflow Graphs*. Boston/Dordrecht/London: Kluwer Academic Publishers.

Bilsen, G., M. Engels, R. Lauwereins, and J. Peperstraete. 1996. "Cyclo-static Dataflow." *IEEE Transactions on Signal Processing* 44 (2): 397–408.

Binkert, Nathan L., Ronald G. Dreslinski, Lisa R. Hsu, Kevin T. Lim, Ali G. Saidi, and Steven K. Reinhardt. 2006. "The M5 Simulator: Modeling Networked Systems." *IEEE Micro* 26 (4): 52–60. doi:10.1109/MM.2006.82.

Birrell, Andrew D., and Bruce Jay Nelson. 1984. "Implementing Remote Procedure Calls." *ACM Transactions on Computer Systems* 2 (1): 39–59. doi:10.1145/2080.357392.

Bjerregaard, T., and J. Sparso. 2005. "A Router Architecture for Connection-Oriented Service Guarantees in the MANGO Clockless Network-on-Chip." In *Proceedings of the Conference on Design, Automation and Test in Europe*, 2:1226–1231.

Bobda, Christophe, Ali Ahmadinia, Mateusz Majer, Jurgen Teich, Sandor Fekete, and Jan van der Veen. 2005. "DyNoC: A Dynamic Infrastructure for Communication in Dynamically Reconfigurable Devices." In *Proceedings of the IEEE International Conference on Field Programmable Logic and Applications*, 153–158. IEEE.

Bolotin, Evgeny, Israel Cidon, Ran Ginosar, and Avinoam Kolodny. 2004. "QNoC: QoS Architecture and Design Process for Network on Chip." *Journal of Systems Architecture* 50, nos. 2–3 (February): 105–128.

Bonfietti, A., M. Lombardi, M. Milano, and L. Benini. 2010. "An Efficient and Complete Approach for Throughput-maximal SDF Allocation and Scheduling on Multi-Core Platforms." In *Proceedings of International Conference on Design, Automation and Test in Europe, DATE'10*, 897–902. IEEE.

Borkar, Shekhar. 2007. "Thousand Core Chips: A Technology Perspective." In *Proceedings of the 44th Annual Design Automation Conference*, 746–749. ACM.

Boukhechem, S., and E.-B. Bouernnane. 2008. "TLM Platform Based on SystemC for STARSoC Design Space Exploration." In *Proceedings of NASA/ESA Conference on Adaptive Hardware and Systems*, 354–361. Noordwijk, The Netherlands, June.

Boyan, Justin A., and Michael L. Littman. 1994. "Packet Routing in Dynamically Changing Networks: A Reinforcement Learning Approach." *Advances in Neural Information Processing Systems* 6:671.

Boyd-Wickizer, Silas, Robert Morris, and M. Frans Kaashoek. 2009. "Reinventing Scheduling for Multicore Systems." In *Proceedings of the 12th Workshop on Hot Topics in Operating Systems (HotOS-XII)*. Monte Verita, Switzerland.

Brown, Jeffery A., Rakesh Kumar, and Dean Tullsen. 2007. "Proximity-Aware Directory-Based Coherence for Multi-Core Processor Architectures." In *Proceedings of the Nineteenth Annual ACM Symposium on Parallel Algorithms and Architectures*, 126–134. ACM.

Bukhari, K. Z., G. K. Kuzmanov, and S. Vassiliadis. 2002. "DCT and IDCT Implementations on Different FPGA Technologies." In *Proceedings of the 13th Annual Workshop on Circuits, Systems and Signal Processing (ProRISC02)*, 232–235. Veldhoven, The Netherlands.

Burtscher, M., I. Ganusov, S. J. Jackson, J. Ke, P. Ratanaworabhan, and N. B. Sam. 2005. "The VPC Trace-Compression Algorithms." *IEEE Transactions on Computers* 54 (11): 1329–1344. doi:10.1109/TC.2005.186.

Buyukkurt, B., Z. Guo, and W. Najjar. 2006. "Impact of Loop Unrolling on Area, Throughput and Clock Frequency in ROCCC: C to VHDL Compiler for FPGAs." In *Reconfigurable Computing: Architectures and Applications*, 3985:401–412. Lecture Notes in Computer Science. Springer.

Calandrino, John M., and James H. Anderson. 2008. "Cache-Aware Real-Time Scheduling on Multicore Platforms: Heuristics and a Case Study." In *EuroMicro Conference on Real-Time Systems (ECRTS '08)*, 299–308. July.

Casper, Jared, Tayo Oguntebi, Sungpack Hong, Nathan G. Bronson, Christos Kozyrakis, and Kunle Olukotun. 2011. "Hardware Acceleration of Transactional Memory on Commodity Systems." *ACM SIGARCH Computer Architecture News* 39 (1): 27–38.

Chafi, Hassan, Jared Casper, Brian D. Carlstrom, Austen McDonald, Chi Cao Minh, Woongki Baek, Christos Kozyrakis, and Kunle Olukotun. 2007. "A Scalable, Non-Blocking Approach to Transactional Memory." In *Proceedings of IEEE 13th International Symposium on High Performance Computer Architecture, 2007. HPCA 2007.* 97–108.

Chakraborty, Koushik, Philip M. Wells, and Gurindar S. Sohi. 2006. "Computation Spreading: Employing Hardware Migration to Specialize CMP Cores on-the-Fly." *ACM SIGOPS Operating Systems Review* 40, no. 5 (October): 283–292. doi:10.1145/1168917.1168893.

Chalamalasetti, S. R., W. Vanderbauwhede, S. Purohit, and M. Margala. 2009. "A Low Cost Reconfigurable Soft Processor for Multimedia Applications: Design Synthesis and Programming Model." In *Proceedings of the International Conference on Field Programmable Logic and Applications, 2009. FPL 2009*. 534–538. IEEE.

Chang, Jichuan, and Gurindar S. Sohi. 2007. "Cooperative Cache Partitioning for Chip Multiprocessors." In *Proceedings of the International Conference on Supercomputing (ICS '07)*, 242–252. June.

Chapiro, Daniel Marcos. 1985. "Globally-Asynchronous Locally-Synchronous Systems (Performance, Reliability, Digital)." AAI8506166. PhD diss., Stanford University.

Charest, L., E. M. Aboulhamid, C. Pilkington, and P. Paulin. 2002. "SystemC Performance Evaluation Using a Pipelined DLX Multiprocessor." In *IEEE Design Automation and Test in Europe*, 3.

Chaudhuri, M. 2009. "PageNUCA: Selected Policies for Page-Grain Locality Management in Large Shared Chip-Multiprocessor Caches." In *Proceedings of the IEEE 15th International Symposium on High Performance Computer Architecture. HPCA 2009*. 227–238. IEEE.

Chen, Kuan-Ju, Chin-Hung Peng, and Feipei Lai. 2010. "Star-Type Architecture with Low Transmission Latency for a 2D Mesh NOC." In *Proceedings of the IEEE Asia Pacific Conference on Circuits and Systems (APCCAS)*, 919–922. IEEE.

Chiou, D., H. Sunjeliwala, H. Sunwoo, J. Dam Xu, and N. Patil. 2006. "FPGA-based Fast, Cycle-Accurate, Full-System Simulators." In *UTFAST-2006-01*, 15:795–825. 5. Austin, TX, USA, November.

Chiu, Ge-Ming. 2000. "The Odd-Even Turn Model for Adaptive Routing." *IEEE Transactions On Parallel And Distributed Systems* 11, no. 7 (July): 729–738.

Cho, Myong Hyon, Keun Sup Shim, Mieszko Lis, Omer Khan, and Srinivas Devadas. 2011. "Deadlock-Free Fine-Grained Thread Migration." In *Proceedings of the Fifth ACM/IEEE International Symposium on Networks-on-Chip*, 33–40. ACM.

Cho, Sangyeun, and Lei Jin. 2006. "Managing Distributed, Shared L2 Caches through OS-Level Page Allocation." In *Proceedings of the 39th Annual IEEE/ACM International Symposium on Microarchitecture*, 455–468.

Christie, Dave, Jae-Woong Chung, Stephan Diestelhorst, Michael Hohmuth, Martin Pohlack, Christof Fetzer, Martin Nowack, et al. 2010. "Evaluation of AMD's Advanced Synchronization Facility within a Complete Transactional Memory Stack." In *Proceedings of the 5th European Conference on Computer Systems (EuroSys'10)*, 27–40. Paris, France.

Chung, E. S., and J. C. Hoe. 2010. "High-Level Design and Validation of the BlueSPARC Multithreaded Processor." *IEEE Transactions on CAD* 29 (10): 1459–1470. doi:10.1109/TCAD.2010.2057870.

Chung, E. S., E. Nurvitadhi, J. C. Hoe, B. Falsafi, and K. Mai. 2008. "A Complexity-Effective Architecture for Accelerating Full-System Multi-processor Simulations Using FPGAs." In *Proceedings of the International Symposium on FPGAs*, 77–86.

Cong, J. 2008. "A New Generation of C-Base Synthesis Tool and Domain-Specific Computing." In *Proceedings of the IEEE International SOC Conference, 2008*, 386–386. IEEE.

Cong, J., K. Gururaj, G. Han, A. Kaplan, M. Naik, and G. Reinman. 2008. "MC-Sim: An Efficient Simulation Tool for MPSoC Designs." In *Proceedings of the IEEE/ACM International Conference on Computer-Aided Design (ICCAD)*, 364–371. San Jose, CA, USA, November. doi:10.1109/ICCAD.2008.4681599.

Coppola, M., S. Curaba, M. Grammatikakis, and G. Maruccia. 2003. "IP-SIM: SystemC 3.0 Enhancements for Communication Refinement." In *Proceedings of the Design, Automation and Test in Europe Conference and Exhibition*, 106–111. December 19.

Coppola, M., R. Locatelli, G. Maruccia, L. Pieralisi, and A. Scandurra. 2004. "Spidergon: A Novel on-Chip Communication Network." In *Proceedings of International Symposium on System-on-Chip*, 15. IEEE.

Cormen, Thomas H., Charles E. Leiserson, Ronald L. Rivest, and Clifford Stein. 2010. *Introduction to Algorithms.* Third edition. Cambridge, MA, USA: MIT Press.

Craven, S., C. Patterson, and P. Athanas. 2006. "A Methodology for Generating Application-Specific Heterogeneous Processor Arrays." In *Proceedings of the Hawaii International Conference on System Sciences*, 39:251.

Culler, David E., Jaswinder Pal Singh, and Anoop Gupta. 1999. *Parallel Computer Architecture: A Hardware/Software Approach.* San Francisco: Morgan Kaufmann.

Cytron, R., J. Ferrante, B. K. Rosen, M. N. Wegman, and F. K. Zadeck. 1991. "Efficiently Computing Static Single Assignment Form and the Control Dependence Graph." *ACM Transactions on Programming Languages and Systems* 13 (4): 490.

Dall'Osso, M., G. Biccari, L. Giovannini, D. Bertozzi, and L. Benini. 2003. "Xpipes: A Latency Insensitive Parameterized Network-on-Chip Architecture for Multi-Processor SoCs." In *Proceedings of the 21st International Conference on Computer Design*, 536–539.

Dally, W. J. 1990. "Performance Analysis of k-ary n-cube Interconnection Networks." *IEEE Transactions on Computers* 39, no. 6 (June): 775–785. doi:10.1109/12.53599.

―――. 1992. "Virtual-Channel Flow Control." *IEEE Transactions on Parallel and Distributed Systems* 3, no. 2 (March): 194–205.

Dally, William J., and Brian Towles. 2001. "Route Packets, not Wires: on-Chip Interconnection Networks." In *Proceedings of the 38th Design Automation Conference,* 684–689. Las Vegas, NV, USA, June.

―――. 2004. *Principles and Practices of Interconnection Networks.* Amsterdam/London: Morgan Kaufmann.

Dave, Nirav, Michael Pellauer, and Joel Emer. 2006. "Implementing a Functional/Timing Partitioned Microprocessor Simulator with an FPGA." In *Proceedings of the 2nd Workshop on Architecture Research Using FPGA Platforms (WARFP 2006).*

Dean, Jeffrey, and Sanjay Ghcmawat. 2008. "MapReduce: Simplified Data Processing on Large Clusters." *Communications of the ACM* 51 (1): 107–113.

Dinechin, F. de, B. Pasca, O. Cret, and R. Tudoran. 2008. "An FPGA-Spccific Approach to Floating-Point Accumulation and Sum-of-Products." In *Proceedings of the International Conference on ICECE Technology, 2008. FPT 2008.* 33–40. December. doi:10.1109/FPT.2008.4762363.

Dinechin, Florent de, and Bogdan Pasca. 2011. *FloPoCO Compiler.* http://flopoco.gforge.inria.fr/.

Draper, J. T., and J. Ghosh. 1994. "A Comprehensive Analytical Model for Wormhole Routing in Multicomputer Systems." *Journal of Parallel and Distributed Computing* 23, no. 2 (November): 202–214.

Duato, J., S. Yalamanchili, and L. M. Ni. 2003. *Interconnection Networks: An Engineering Approach.* Amsterdam: Morgan Kaufmann.

Dybdahl, Haakon, Per Stenström, and Lasse Natvig. 2006. "A Cache-Partitioning Aware Replacement Policy for Chip Multiprocessors." In *Proceedings High Performance Computing-HiPC 2006,* edited by Y. Robert, M. Parashar, R. Badrinath, and V. K. Prasanna, 4297:22–34. Lecture Notes in Computer Science. Berlin/Heidelberg: Springer.

Elmiligi, H., A. A. Morgan, M. W. El-Kharashi, and F. Gebali. 2007. "Performance Analysis of Networks-on-Chip Routers." In *Proceedings of the International Design and Test Workshop,* 232–236. IEEE, December.

Emulation and Verification Engineering. 2010. *ZeBu™: A Unified Verification Approach for Hardware Designers and Embedded Software Developers.* Technical report. White Paper. San Jose, CA, USA, April. www.eve-team.com.

Fauth, A., J. Van Praet, and M. Freericks. 1995. "Describing Instruction Set Processors Using nML." In *EDTC '95: Proceedings of the 1995 European Conference on Design and Test,* 503. Washington, DC, USA: IEEE Computer Society.

Fedorova, Alexandra, Margo Seltzer, and Michael D. Smith. 2006. *Cache-Fair Thread Scheduling for Multicore Processors.* Technical report TR-17-06. Harvard University.

Feero, B. S., and P. P. Pande. 2009. "Networks-on-Chip in a Three-Dimensional Environment: A Performance Evaluation." *IEEE Transactions on Computers* 58, no. 1 (January): 32–45. doi:10.1109/TC.2008.142.

Felber, Pascal, Christof Fetzer, and Torvald Riegel. 2008. "Dynamic Performance Tuning of Word-Based Software Transactional Memory." In *Proceedings of the 13th ACM SIGPLAN Symposium on Principles and Practice of Parallel Programming,* 237–246.

Feliciian, F., and S. B. Furber. 2004. "An Asynchronous on-Chip Network Router with Quality-of-Service (QoS) Support." In *Proceedings of International IEEE SoC Conference,* 274–277.

Fensch, C., and M. Cintra. 2008. "An OS-Based Alternative to Full Hardware Coherence on Tiled CMPs." In *Proceedings of the IEEE 14th International Symposium on High Performance Computer Architecture, 2008. HPCA 2008.* 355–366. IEEE.

Ferri, Cesare, Samantha Wood, Tali Moreshet, R. Iris Bahar, and Maurice Herlihy. 2010. "Embedded-TM: Energy and Complexity-Effective Hardware Transactional Memory for Embedded Multicore Systems." *Journal of Parallel and Distributed Computing* 70 (10): 1042–1052.

Flanagan, Cormac, and Patrice Godefroid. 2005. "Dynamic Partial-Order Reduction for Model Checking Software." In *Proceedings of the 32nd ACM SIGPLAN-SIGACT Symposium on Principles of Programming Languages (POPL'05),* 110–121. Long Beach, CA, USA: ACM. doi:10.1145/1040305.1040315.

Foroutan, S., Y. Thonnart, R. Hersemeule, and A. Jerraya. 2010. "An Analytical Method for Evaluating Network-on-Chip Performance." In *Proceedings of the Conference on Design, Automation and Test in Europe (DATE'10),* 1629–1632. Dresden: IEEE.

Ganguly, Amlan, Kevin Chang, Sujay Deb, Partha Pande, Benjamin Belzer, and Christof Teuscher. 2011. "Scalable Hybrid Wireless Network-on-Chip Architectures for Multi-Core Systems." *IEEE Transactions on Computers* 60, no. 10 (September): 1485–1502.

Ganguly, Amlan, Kevin Chang, Partha Pratim Pande, Benjamin Belzer, and Alireza Nojeh. 2009. "Performance Evaluation of Wireless Networks on Chip Architectures." In *Proceedings of the IEEE International Symposium on Quality Electronic Design (ISQED)*, 350–355. March 16–18.

Ganguly, Amlan, Partha Pande, and Benjamin Belzer. 2009. "Crosstalk-Aware Channel Coding Schemes for Energy Efficient and Reliable NoC Interconnects." *IEEE Transactions on VLSI* 17, no. 11 (November): 1626–1639.

Ganguly, Amlan, Partha Pande, Benjamin Belzer, and Cristian Grecu. 2008. "Design of Low Power & Reliable Networks on Chip through Joint Crosstalk Avoidance and Multiple Error Correction Coding." Special Issue on Defect and Fault Tolerance, *Journal of Electronic Testing: Theory and Applications (JETTA)* 24 (1): 67–81.

Gangwal, O. P., A. Rădulescu, K. Goossens, S. González Pestana, and E. Rijpkema. 2005. "Building Predictable Systems on Chip: An Analysis of Guaranteed Communication in the AEthereal Network on Chip." In *Dynamic and Robust Streaming in and Between Connected Consumer-Electronics Devices,* edited by Peter van der Stok, 3:1–36. Philips Research Book Series. Dordrecht: Springer.

Garcia-Molina, H., R. J. Lipton, and J. Valdes. 1984. "A Massive Memory Machine." *IEEE Transactions on Computers* 100 (5): 391–399.

Gebali, F., H. Elmiligi, and M. W. El-Kharashi, eds. 2009. *Networks-on-Chips: Theory and Practice.* Boca Raton, FL, USA: Taylor & Francis Group, LLC/CRC Press.

Geilen, M. C. W., and T. Basten. 2003. "Requirements on the Execution of Kahn Process Networks." In *12th European Symposium on Programming, ESOP 2003, held as part of the Joint European Conferences on Theory and Practice of Software, ETAPS 2003, Warsaw, Poland, April 7-11, 2003,* edited by Pierpaolo Degano, 2618:319–334. Lecture Notes in Computer Science. Berlin/Heidelberg/New York: Springer-Verlag.

Geilen, M., T. Basten, and S. Stuijk. 2005. "Minimising Buffer Requirements of Synchronous Dataflow Graphs with Model Checking." In *Proceedings of the 42nd Annual Design Automation Conference,* 819–824. DAC '05. Anaheim, CA, USA: ACM. doi:10.1145/1065579.1065796.

Ghamarian, A. H., M. C. W. Geilen, T. Basten, B. D. Theelen, M. R. Mousavi, and S. Stuijk. 2006. "Liveness and Boundedness of Synchronous Data Flow Graphs." In *Proceedings of the International Conference on Formal Methods in Computer Aided Design, FMCAD'06*, 68–75. IEEE.

Ghamarian, A. H., M. C. W. Geilen, S. Stuijk, T. Basten, A. J. M. Moonen, M. J. G. Bekooij, B. D. Theelen, and M. R. Mousavi. 2006. "Throughput Analysis of Synchronous Data Flow Graphs." In *Proceedings of the International Conference on Application of Concurrency to System Design, ACSD'06*, 25–36. IEEE. doi:10.1109/ACSD.2006.33.

Ghamarian, A. H., S. Stuijk, T. Basten, M. C. W. Geilen, and B. D. Theelen. 2007. "Latency Minimization for Synchronous Data Flow Graphs." In *Proceedings of the Conference on Digital System Design, DSD'07*, 189–196. IEEE.

Gheorghita, S. V., S. Stuijk, T. Basten, and H. Corporaal. 2005. "Automatic Scenario Detection for Improved WCET Estimation." In *Proceedings of the Design Automation Conference, DAC 05*, 101–104. ACM.

Ghosal, Prasun, and Tuhin Subhra Das. 2012a. "Network-on-Chip Routing Using Structural Diametrical 2D Mesh Architecture." In *Proceedings of Third International Conference on Emerging Applications of Information Technology (EAIT 2012)*.

———. 2012b. "SD2D: A Novel Routing Architecture For Network-on-Chip." In *Proceedings of 3rd International Symposium on Electronic System Design (ISED 2012)*.

———. 2013. "A Novel Routing Algorithm for on-Chip Communication in NoC on Diametrical 2D Mesh Interconnection Architecture." In *Advances in Computing and Information Technology. Proceedings of the Second International Conference on Advances in Computing and Information Technology (ACITY), Volume 3*, July 13–15, 2012, Chennai, India, edited by Natarajan Meghanathan, Dhinaharan Nagamalai, and Nabendu Chaki, 178:667–676. Advances in Intelligent Systems and Computing. Berlin/Heidelberg: Springer-Verlag.

Ghosal, Prasun, and Sankar Karmakar. 2012. "Diametrical Mesh of Tree (D2D-MoT) Routing Architecture for Network-on-Chip." *International Journal of Advanced Engineering Technology* III, no. I (January): 243–247.

Gibson, J., R. Kunz, D. Ofelt, M. Horowitz, J. Hennessy, and M. Heinrich. 2000. "FLASH vs. (simulated) FLASH: Closing the Simulation Loop." *ACM SIGOPS Operating Systems Review* 34, no. 5 (March): 49–58.

GNU Project. 2010. *GDB: The GNU Project Debugger*, March. http://www.gnu.org/software/gdb/.

Godefroid, Patrice, ed. 1996. *Partial-Order Methods for the Verification of Concurrent Systems: An Approach to the State-Explosion Problem.* Lecture Notes in Computer Science. Berlin/Heidelberg/New York: Springer-Verlag. doi:10.1007/3-540-60761-7.

Goel, A. K. 2001. "Nanotechnology Circuit Design - The 'Interconnect Problem'." In *Proceedings of the 2001 1st IEEE Conference on Nanotechnology (IEEE-NANO'01)*, 123–127. IEEE. doi:10.1109/NANO.2001.966405.

———. 2007. *High-Speed VLSI Interconnections.* Second Edition. Microwave and Optical Engineering. Hoboken, NJ, USA: Wiley-IEEE Press.

Gokhale, M. B., J. M. Stone, J. Arnold, and M. Kalinowski. 2000. "Stream-oriented FPGA Computing in the Streams-C High Level Language." In *Proceedings of the 2000 IEEE Symposium on Field-Programmable Custom Computing Machines,* 49–56. Washington, DC, USA: IEEE Computer Society.

Goossens, K., J. Dielissen, O. P. Gangwal, S. G. Pestana, A. Radulescu, and E. Rijpkema. 2005. "A Design Flow for Application-Specific Networks on Chip with Guaranteed Performance to Accelerate SOC Design and Verification." In *Proceedings of the Conference on Design, Automation and Test in Europe,* 2:1182–1187. ACM.

Goossens, Kees, John Dielissen, and Andrei Radulescu. 2005. "Æthereal Network on Chip: Concepts, Architectures, and Implementations." *IEEE Design & Test of Computers* 22, no. 5 (September): 414–421.

Goyal, Pawan, Harrick M. Vin, and Haichen Cheng. 1996. "Start-Time Fair Queueing: A Scheduling Algorithm for Integrated Services Packet Switching Networks." *ACM SIGCOMM Computer Communication Review* 26 (4): 157–168.

Gray, J. 1986. "Why Do Computers Stop and What Can Be Done About It?" In *Symposium on Reliability in Distributed Software and Database Systems,* 3–12.

Gray, Jan. 1998. *The Myriad Uses of Block RAM,* October 27. http://www.fpgacpu.org/usenet/bb.html.

———. 2000. "Hands-on Computer Architecture - Teaching Processor and Integrated Systems Design with FPGAs." In *Proceedings of the 2000 Workshop on Computer Architecture Education,* 17. ACM.

Grecu, C., P. P. Pande, A. Ivanov, and R. Saleh. 2004. "A Scalable Communication-Centric SoC Interconnect Architecture." In *Proceedings of the 5th International Symposium on Quality Electronic Design, 2004.* 343–348. doi:10.1109/ISQED.2004.1283698.

Greenberg, R. I., and Lee Guan. 1997. "An Improved Analytical Model for Wormhole Routed Networks with Application to Butterfly Fat-Trees." In *Proceedings of the 1997 International Conference on Parallel Processing, 1997,* 44–48. August. doi:10.1109/ICPP.1997.622554.

Grottke, M., and K. S. Trivedi. 2005. "A Classification of Software Faults." In *Proceedings of the International Symposium on Software Reliability Engineering,* 4–19.

Gschwind, M., B. D'Amora, and A. Eichenberger. 2006. "Cell Broadband Engine – Enabling Density Computing for Data-Rich Environment." In *Proceedings of the International Symposium on Computer Architecture.* Slides from conference presentation. June. http : / / researcher . watson . ibm . com / researcher / files / us – mkg / cell _ isca2006.pdf.

Gu, Huaxi, Jiang Xu, and Wei Zhang. 2009. "A Low Power Fat Tree Based Optical Network-on-Chip for Multiprocessor System-on-Chip." In *Proceedings of the Design, Automation and Test in Europe Conference and Exhibition,* 3–8.

Guan, W. J., W. K. Tsai, and D. Blough. 1993. "An Analytical Model for Wormhole Routing in Multicomputer Interconnection Networks." In *Proceedings of Seventh International Parallel Processing Symposium,* 650–654. Newport, CA: IEEE.

Guerre, A., N. Ventroux, R. David, and A. Merigot. 2009. "Approximate-Timed Transactional Level Modeling for MPSoC Exploration: A Network-on-Chip Case Study." In *Proceedings of the IEEE EUROMICRO Conference on Digital System Design (DSD),* 390–397. Patras, Greece, August.

———. 2010. "Hierarchical Network-on-Chip for Embedded Many-Core Architectures." In *Proceedings of the ACM/IEEE International Symposium on Networks-on-Chip (NOCS),* 189–196. Grenoble, France, May.

Guerrier, P., and A. Greiner. 2000. "A Generic Architecture for on-Chip Packet-Switched Interconnections." In *Proceedings of the Conference on Design, Automation and Test in Europe,* 250–256. Paris, France.

Gupta, Anoop, Wolf-Dietrich Weber, and Todd Mowry. 1992. "Reducing Memory and Traffic Requirements for Scalable Directory-Based Cache Coherence Schemes." In *Scalable Shared Memory Multiprocessors,* edited by Michel Dubois and Shreekant Thakkar, 167–192. Originally published in: *Proceedings of the International Conference on Parallel Processing,* 1990. New York: Springer Science + Business Media, LLC. doi:10.1007/ 978-1-4615-3604-8_9.

Gupta, T., C. Bertolini, O. Heron, N. Ventroux, T. Zimmer, and F. Marc. 2010. "High Level Power and Energy Exploration Using ArchC." In *Proceedings of the IEEE International Symposium on Computer Architecture and High Performance Computing (SBAC-PAD)*, 25–32. Petrópolis, Brazil, October.

Gustafsson, J. 2006. "The Worst Case Execution Time Tool Challenge 2006." In *Proceedings of the International Symposium on Leveraging Applications of Formal Methods, Verification and Validation*, 233–240.

Guz, Zvika, Isask'har Walter, Evgeny Bolotin, Israel Cidon, Ran Ginosar, and Avinoam Kolodny. 2006. "Efficient Link Capacity and QoS Design for Network-on-Chip." In *Proceedings of the Conference on Design, Automation and Test in Europe*, 9–14. Leuven, Belgium: European Design / Automation Association. http://dl.acm.org/citation.cfm?id= 1131481.1131487.

———. 2007. "Network Delays and Link Capacities in Application-Specific Wormhole NoCs." *VLSI Design* 2007. doi:10.1155/2007/90941.

Hadjiyiannis, G., S. Hanono, and S. Devadas. 1997. "ISDL: An Instruction Set Description Language For Retargetability." In *Proceedings of the 34th Design Automation Conference, 1997*, 299–302. June.

Haid, W., Kai Huang, I. Bacivarov, and L. Thiele. 2009. "Multiprocessor SoC Software Design Flows." *Signal Processing Magazine* 26 (6): 64–71.

Hailpern, B., and P. Santhanam. 2002. "Software Debugging, Testing, and Verification." *IBM Systems Journal* 41, no. 1 (January): 4–12. doi:10.1147/sj.411.0004.

Halambi, Ashok, Peter Grun, Vijay Ganesh, Asheesh Khare, Nikil Dutt, and Alex Nicolau. 1999. "EXPRESSION: A Language for Architecture Exploration through Compiler/Simulator Retargetability." In *Proceedings of the Conference on Design, Automation and Test in Europe*. 100. Munich, Germany: ACM. doi:10.1145/307418.307549.

Hardavellas, Nikos, Michael Ferdman, Babak Falsafi, and Anastasia Ailamaki. 2009. "Reactive NUCA: Near-Optimal Block Placement and Replication in Distributed Caches." *SIGARCH Computer Architecture News* 37, no. 3 (June): 184–195. doi:10.1145/1555815.1555779.

Heinrich, Joe. 1994. *MIPS R4000 Microprocessor User's Manual*. MIPS Technologies, Inc.

Hennessy, John L., and D. Patterson. 2003. *Computer Architecture: A Quantitive Approach*. Third Edition. Amsterdam: Morgan Kaufmann.

Henriksson, T., and P. van der Wolf. 2006. "TTL Hardware Interface: A High-Level Interface for Streaming Multiprocessor Architectures." In *Proceedings of the IEEE/ACM/IFIP Workshop on Embedded Systems for Real Time Multimedia (ESTIMedia)*, 107–112. Seoul, Korea: IEEE Computer Society, October.

Herlihy, Maurice, and J. Moss. 1993. "Transactional Memory: Architectural Support for Lock-Free Data Structures." In *Proceedings of the 20th Annual International Symposium on Computer Architecture (ISCA-20)*, 289–300.

Hill, Mark, and Michael Marty. 2008. "Amdahl's Law in the Multicore Era." *IEEE Computer* 41, no. 7 (July): 33–38. doi:10.1109/MC.2008.209.

Hoare, C. A. R. 1978. "Communicating Sequential Processes." *Communications of the ACM* 21 (8): 666–677.

Hofstee, H. Peter. 2005. "Power Efficient Processor Architecture and the Cell Processor." In *Proceedings of the 11th International Symposium on High-Performance Computer Architecture, 2005 (HPCA'11)*, 258–262. Washington, DC, USA: IEEE Computer Society. doi:10.1109/HPCA.2005.26.

Holsti, Niklas, Jan Gustafsson, Guillem Bernat, Clément Ballabriga, Armelle Bonenfant, Roman Bourgade, Hugues Cassé, et al. 2008. "WCET 2008 – Report from the Tool Challenge 2008." In *Proceedings of the 8th International Workshop on Worst-Case Execution Time (WCET) Analysis*, edited by Raimund Kirner, 149–171. Also published in print by Austrian Computer Society (OCG) under ISBN 978-3-85403-237-3. Dagstuhl, Germany: Schloss Dagstuhl - Leibniz-Zentrum fuer Informatik, Germany. doi:10.4230/OASIcs.WCET.2008.1663.

Hong, Sungpack, Tayo Oguntebi, Jared Casper, Nathan Bronson, Christos Kozyrakis, and Kunle Olukotun. 2010. "EigenBench: A Simple Exploration Tool for Orthogonal TM Characteristics." In *Proceedings of the IEEE International Symposium on Workload Characterization (IISWC), 2010*, 1–11. IEEE.

Hopkins, A. B. T., and K. D. McDonald-Maier. 2006. "Debug Support Strategy for Systems-on-Chips with Multiple Processor Cores." *IEEE Transactions on Computers* 55 (2): 174–184. doi:10.1109/TC.2006.22.

Howes, L. W., O. Pell, O. Mencer, and O. Beckmann. 2006. "Accelerating the Development of Hardware Accelerators." In *Workshop on Edge Computing, North Carolina, USA.* May. http://gamma.cs.unc.edu/events/workshops/edge-06/.

HPC Project. n.d. "Par4All, An Automatic Parallelizing and Optimizing Compiler for C and Fortran Sequential Programs." http://www.par4all.org.

Hsieh, Wilson C., Paul Wang, and William E. Weihl. 1993. "Computation Migration: Enhancing Locality for Distributed-Memory Parallel Systems." Also published in: Proceedings of the Fourth ACM SIGPLAN Symposium on Principles and Practice of Parallel Programming: San Diego, CA, USA, *SIGPLAN Notices* (New York, NY, USA) 28, no. 7 (July): 239–248. doi:10.1145/173284.155357.

Hu, Jingcao, and Radu Marculescu. 2004. "DyAD: Smart Routing for Networks-on-Chip." In *Proceedings of the 41st Annual Design Automation Conference (DAC'04)*, 260–263. New York, NY, USA: ACM. doi:10.1145/996566.996638.

Hu, P.-C., and L. Kleinrock. 1997. "An Analytical Model for Wormhole Routing with Finite Size Input Buffers." In *Proceedings of 15th International Teletraffic Congress*, 549–560. Washington, DC, June 23–27.

IEEE Standard Test Access Port and Boundary-Scan Architecture. 2001. IEEE Standard 1149.1-2001. doi:10.1109/IEEESTD.2001.92950.

IEEE Standard for Reduced-Pin and Enhanced-Functionality Test Access Port and Boundary-Scan Architecture. 2009. IEEE Standard 1149.7-2009. doi:10.1109/IEEESTD.2010.5412866.

Infineon Technologies. 2010. *MCDS - Multi-Core Debug Solution*, March. http://www.ip-extreme.com/IP/mcds.shtml.

Intel Corporation. 2005. *Intel PXA27x Processor Family, Electrical, Mechanical, and Thermal Specification Datasheet.*

Intel Corporation. 2009. *Intel 64 and IA-32 Architectures Software Developer's Manual, Volume 3: System Programming Guide.* Intel Corporation, June.

———. 2009. *Intel Multi-Core Technology.* Intel Corporation. http://www.intel.com/multi-core/.

Intel Corporation. 2010a. *Intel Itanium Architecture Software Developer's Manual.* http://www.intel.com/design/itanium/manuals/iiasdmanual.htm.

———. 2010b. *Multi-Core Debugging for Intel Processors.* http://www.intel.com/intelpress/articles/ms2a_2.pdf.

International Business Machines Corporation and Sony Computer Entertainment Incorporated and Toshiba Corporation. 2008. *Cell Broadband Engine Programming Handbook.* 1.0. Hopewell Junction, NY: IBM Systems and Technology Group, August.

International Telecommunications Union. 2005. "National Spectrum Management." http://www.itu.int/.

Ipek, E., M. Kirman, N. Kirman, and J. F. Martinez. 2007. "Core Fusion: Accommodating Software Diversity in Chip Multiprocessors." *ACM SIGARCH Computer Architecture News* 35 (2): 186–197.

The International Technology Roadmap for Semiconductors: Assembly and Packaging. 2007. http://www.itrs.net/Links/2007ITRS/2007_Chapters/2007_Assembly.pdf.

The International Technology Roadmap for Semiconductors 2009 for Interconnects. 2009. http://public.itrs.net/links/2009ITRS/Home2009.htm.

The International Technology Roadmap for Semiconductors. 2008. http://www.itrs.net/Links/2008ITRS/Update/2008_Update.pdf.

Iyer, Ravi. 2004. "CQoS: A Framework for Enabling QoS in Shared Caches of CMP Platforms." In *Proceedings of the 18th Annual International Conference on Supercomputing*, 257–266.

Jantsch, A. 2003. "Communication Performance in Network-on-Chips." Presentation at the Swedish INTELECT Summer School on Multiprocessor Systems on Chip, Stockholm.

Jayadevappa, Suryaprasad, Ravi Shankar, and Imad Mahgoub. 2004. "A Comparative Study of Modeling at Different Levels of Abstraction in System on Chip Designs: A Case Study." In *Proceedings of the Annual Symposium on VLSI*, 52–58. Los Alamitos, CA, USA: IEEE Computer Society, February. doi:10.1109/ISVLSI.2004.1339508.

Jenks, Stephen, and Jean-Luc Gaudiot. 1996. "Nomadic Threads: A Migrating Multithreaded Approach to Remote Memory Accesses in Multiprocessors." In *Proceedings of the 1996 Conference on Parallel Architectures and Compilation Techniques, 1996*, 2–11. IEEE.

———. 2002. "An Evaluation of Thread Migration for Exploiting Distributed Array Locality." In *Proceedings of the 16th Annual International Symposium on High Performance Computing Systems and Applications, 2002*, 190–195. IEEE.

Jerraya, A. A., and W. Wolf, eds. 2005. *Multiprocessor Systems-on-Chips*. Amsterdam, London: Morgan Kaufmann.

Jordans, R., F. Siyoum, S. Stuijk, A. Kumar, and H. Corporaal. 2011. "An Automated Flow to Map Throughput Constrained Applications to a MP-SoC." In *Bringing Theory to Practice: Predictability and Performance in Embedded Systems, DATE Workshop PPES 2011,* Grenoble, France, 47–58. Leibniz-Zentrum fuer Informatik, Germany: Schloss Dagstuhl, March. doi:10.4230/OASIcs.PPES.2011.47.

Joshi, Ajay, Christopher Batten, Yong-Jin Kwon, Scott Beamer, Imran Shamim, Krste Asanovic, and Vladimir Stojanovic. 2009. "Silicon-Photonic Clos Networks for Global on-Chip Communication." In *Proceedings of the 3rd ACM/IEEE International Symposium on Networks-on-Chip (NoCS),* 124–133. San Diego, CA, USA, May.

Kachris, Christoforos, and Chidamber Kulkarni. 2007. "Configurable Transactional Memory." In *Proceedings of the 15th Annual IEEE Symposium on Field-Programmable Custom Computing Machines, 2007 (FCCM 2007),* 65–72. IEEE.

Kahn, G. 1974. "The Semantics of a Simple Language for Parallel Programming." In *Proceedings of the IFIP Congress on Information Processing '74,* edited by J. L. Rosenfeld, 471–475. New York, NY: North-Holland.

Kandemir, M., Feihui Li, M. J. Irwin, and Seung Woo Son. 2008. "A Novel Migration-Based NUCA Design for Chip Multiprocessors." In *Proceedings of the International Conference for High Performance Computing, Networking, Storage and Analysis, 2008 (SC 2008),* 1–12. IEEE.

Kao, C.-F., S.-M. Huang, and I.-J. Huang. 2007. "A Hardware Approach to Real-Time Program Trace Compression for Embedded Processors." *IEEE Transactions on Circuits and Systems* 54 (3): 530–543. doi:10.1109/TCSI.2006.887613.

Karim, F., A. Nguyen, and S. Dey. 2002. "An Interconnect Architecture for Networking Systems on Chips." *IEEE Micro* 22 (5): 36–45.

Kavvadias, N., and S. Nikolaidis. 2008. "Elimination of Overhead Operations in Complex Loop Structures for Embedded Microprocessors." *IEEE Transactions on Computers* 57, no. 2 (February): 200–214. doi:10.1109/TC.2007.70790.

Kermani, Parviz, and Leonard Kleinrock. 1979. "Virtual Cut-Through: A New Computer Communication Switching Technique." *Computer Networks* 3 (4): 267–286.

Khan, M. A., and A. Q. Ansari. 2011a. "128-Bit High-Speed FIFO for Network-on-Chip." In *Proceedings of the IEEE International Conference on Emerging Trends in Computing,* 116–121. March.

Khan, M. A., and A. Q. Ansari. 2011b. "A Quadrant-XYZ Routing Algorithm for 3-D Asymmetric Torus Routing Chip." *International Journal of ACM Jordan (IJJ): The Research Bulletin of Jordan ACM-ISWSA* 2 (2): 18–26.

———. 2011c. "An Efficient Tree-Based Topology for Network-on-Chip." In *Proceedings of the IEEE World Congress on Information Technology,* edited by Ajith Abraham, 11–14. Mumbai: University of Mumbai, IEEE, December.

Khonsari, A., M. Ould-Khaoua, and J. D. Ferguson. 2003. "A General Analytical Model of Adaptive Wormhole Routing in k-ary n-cubes." In *International Symposium on Performance Evaluation of Computer and Telecommunication Systems.* Montreal, Canada, July 20–24.

Ki, Woo-Seo, Hyeong-Ok Lee, and Jae-Cheol Oh. 2009. "The New Torus Network Design Based on 3-Dimensional Hypercube." In *Proceedings of the 11th International Conference on Advanced Communication Technology, 2009 (ICACT 2009),* 1:615–620. February.

Kiasari, A. E., D. Rahmati, H. Sarbazi-Azad, and S. Hessabi. 2008. "A Markovian Performance Model for Networks-on-Chip." In *Proceedings of the 16th Euromicro Conference on Parallel, Distributed and Network-Based Processing (PDP 2008),* 157–164. Washington, DC, USA: IEEE Computer Society. doi:10.1109/PDP.2008.83.

Kim, Changkyu, Doug Burger, and Stephen W. Keckler. 2002. "An Adaptive, Non-Uniform Cache Structure for Wire-Delay Dominated on-Chip Caches." *ACM Sigplan Notices* 37 (10): 211–222.

Kim, Kwanho, Se-Joong Lee, Kangmin Lee, and Hoi-Jun Yoo. 2005. "An Arbitration Look-Ahead Scheme for Reducing End-to-End Latency in Networks on Chip." In *Proceedings of the IEEE International Symposium Circuits and Systems (ISCAS'05),* 3:2357–2360.

Kim, Seongbeom, Dhruba Chandra, and Yan Solihin. 2004. "Fair Cache Sharing and Partitioning in a Chip Multiprocessor Architecture." In *Proceedings of the 13th International Conference on Parallel Architectures and Compilation Techniques,* 111–122. IEEE Computer Society, October.

Kindratenko, Volodymyr V., Robert J. Brunner, and Adam D. Myers. 2007. "Mitrion-C Application Development on SGI Altix 350/RC100." In *Proceedings of the 15th Annual IEEE Symposium on Field-Programmable Custom Computing Machines (FCCM '07),* 239–250. Washington, DC, USA: IEEE Computer Society. doi:10.1109/FCCM.2007.45.

Kini, N. G., M. S. Kumar, and H. S. Mruthyunjaya. 2009. "A Torus Embedded Hypercube Scalable Interconnection Network for Parallel Architecture." In *Proceedings of the IEEE International Advance Computing Conference, 2009 (IACC 2009)*, 858–861. March. doi:10.1109/IADCC.2009. 4809127.

Kock, E. A. de, W. J. M. Smits, P. van der Wolf, J.-Y. Brunel, W. M. Kruijtzer, P. Lieverse, K. A. Vissers, and G. Essink. 2000. "YAPI: Application Modeling for Signal Processing Systems." In *Proceedings of the Design Automation Conference, DAC'00*, 402–405. ACM.

Konstantakopulos, Theodoros, Jonathan Eastep, James Psota, and Anant Agarwal. 2008. *Energy Scalability of on-Chip Interconnection Networks in Multicore Architectures*. Technical report MIT-CSAIL-TR-2008-066. MIT.

Koohi, Somayyeh, Meisam Abdollahi, and Shaahin Hessabi. 2011. "All-Optical Wavelength-Routed NoC Based on a Novel Hierarchical Topology." In *Proceedings of the Fifth ACM/IEEE International Symposium on Networks-on-Chip NOCS'11*, 97–104.

Koohi, Somayyeh, and Shaahin Hessabi. 2009. "Contention-Free on-Chip Routing of Optical Packets." In *Proceedings of the 3rd ACM/IEEE International Symposium on Networks-on-Chip*, 134–143.

Kopp, C., S. Bernabe, B. B. Bakir, J. Fedeli, R. Orobtchouk, F. Schrank, H. Porte, L. Zimmermann, and T. Tekin. 2011. "Silicon Photonic Circuits: On-CMOS Integration, Fiber Optical Coupling, and Packaging." *IEEE Journal of Selected Topics in Quantum Electronics* 17 (3): 498–509.

Krasnov, Alex, Andrew Schultz, John Wawrzynek, Greg Gibeling, and Pierre-Yves Droz. 2007. "RAMP Blue: A Message-Passing Manycore System in FPGAs." In *Proceedings of the International Conference on Field Programmable Logic and Applications, 2007 (FPL 2007)*, 54–61. IEEE.

Kreupl, F., A. P. Graham, G. S. Duesberg, W. Steinhögl, M. Liebau, E. Unger, and W. Hönlein. 2002. "Carbon Nanotubes in Interconnect Applications." *Microelectronic Engineering* 64 (1): 399–408.

Kumar, A., S. Fernando, Y. Ha, B. Mesman, and H. Corporaal. 2008. "Multiprocessor Systems Synthesis for Multiple Use-Cases of Multiple Applications on FPGA." *ACM Transactions on Design Automation of Electronic Systems* 13 (3): 1–27.

Kumar, Amit, Partha Kundu, Arvind P. Singh, Li-Shiuan Peh, and Niraj K. Jha. 2007. "A 4.6 Tbits/s 3.6 GHz Single-Cycle NoC Router with a Novel Switch Allocator." In *Proceedings of the 25th International Conference on Computer Design, 2007 (ICCD 2007)*, 63–70. IEEE.

Kumar, R., D. M. Tullsen, N. P. Jouppi, and P. Ranganathan. 2005. "Hetero-geneous Chip Multiprocessors." *IEEE Computer* 38 (11): 32–38.

Kumar, S., A. Jantsch, J.-P. Soininen, M. Forsell, M. Millberg, J. Oberg, K. Tiensyrja, and A. Hemani. 2002. "A Network on Chip Architecture and Design Methodology." In *Proceedings of the IEEE Computer Society Annual Symposium on VLSI*, 105–112. Pittsburgh, PA.

Kundu, S., R. P. Dasari, S. Chattopadhyay, and K. Manna. 2008. "Mesh-of-Tree Based Scalable Network-on-Chip Architecture." In *Proceedings of the IEEE Region 10 Third international Conference on Industrial and Information Systems, 2008 (ICIIS 2008)*, 1–6. IEEE.

Kurian, G., J. E. Miller, J. Psota, J. Eastep, J. Liu, J. Michel, L. C. Kimerling, and A. Agarwal. 2010. "ATAC: A 1000-Core Cache-Coherent Processor with on-Chip Optical Network." In *Proceedings of the 19th International Conference on Parallel Architectures and Compilation Techniques*, 477–488. ACM.

Labrecque, Martin, Mark Jeffrey, and J. Gregory Steffan. 2010. "Application-Specific Signatures for Transactional Memory in Soft Processors." In *Proceedings of the 6th International Conference on Reconfigurable Computing: Architectures, Tools and Applications*, 42–54. ARC'10. Bangkok, Thailand: Springer-Verlag. doi:10.1007/978-3-642-12133-3_7.

Lam, M. 1988. "Software Pipelining: An Effective Scheduling Technique for VLIW Machines." *ACM SIGPLAN Notices* 23 (7): 318–328. doi:10.1145/960116.54022.

Lawler, E. L., and D. E. Wood. 1966. "Branch-And-Bound Methods: A Survey." *Operations Research* 14 (4): 699–719.

Lee, E. A., and D. G. Messerschmitt. 1987a. "Static Scheduling of Synchronous Data Flow Programs for Digital Signal Processing." *IEEE Transactions on Computers* 36 (1): 24–35.

———. 1987b. "Synchronous Data Flow." *Proceedings of the IEEE* 75, no. 9 (September): 1235–1245.

Lee, S. B., S. W. Tam, I. Pefkianakis, S. Lu, M. F. Chang, C. Guo, G. Reinman, C. Peng, M. Naik, L. Zhang, et al. 2009. "A Scalable Micro Wireless Interconnect Structure for CMPs." In *Proceedings of the 15th Annual International Conference on Mobile Computing and Networking*, 217–228. ACM.

Lee, S.-J., Seong-Jun Song, Kangmin Lee, Jeong-Ho Woo, and Sung-Eun. 2003. "An 800 MHz Star-Connected on-Chip Network for Application to Systems on a Chip." In *Digest of Technical Papers. 2003 IEEE International Solid-State Circuits Conference, 2003 (ISSCC)*, 1:468–469. ISSCC.

Lelewer, D., and D. Hirschberg. 1987. "Data Compression." *ACM Computing Surveys* 19, no. 3 (September): 261–296. doi:10.1145/45072.45074.

Leupers, R., and P. Marwedel. 1998. "Retargetable Code Generation Based on Structural Processor Description." *Design Automation for Embedded Systems* 3 (1): 75–108.

Li, Yonghui, and Huaxi Gu. 2009. "XY-Turn Model for Deadlock Free Routing in Honeycomb Networks-on-Chip." In *Proceedings of the 15th Asia-Pacific Conference on Communications, (APCC'09)*, 900–903. August. doi:10.1109/APCC.2009.5375521.

Li, Z., D. Fay, A. Mickelson, L. Shang, M. Vachharajani, D. Filipovic, W. Park, and Y. Sun. 2009. "Spectrum: A Hybrid Nanophotonic-Electric on-Chip Network." In *Proceedings of the 46th ACM/IEEE Design Automation Conference, 2009 (DAC'09)*, 575–580. IEEE.

Liang, J., S. Swaminathan, and R. Tessier. 2000. "aSOC: A Scalable, Single-Chip Communications Architecture." In *Proceedings of the 2000 International Conference on Parallel Architectures and Compilation Techniques*, 37–46. Philadelphia, PA.

Liedtke, Jochen, Hermann Härtig, and Michael Hohmuth. 1997. "OS-Controlled Cache Predictability for Real-Time Systems." In *Proceedings of the Third IEEE Real-Time Technology and Applications Symposium*, 213–224. IEEE.

Lin, Jiang, Qingda Lu, Xiaoning Ding, Zhao Zhang, Xiaodong Zhang, and P. Sadayappan. 2008. "Gaining Insights into Multicore Cache Partitioning: Bridging the Gap between Simulation and Real Systems." In *Procedings of the 14th IEEE International Symposium on High Performance Computer Architecture*, 367–378.

Lines, A. 2004. "Asynchronous Interconnect for Synchronous SoC Design." *IEEE Micro* 24, no. 1 (January): 32–41.

Liu, Chun, Anand Sivasubramaniam, and Mahmut Kandemir. 2004. "Organizing the Last Line of Defense before Hitting the Memory Wall for CMPs." In *Proceedings of the International Symposium on High-Performance Computer Architecture*, 176–185.

Liu, W., W. Yuan, X. He, Z. Gu, and X. Liu. 2008. "Efficient SAT-Based Mapping and Scheduling of Homogeneous Synchronous Dataflow Graphs for Throughput Optimization." In *Proceedings of the Real-Time Systems Symposium, RTSS'08*, 492–504. IEEE.

Loo, S. M., B. E. Wells, N. Freije, and J. Kulick. 2002. "Handel C for Rapid Prototyping of VLSI Coprocessors for Real Time Systems." In *Proceedings of the Thirty-Fourth Southeastern Symposium on System Theory, 2002*, 34:6–10.

Loucif, S., and M. Ould-Khaoua. 2004. "Modeling Latency in Deterministic Wormhole-Routed Hypercubes under Hot-Spot Traffic." *The Journal of Supercomputing* 27 (3): 265–278.

Loucif, S., M. Ould-Khaoua, and G. Min. 2005. "Analytical Modelling of Hot-Spot Traffic in Deterministically-Routed *k*-ary *n*-cubes." In *Proceedings of the 19th IEEE International Parallel and Distributed Processing Symposium (IPDPS'05) - Workshop 15,* vol. 16, 8–pp. IPDPS '05. Washington, DC, USA: IEEE Computer Society. doi:10.1109/IPDPS.2005.108.

Lu, Z. 2007. "Design and Analysis of on-Chip Communication for Network-on-Chip Platforms." PhD diss., KTH Royal Institute of Technology.

Lu, Zhonghai, Axel Jantsch, and Ingo Sander. 2005. "Feasibility Analysis of Messages for on-Chip Networks Using Wormhole Routing." In *Proceedings of the 2005 Asia and South Pacific Design Automation Conference,* 960–964. ASP-DAC'05. Shanghai, China: ACM. doi:10.1145/1120725.1120767.

Magnusson, Peter S., Magnus Christensson, Jesper Eskilson, Daniel Forsgren, Gustav Hållberg, Johan Högberg, Fredrik Larsson, Andreas Moestedt, and Bengt Werner. 2002. "Simics: A Full System Simulation Platform." *Computer* 35 (2): 50–58. doi:10.1109/2.982916.

Majer, M., C. Bobda, A. Ahmadinia, and J. Teich. 2005. "Packet Routing in Dynamically Changing Networks on Chip." In *Proceedings of the 19th IEEE International Parallel and Distributed Processing Symposium,* 154b. IEEE, April. doi:10.1109/IPDPS.2005.323.

Manna, K., S. Chattopadhyay, and I. S. Gupta. 2010. "Energy and Performance Evaluation of a Dimension Order Routing Algorithm for Mesh-of-Tree Based Network-on-Chip Architecture." In *Proceedings of the Annual IEEE India Conference (INDICON),* 1–4. December 17–19. doi:10.1109/INDCON.2010.5712666.

Manzke, Michael, and Ross Brennan. 2004. "Extending FPGA Based Teaching Boards into the Area of Distributed Memory Multiprocessors." In *Proceedings of the 2004 Workshop on Computer Architecture Education: held in conjunction with the 31st International Symposium on Computer Architecture (WCAE '04).* New York, NY, USA.

Marchetti, M., L. Kontothanassis, R. Bianchini, and M. L. Scott. 1995. "Using Simple Page Placement Policies to Reduce the Cost of Cache Fills in Coherent Shared-Memory Systems." In *Proceedings of the 9th International Parallel Processing Symposium,* 480–485. IEEE.

Martin, G. 2006. "Overview of the MPSoC Design Challenge." In *Proceedings of the 43rd Annual Design Automation Conference, DAC'06,* 274–279. ACM.

Martínez, José F., and Josep Torrellas. 2002. "Speculative Synchronization: Applying Thread-Level Speculation to Explicitly Parallel Applications." *ACM SIGOPS Operating Systems Review* 36 (5): 18–29.

Mattson, Richard L., Jan Gecsei, Donald R. Slutz, and Irving L. Traiger. 1970. "Evaluation Techniques for Storage Hierarchies." *IBM Systems Journal* 9 (2): 78–117.

Mazurkiewicz, Antoni. 1987. "Trace Theory." In *Petri Nets: Applications and Relationships to Other Models of Concurrency,* edited by W. Brauer, W. Reisig, and G. Rozenberg, 255:278–324. Lecture Notes in Computer Science. Berlin, Heidelberg: Springer. doi:10.1007/3-540-17906-2_30.

Meincke, T., A. Hemani, S. Kumar, P. Ellervee, J. Oberg, T. Olsson, P. Nilsson, D. Lindqvist, and H. Tenhunen. 1999. "Globally Asynchronous Locally Synchronous Architecture for Large High-Performance ASICs." In *Proceedings of the 1999 IEEE International Symposium on Circuits and Systems, 1999 (ISCAS'99),* 2:512–515. IEEE.

Michaud, P. 2004. "Exploiting the Cache Capacity of a Single-Chip Multi-Core Processor with Execution Migration." In *Proceedings of the 10th International Symposium on High Performance Computer Architecture,* 186–195. IEEE Computer Society. doi:10.1109/HPCA.2004.10026.

Mihajlovic, B., M. H. Neishaburi, J. G. Tong, N. Azuelos, Z. Zilic, and W. J. Gross. 2009. "Providing Infrastructure Support to Assist NoC Software Development." In *Proceedings of the Workshop on Diagnostic Services in Network-on-Chips.* Acropolis, Nice, France, April. http://www.dsnoc.org.

Mihajlovic, B., and Z. Zilic. 2011. "Real-Time Address Trace Compression for Emulated and Real System-on-Chip Processor Core Debugging." In *Proceedings of the ACM Great Lakes Symposium on VLSI,* 331–336. Lausanne, Switzerland. doi:10.1145/1973009.1973075.

Mihajlovic, B., Z. Zilic, and K. Radecka. 2007. "Compression and Encryption of Self-Test Programs for Wireless Sensor Network Nodes." In *Proceedings of the IEEE Midwest Symposium on Circuits and Systems,* 1344–1347.

———. 2010. "Infrastructure for Testing Nodes of a Wireless Sensor Network." Chap. 5 in *Handbook of Research on Developments and Trends in Wireless Sensor Networks,* edited by H. Jin and W. Jiang, 79–107. Hershey, PA, USA: IGI Global. doi:10.4018/978-1-61520-701-5.ch005.

Milenkovic, A., V. Uzelac, M. Milenkovic, and M. Burtscher. 2011. "Caches and Predictors for Real-Time, Unobtrusive, and Cost-Effective Program Tracing in Embedded Systems." *IEEE Transactions on Computers* 60, no. 7 (July): 992–1005. doi:10.1109/TC.2010.146.

Milenkovic, M., and M. Burtscher. 2007. "Algorithms and Hardware Structures for Unobtrusive Real-Time Compression of Instruction and Data Address Traces." In *Proceedings of the Data Compression Conference*, 283–292. doi:10.1109/DCC.2007.10.

Millberg, M., E. Nilsson, T. Thid, and A. Jantsch. 2004. "Guaranteed Bandwidth Using Looped Containers in Temporally Disjoint Networks within the Nostrum Network on Chip." In *Proceedings of the Conference on Design, Automation and Test in Europe*, 2:890–895.

Miller, Jason E., Harshad Kasture, George Kurian, Charles Gruenwald, Nathan Beckmann, Christopher Celio, Jonathan Eastep, and Anant Agarwal. 2010. "Graphite: A Distributed Parallel Simulator for Multicores." In *Proceedings of the 2010 IEEE 16th International Symposium on High Performance Computer Architecture*, 1–12. IEEE.

Minh, Chi Cao, Jae Woong Chung, Christos Kozyrakis, and Kunle Olukotun. 2008. "STAMP: Stanford Transactional Applications for Multi-Processing." In *Proceedings of the IEEE International Symposium on Workload Characterization, 2008 (IISWC 2008)*, 35–46. IEEE.

ModelSim. *http://www.model.com/*.

Mookherjea, S., and A. Melloni. 2008. *Microring Resonators in Integrated Optics.* http://mnp.ucsd.edu/ece240a_2009/chapter_microring.pdf.

Moore, G. E. 1998. "Cramming More Components onto Integrated Circuits." *Proceedings of the IEEE* 86 (1): 82–85.

Moore, Kevin E., Jayaram Bobba, Michelle J. Moravan, Mark D. Hill, and David A. Wood. 2006. "LogTM: Log-Based Transactional Memory." In *Proceedings of the 12th International Symposium on High-Performance Computer Architecture (HPCA-'06)*, 254–265. Austin: IEEE Computer Society.

Morad, Tomer Y., U. C. Weiser, A. Kolodny, M. Valero, and E. Ayguade. 2006. "Performance, Power Efficiency and Scalability of Asymmetric Cluster Chip Multiprocessors." *Computer Architecture Letters* 5 (1): 14–17.

Moraes, F., N. Calazans, A. Mello, L. Möller, and L. Ost. 2004. "HERMES: An Infrastructure for Low Area Overhead Packet-Switching Networks on Chip." *Integration, the VLSI Journal* 38, no. 1 (October): 69–93. doi:10.1016/j.vlsi.2004.03.003.

Moreira, O., J.-D. Mol, M. Bekooij, and J. van Meerbergen. 2005. "Multiprocessor Resource Allocation for Hard-Real-Time Streaming with a Dynamic Job-Mix." In *Proceedings of the Real Time and Embedded Technology and Applications Symposium*, 332–341. IEEE.

Moses, J., K. Aisopos, A. Jaleel, R. Iyer, R. Illikkal, D. Newell, and S. Makineni. 2009. "CMPSched$im: Evaluating OS/CMP Interaction on Shared Cache Management." In *Proceedings of the IEEE International Symposium on Performance Analysis of Systems and Software (ISPASS '09)*, 113–122. April.

Najaf-Abadi, H. H., and H. Sarbazi-Azad. 2004. "An Accurate Combinatorial Model for Performance Prediction of Deterministic Wormhole Routing in Torus Multicomputer Systems." In *Proceedings of the IEEE International Conference on Computer Design (ICCD'04)*, 548–553. Washington, DC, USA: IEEE Computer Society. http://dl.acm.org/citation.cfm?id=1032648.1033415.

Naveh, A., E. Rotem, A. Mendelson, S. Gochman, R. Chabukswar, K. Krishnan, and A. Kumar. 2006. "Power and Thermal Management in the Intel Core Duo Processor." *Intel Technology Journal* 10 (2): 109–122.

Nexus 5001™ Forum. 2012. *The Nexus 5001™ Forum Standard for a Global Embedded Processor Debug Interface*. IEEE-ISTO 5001-2012. IEEE Industry Standards and Technology Organization. http://www.nexus5001.org/standard.

Njoroge, Njuguna, Jared Casper, Sewook Wee, Yuriy Teslyar, Daxia Ge, Christos Kozyrakis, and Kunle Olukotun. 2007. "ATLAS: A Chip-Multiprocessor with TM Support." In *Proceedings of the Conference on Design, Automation and Test in Europe*, 3–8.

Noakes, Michael D., Deborah A. Wallach, and William J. Dally. 1993. "The J-Machine Multicomputer: An Architectural Evaluation." *ACM SIGARCH Computer Architecture News* 21 (2): 224–235.

Nurmi, J., H. Tenhunen, J. Isoaho, and A. Jantsch, eds. 2004. *Interconnect-Centric Design for Advanced SoC and NoC*. Dordrecht: Kluwer.

Ogras, U. 2007. "Modeling, Analysis and Optimization of Network-on-Chip Communication Architectures." PhD diss., Carnegie Mellon University.

Ogras, U. Y., and R. Marculescu. 2006. "It's a Small World After All: NoC Performance Optimization via Long-Range Link Insertion." *IEEE Transactions on Very Large Scale Integration Systems* 14, no. 7 (July): 693–706.

———. 2007. "Analytical Router Modeling for Networks-on-Chip Performance Analysis." In *Proceedings of the Conference on Design, Automation and Test in Europe*, 1096–1101. San Jose, CA, USA: EDA Consortium. http://dl.acm.org/citation.cfm?id=1266366.1266602.

Oi, Hitoshi, and N. Ranganathan. 1999. "A Cache Coherence Protocol for the Bidirectional Ring Based Multiprocessor." In *Proceedings of the International Conference on Parallel and Distributed Computing and Systems, PDCS'99*, 3–6.

Open SystemC Initiative. http://www.systemc.org.

Ould-Khaoua, M., and H. Sarbazi-Azad. 2001. "An Analytical Model of Adaptive Wormhole Routing in Hypercubes in the Presence of Hot Spot Traffic." *IEEE Transactions on Parallel and Distributed Systems* 12, no. 3 (March): 283–292.

Pan, Y., P. Kumar, J. Kim, G. Memik, Y. Zhang, and A. Choudhary. 2009. "Firefly: Illuminating Future Network-on-Chip with Nanophotonics." *ACM SIGARCH Computer Architecture News* 37 (3): 429–440.

Panainte, E. Moscu, K. L. M. Bertels, and S. Vassiliadis. 2007. "The Molen Compiler for Reconfigurable Processors." *ACM Transactions on Embedded Computing Systems (TECS)* 6, no. 1 (February): 6.

Pande, P. P., C. Grecu, A. Ivanov, and R. Saleh. 2003. "Design of a Switch for Network on Chip Applications." In *Proceedings of the International Symposium on Circuits and Systems*, 5:217–220.

———. 2005. "Timing Analysis of Network on Chip Architectures for MP-SoC Platforms." *Microelectronics Journal* 36 (September): 833–45. doi:10.1016/j.mejo.2005.03.006.

Pande, P. P., C. Grecu, M. Jones, A. Ivanov, and R. Saleh. 2005. "Performance Evaluation and Design Trade-Offs for Network-on-Chip Interconnect Architectures." *IEEE Transactions on Computers* 54, no. 8 (August): 1025–1040.

Pande, Partha Pratim, Amlan Ganguly, Sujay Deb, and Kevin Chang. 2012. "Energy-Efficient Network-on-Chip Architectures for Multicore Systems." In *Handbook of Energy-Aware and Green Computing*, edited by Ishfaq Ahmad and Sanjay Ranka. Chapman & Hall/CRC Computer & Information Science Series. Boca Raton, FL, USA; London: Chapman & Hall/CRC Press.

Pande, Partha Pratim, Resve Saleh, Amlan Ganguly, Andre Ivanov, and Cristian Grecu. 2009. "Test and Fault Tolerance for Networks-on-Chip Infrastructures." In Gebali, Elmiligi, and El-Kharashi 2009, 191–222.

Papadopoulos, Gregory M., and David E. Culler. 1990. "Monsoon: An Explicit Token-Store Architecture." *ACM SIGARCH Computer Architecture News* 18 (3a): 82–91.

Parekh, A. 1992. "A Generalized Processor Sharing Approach to Flow Control in Integrated Services Networks." PhD diss., Massachusetts Institute of Technology.

Parhami, Behrooz. 1999. *Introduction to Parallel Processing: Algorithms and Architectures*. Plenum Series in Computer Science. New York, London: Plenum Press.

Parks, T. M. 1995. "Bounded Scheduling of Process Networks." PhD diss., University of California, EECS Department.

Paulin, P., C. Pilkington, and E. Bensoudane. 2002. "StepNP: A System-Level Exploration Platform for Network Processors." *IEEE Design & Test* 19, no. 6 (November): 17–26.

Pavlidis, Vasilis F., and Eby G. Friedman. 2007. "3-D Topologies for Networks-on-Chip." *IEEE Transactions on Very Large Scale Integration (VLSI) Systems* 15, no. 10 (October): 1081–1090. doi:10.1109/TVLSI.2007.893649.

Pees, Stefan, Andreas Hoffmann, Vojin Zivojnovic, and Heinrich Meyr. 1999. "LISA – Machine Description Language for Cycle-Accurate Models of Programmable DSP Architectures." In *Proceedings of the 36th annual ACM/IEEE Design Automation Conference, (AC '99)*, 933–938. New Orleans, LA, USA: ACM. doi:10.1145/309847.310101.

Petracca, Michele, Benjamin G. Lee, Keren Bergman, and Luca P. Carloni. 2009. "Photonic NoCs: System-Level Design Exploration." *IEEE Micro* 29 (4): 74–85.

Peyton Jones, Simon, ed. 2003. *Haskell 98 Language and Libraries: The Revised Report*. Cambridge: Cambridge University Press.

Pfister, G. F., and V. Norton. 1985. "Hot Spot Contention and Combining in Multistage Interconnection Networks." *IEEE Transactions on Computers* 34 (10): 943–948.

Pimentel, A. D. 2008. "The Artemis Workbench for System-Level Performance Evaluation of Embedded Systems." *International Journal of Embedded Systems* 3 (3): 181–196.

Pimentel, A. D., C. Erbas, and S. Polstra. 2006. "A Systematic Approach to Exploring Embedded System Architectures at Multiple Abstraction Levels." *IEEE Transactions on Computers* 55, no. 2 (February): 99–112.

Pinkston, Timothy Mark, and Jose Duato. 2006. "Appendix E: Interconnection Networks." In John L. Hennessy and David A. Patterson, *Computer Architecture: A Quantitative Approach,* Fourth Edition. San Francisco, CA, USA: Morgan Kaufmann.

Pisinger, David, and Mikkel Sigurd. 2007. "Using Decomposition Techniques and Constraint Programming for Solving the Two-Dimensional Bin-Packing Problem." *INFORMS Journal on Computing* (Linthicum, MD, USA) 19 (1): 36–51. doi:10.1287/ijoc.1060.0181.

Plattner, B. 1984. "Real-Time Execution Monitoring." *IEEE Transactions on Software Engineering* 10 (6): 756–764.

Puente, V., J. Gregorio, and R. Beivide. 2002. "SICOSYS: An Integrated Framework for Studying Interconnection Network Performance in Multiprocessor Systems." In *Proceedings of the Euromicro Workshop on Parallel, Distributed and Network-Based Processing*, 15–22. Canary Islands, Spain: IEEE, January.

Purohit, Sohan, Sai Rahul Chalamalasetti, Martin Margala, and Pasquale Corsonello. 2008. "Power-Efficient High Throughput Reconfigurable Datapath Design for Portable Multimedia Devices." In *Proceedings of the International Conference on Reconfigurable Computing and FPGAs (Reconfig08)*, 217–222.

Pusceddu, Matteo, Simone Ceccolini, Gianluca Palermo, Donatella Sciuto, and Antonino Tumeo. 2010. "A Compact TM Multiprocessor System on FPGA." In *Proceedings of the International Conference on Field Programmable Logic*, 578–581. Milano, Italy, August.

Qin, Wei, Subramanian Rajagopalan, and Sharad Malik. 2004. "A Formal Concurrency Model Based Architecture Description Language for Synthesis of Software Development Tools." In *Proceedings of the 2004 ACM SIGPLAN/SIGBED Conference on Languages, Compilers, and Tools for Embedded Systems (LCTES '04)*, 47–56. Washington, DC, USA: ACM. doi:10.1145/997163.997171.

Qureshi, Moinuddin K., and Yale N. Patt. 2006. "Utility-Based Cache Partitioning: A Low-Overhead, High-Performance, Runtime Mechanism to Partition Shared Caches." In *Proceedings of the 39th Annual IEEE/ACM International Symposium on Microarchitecture*, 423–432.

Rabaey, J. M., A. P. Chandrakasan, and B. Nikolic. 2003. "Coping with Interconnect." Chap. 9 in *Digital Integrated Circuits: a Design Perspective*, Second Edition, 445–490. Prentice Hall electronics and VLSI series. Upper Saddle River, NJ, USA: Pearson Education.

Rafique, Nauman, Won-Taek Lim, and Mithuna Thottethodi. 2006. "Architectural Support for Operating System-Driven CMP Cache Management." In *Proceedings of the Fifteenth International Conference on Parallel Architectures and Compilation Techniques (PACT '06)*, 2–12. September.

Rahman, M. M. H., and S. Horiguchi. 2004. "High Performance Hierarchical Torus Network under Matrix Transpose Traffic Patterns." In *Proceedings of the 7th International Symposium on Parallel Architectures, Algorithms and Networks*, 111–116. May. doi:10.1109/ISPAN.2004.1300467.

Rajwar, Ravi, and James R. Goodman. 2001. "Speculative Lock Elision: Enabling Highly Concurrent Multithreaded Execution." In *Proceedings of the 34th Annual ACM/IEEE International Symposium on Microarchitecture*, 294–305. IEEE Computer Society.

———. 2002. "Transactional Lock-Free Execution of Lock-Based Programs." In *Proceedings of the 10th International Conference on Architectural Support for Programming Languages and Operating Systems*, 5–17. ASPLOS-X. San Jose, CA, USA: ACM. doi:10.1145/605397.605399.

Ramanujam, Rohit Sunkam, and Bill Lin. 2009. "A Layer-Multiplexed 3D on-Chip Network Architecture." *IEEE Embedded Systems Letters* 1 (2): 50–55.

Rangan, Krishna K., Gu-Yeon Wei, and David Brooks. 2009. "Thread Motion: Fine-Grained Power Management for Multi-Core Systems." *ACM SIGARCH Computer Architecture News* 37 (3): 302–313.

Ranganathan, P., V. S. Pai, H. Abdel-Shafi, and S. V. Adve. 1997. "The Interaction of Software Prefetching with ILP Processors in Shared-Memory Systems." *ACM SIGARCH Computer Architecture News* 25 (2): 144–156.

Ranganathan, Parthasarathy, Sarita V. Adve, and Norman P. Jouppi. 2000. "Reconfigurable Caches and Their Application to Media Processing." In *Proceedings of the 27th Annual International Symposium on Computer Architecture*, 214–224. June.

Rantala, Ville, Teijo Lehtonen, and Juha Plosila. 2006. *Network on Chip Routing Algorithms*. Technical report, TUCS 779. Turku, Finland: Turku Centre for Computer Science. http://tucs.fi/.

Reshadi, Midia, Ahmad Khademzadeh, Akram Reza, and Maryam Bahmani. n.d. *A Novel Mesh Architecture for on-Chip Networks*. D & R Industry Articles. http://www.design-reuse.com/articles/23347/on-chip-network.html.

Rhoads, Steve. 2001. *Plasma Soft Core*. Open Source Soft Core. http://opencores.org/project,plasma.

Richardson, A. 2006. *WCDMA Design Handbook*. Cambridge: Cambridge University Press.

Rigo, S., G. Araujo, M. Bartholomeu, and R. Azevedo. 2004. "ArchC: A SystemC-Based Architecture Description Language." In *Proceedings of the 16th Symposium on Computer Architecture and High Performance Computing SBAC-PAD 2004*, 66–73. doi:10.1109/SBAC-PAD.2004. 8.

Rigo, Sandro, Marcio Juliato, Rodolfo Azevedo, Guido Araújo, and Paulo Centoducatte. 2004. "Teaching Computer Architecture Using an Architecture Description Language." In *Proceedings of the 2004 Workshop on Computer Architecture Education (WCAE '04)*, 6. Munich, Germany: ACM. doi:10.1145/1275571.1275580.

Ritchie, C. 2008. *Database Principles and Design*. Third Edition. London: Cengage Learning EMEA.

Rusu, S., S. Tam, H. Muljono, D. Ayers, J. Chang, R. Varada, M. Ratta, and S. Vora. 2010. "A 45 nm 8-Core Enterprise Xeon® Processor." *IEEE Journal of Solid-State Circuits* 45 (1): 7–14.

Saastamoinen, I., D. Siguenza-Tortosa, and J. Nurmi. 2002. "Interconnect IP Node for Future System-on-Chip Designs." In *Proceedings of the the First IEEE International Workshop on Electronic Design, Test and Applications (DELTA '02)*, 116–120.

Salminen, E., T. Kangasb, V. Lahtinenb, J. Riihim'kib, K. Kuusilinnac, and T. D. Hämäläinen. 2007. "Benchmarking Mesh and Hierarchical Bus Networks in System-on-Chip Context." *Journal of Systems Architecture* 53, no. 8 (August): 477–488.

Salminen, E., A. Kulmala, and T. D. Hamalainen. 2007. "On Network-on-Chip Comparison." In *Euromicro Symposium on Digital Systems Design*, 503–510. Los Alamitos, CA, USA: IEEE Computer Society. doi:10.1109/DSD.2007.80.

———. 2008. "On the Credibility of Load-Latency Measurement of Network-on-Chips." In *Proceedings of the International Symposium on System-on-Chip (SOC 2008)*, 1–7. Tampere, November.

Sangiovanni-Vincentelli, A., and G. Martin. 2001. "Platform-Based Design and Software Design Methodology for Embedded Systems." *IEEE Design and Test of Computers* 18 (6): 23–33.

Santambrogio, A., M. D. Fracassi, M. Gotti, P. Sandionigi, and C. Antola. 2007. "A Novel Hardware/Software Codesign Methodology Based on Dynamic Reconfiguration with Impulse C and Codeveloper." In *Proceedings of the 3rd Southern Conference on Programmable Logic (SPL '07)*, 221–224. IEEE, February.

Sarbazi-Azad, H., A. Khonsari, and M. Ould-Khaoua. 2002. "Analysis of Deterministic Routing in k-ary n-cubes with Virtual Channels." *Journal of Interconnection Networks* 3 (August): 85–101.

Sarbazi-Azad, H., M. Ould-Khaoua, and L. M. Mackenzie. 2001. "Communication Delay in Hypercubes in the Presence of Bit-Reversal Traffic." *Parallel Computing* 27, no. 13 (December): 1801–1816.

Sassolas, T., N. Ventroux, N. Boudouani, and G. Blanc. 2011. "A Power-Aware Online Scheduling Algorithm for Streaming Applications in Embedded MPSoC." In *Proceedings of the IEEE International Workshop on Power and Timing Modeling, Optimization and Simulation (PATMOS)*, 1–10. Grenoble, France: Springer, September.

Sazeides, Y., and J. E. Smith. 1997. "The Predictability of Data Values." In *Proceedings of the IEEE/ACM International Symposium on Microarchitecture*, 248–258. doi:10.1109/MICRO.1997.645815.

Schultz, M. R. de, A. K. I. Mendonca, F. G. Carvalho, O. J. V. Furtado, and L. C. V. Santos. 2007. "Automatically-Retargetable Model-Driven Tools for Embedded Code Inspection in SoCs." In *Proceedings of the 50th Midwest Symposium on Circuits and Systems (MWSCAS'07)*, 245–248. May. doi:10.1109/MWSCAS.2007.4488580.

Seiler, L., D. Carmean, E. Sprangle, T. Forsyth, P. Dubey, S. Junkins, A. Lake, et al. 2009. "Larrabee: A Many-Core x86 Architecture for Visual Computing." *IEEE Micro* 29, no. 1 (January): 10–21. doi:10.1109/MM.2009.9.

Shabbir, A., A. Kumar, S. Stuijk, B. Mesman, and H. Corporaal. 2010. "CA-MPSoC: An Automated Design Flow for Predictable Multi-Processor Architectures for Multiple Applications." Special Issue on HW/SW Co-Design: Systems and Networks on Chip, *Journal of Systems Architecture* 56 (7): 265–277.

Shacham, A., K. Bergman, and L. P. Carloni. 2008. "Photonic Networks-on-Chip for Future Generations of Chip Multiprocessors." *IEEE Transactions on Computers* 57, no. 9 (September): 1246–1260. doi:10.1109/TC.2008.78.

Sheibanyrad, A. 2008. "Implémentation Asynchrone d'un Réseau-sur-Puce Distribué (Asynchronous Implementation of a Distributed Network-on-Chip)." PhD diss., Université de Pierre et Marie Curie, Paris.

Sheibanyrad, A., A. Greiner, and I. Miro-Panades. 2008. "Multisynchronous and Fully Asynchronous NoCs for GALS Architectures." *IEEE Design & Test of Computers* 25, no. 6 (December): 572–580.

Sheibanyrad, A., I. Miro Panades, and A. Greiner. 2007. "Systematic Comparison between the Asynchronous and the Multi-Synchronous Implementations of a Network on Chip Architecture." In *Proceedings of the Conference on Design, Automation and Test in Europe,* 1090–1095. Nice, France: IEEE.

Shen, H., P. Gerin, and F. Pétrot. 2008. "Configurable Heterogeneous MPSoC Architecture Exploration Using Abstraction Levels." In *Proceedings of the IEEE/IFIP International Symposium on Rapid System Prototyping,* 51–57. Paris, France: IEEE, June.

Sherwood, Timothy, Brad Calder, and Joel S. Emer. 1999. "Reducing Cache Misses Using Hardware and Software Page Placement." In *Proceedings of the 13th International Conference on Supercomputing,* 155–164. ACM, June.

Shim, Keun Sup, Mieszko Lis, Myong Hyon Cho, Omer Khan, and Srinivas Devadas. 2011. "System-Level Optimizations for Memory Access in the Execution Migration Machine (EM^2)." In *Proceedings of the International Workshop on Computer Architecture and Operating System Co-Design.* Heraklion, Crete, Greece, January 24–26. http://projects.csail.mit.edu/caos/caos2011.html.

Shojaei, H., A. H. Ghamarian, T. Basten, M. C. W. Geilen, S. Stuijk, and R. Hoes. 2009. "A Parameterized Compositional Multi-Dimensional Multiple-Choice Knapsack Heuristic for CMP Run-Time Management." In *Proceedings of the Design Automation Conference, DAC'09,* 917–922. ACM.

SoClib (Open Platform for Virtual Prototyping of Multi-Processor Systems-on-Chip). http://www.soclib.fr/.

Soininen, J. P., and H. Hensala. 2003. "A Design Methodology for NoC Based Systems." In *Networks on Chips,* edited by A. Jantsch and H. Tenhunen. Boston: Kluwer.

Song, Zhaohui, Guangsheng Ma, and Dalei Song. 2008. "A NoC-Based High Performance Deadlock Avoidance Routing Algorithm." In *Proceedings of the International Multisymposiums Computer and Computational Sciences, 2008 (IMSCCS'08),* 140–143.

Sonmez, Nehir, Oriol Arcas, Gokhan Sayilar, Osman S. Unsal, Adrian Cristal, Ibrahim Hur, Satnam Singh, and Mateo Valero. 2011. "From Plasma to BeeFarm: Design Experience of an FPGA-Based Multicore Prototype." In *Proceedings of the 7th International Conference on Reconfigurable Computing: Architectures, Tools and Applications,* 350–362. Springer, March 23–25.

Soteriou, V., and Li-Shiuan Peh. 2004. "Design-Space Exploration of Power-Aware on/off Interconnection Networks." In *Proceedings of the IEEE International Conference on Computer Design: VLSI in Computers and Processors, (ICCD'04)*, 510–517. October. doi:10.1109/ICCD.2004.1347970.

Sridharan, Srinivas, Brett Keck, Richard Murphy, Surendar Ch, and Peter Kogge. 2006. "Thread Migration to Improve Synchronization Performance." In *Proceedings of the Workshop on Operating System Interference in High Performance Applications (OSIHPA'06)*.

Srikantaiah, Shekhar, Mahmut Kandemir, and Mary Jane Irwin. 2008. "Adaptive Set Pinning: Managing Shared Caches in CMPs." *ACM SIGARCH Computer Architecture News* 36 (1): 135–144.

Sriram, S., and S. S. Bhattacharyya. 2009. *Embedded Multiprocessors: Scheduling and Synchronization*. Second Edition. Boca Raton, FL, USA: CRC Press.

Stoica, Ion, Hussein Abdel-Wahab, Kevin Jeffay, Sanjoy K. Baruah, Johannes E. Gehrke, and C. Greg Plaxton. 1996. "A Proportional Share Resource Allocation Algorithm for Real-Time, Time-Shared Systems." In *Proceedings of the 17th IEEE Real-Time Systems Symposium*, 288–299. IEEE, December.

Stuijk, S. 2007. "Predictable Mapping of Streaming Applications on Multiprocessors." PhD diss., Eindhoven University of Technology.

Stuijk, S., T. Basten, M. C. W. Geilen, and H. Corporaal. 2007. "Multiprocessor Resource Allocation for Throughput-Constrained Synchronous Dataflow Graphs." In *Proceedings of the Design Automation Conference*, 777–782. ACM.

Stuijk, S., M. C. W. Geilen, and T. Basten. 2006a. "Exploring Trade-Offs in Buffer Requirements and Throughput Constraints for Synchronous Dataflow Graphs." In *Proceedings of the Design Automation Conference, DAC'06*, 899–904. ACM.

———. 2006b. "SDF³: SDF For Free." In *Proceedings of the International Conference on Application of Concurrency to System Design, ACSD'06*, 276–278. IEEE. doi:10.1109/ACSD.2006.23.

———. 2008. "Throughput-Buffering Trade-Off Exploration for Cyclo-Static and Synchronous Dataflow Graphs." *IEEE Transactions on Computers* 57 (10): 1331–1345.

———. 2010. "A Predictable Multiprocessor Design Flow for Streaming Applications with Dynamic Behaviour." In *Proceedings of the Conference on Digital System Design, DSD'10*, 548–555. IEEE. doi:10.1109/DSD.2010.31.

Sudan, Kshitij, Niladrish Chatterjee, David Nellans, Manu Awasthi, Rajeev Balasubramonian, and Al Davis. 2010. "Micro-Pages: Increasing DRAM Efficiency with Locality-Aware Data Placement." *SIGARCH Computer Architecture News* 38:219–230.

Suh, G. E., L. Rudolph, and S. Devadas. 2004. "Dynamic Partitioning of Shared Cache Memory." *Journal of Supercomputing* 28, no. 1 (April): 7–26.

Suh, G. Edward, Srinivas Devadas, and Larry Rudolph. 2001. "Analytical Cache Models with Applications to Cache Partitioning." In *Proceedings of the International Conference on Supercomputing (ICS '01)*, 1–12. June. citeseer.ist.psu.edu/suh01analytical.html.

Suleman, M. Aater, O. Mutlu, M. K. Qureshi, and Y. N. Patt. 2009. "Accelerating Critical Section Execution with Asymmetric Multi-Core Architectures." *ACM Sigplan Notices* 44 (3): 253–264.

Suleman, M. Aater, Yale N. Patt, Eric A. Sprangle, Anwar Rohillah, Anwar Ghuloum, and Doug Carmean. 2007. *ACMP: Balancing Hardware Efficiency and Programmer Efficiency*. Technical report TR-HPS-2007-001. High Performance Substrate/Systems Research Group, University of Texas at Austin, February. http://hps.ece.utexas.edu/.

Sullivan, C., A. Wilson, and S. Chappell. 2004. "Using C Based Logic Synthesis to Bridge the Productivity Gap." In *Proceedings of the 2004 Conference on Asia South Pacific Design Automation: Electronic Design and Solution Fair*, 349–354. Piscataway, NJ, USA: IEEE Press.

Synopsys Inc. n.d. "Design Compiler." http://www.synopsys.com.

———. n.d. "Primetime Power Analysis." http://www.synopsys.com.

Tam, David, Reza Azimi, Livio Soares, and Michael Stumm. 2009. "RapidMRC: Approximating L2 Miss Rate Curves on Commodity Systems for Online Optimizations," *ACM Sigplan Notices* 44 (3): 121–132.

Tam, David, Reza Azimi, and Michael Stumm. 2007. "Thread Clustering: Sharing-Aware Scheduling on SMP-CMP-SMT Multiprocessors." In *Proceedings of EuroSys 2007*, 47–58. Lisbon, Portugal, March.

Tan, Zhangxi, Andrew Waterman, Rimas Avizienis, Yunsup Lee, Henry Cook, David Patterson, and Krste Asanović. 2010. "RAMP Gold: An FPGA-Based Architecture Simulator for Multiprocessors." In *Proceedings of the 47th ACM/IEEE Design Automation Conference (DAC'10)*, 463–468. IEEE.

Thacker, Chuck. 2009. *A DDR2 Controller for BEE3*. Technical report. Microsoft Research.

———. 2010a. *Beehive: A Many-Core Computer for FPGAs (v5)*. http : //projects.csail.mit.edu/beehive/BeehiveV5.pdf.

———. 2010b. *Hardware Transactional Memory for Beehive*. http : / / research . microsoft . com / en – us / um / people / birrell / beehive / hardware % 20transactional % 20memory % 20for % 20beehive3.pdf.

Theelen, B. D., M. C. W. Geilen, T. Basten, J. P. M. Voeten, S. V. Gheorghita, and S. Stuijk. 2006. "A Scenario-Aware Data Flow Model for Combined Long-Run Average and Worst-Case Performance Analysis." In *Proceedings of the International Conference on Formal Methods and Models for Co-Design, MEMOCODE*, 185–194. IEEE.

Thid, Rikard, Ingo Sander, and Axel Jantsch. 2006. "Flexible Bus and NoC Performance Analysis with Configurable Synthetic Workloads." In *Proceedings of the 9th EUROMICRO Conference on Digital System Design*, 681–688. DSD '06. IEEE Computer Society. doi:10.1109/DSD.2006. 52.

Thomas, D., A. Hunt, and C. Fowler. 2001. *Programming Ruby: The Pragmatic Programmer's Guide*. Reading, MA: Addison-Wesley.

Thoziyoor, Shyamkumar, Jung Ho Ahn, Matteo Monchiero, Jay B. Brockman, and Norman P. Jouppi. 2008. "A Comprehensive Memory Modeling Tool and Its Application to the Design and Analysis of Future Memory Hierarchies." In *Proceedings of the 35th International Symposium on Computer Architecture, (ISCA'08)*, 51–62. IEEE.

Trancoso, Pedro, and Josep Torrellas. 1996. "The Impact of Speeding up Critical Sections with Data Prefetching and Forwarding." In *Proceedings of the 1996 International Conference on Parallel Processing, (ICPP'96)*, 3:79–86. IEEE.

Transaction-Level Modeling Working Group. *SystemC*. http : / / www . systemc.org/.

Tripp, J. L., K. D. Peterson, C. Ahrens, J. D. Poznanovic, and M. Gokhale. 2005. "Trident: An FPGA Compiler Framework for Floating-Point Algorithms." In *Proceedings of the 15th International Conference on Field Programmable Logic and Applications (FPL 2005)*, 317–322.

Uzelac, V., and A. Milenkovic. 2009. "A Real-Time Program Trace Compressor Utilizing Double Move-to-Front Method." In *Proceedings of the ACM/IEEE Design Automation Conference*, 738–743.

———. 2010. "Hardware-Based Data Value and Address Trace Filtering Techniques." In *Proceedings of the International Conference on Compilers, Architectures and Synthesis for Embedded Systems*, 117–126. doi:10.1145/ 1878921.1878940.

Uzelac, V., A. Milenkovic, M. Burtscher, and M. Milenkovic. 2010. "Real-Time Unobtrusive Program Execution Trace Compression Using Branch Predictor Events." In *Proceedings of the International Conference on Compilers, Architectures and Synthesis for Embedded Systems*, 97–106. Scottsdale, AZ, USA. doi:10.1145/1878921.1878938.

Vanderbauwhede, W., S. R. Chalamalasetti, S. Purohit, and M. Margala. 2011. "A Few Lines of Code, Thousands of Cores: High-Level FPGA Programming Using Vector Processor Networks." In *Proceedings of the 2011 International Conference on High Performance Computing and Simulation (HPCS)*, 461–467. IEEE.

Vanderbauwhede, W., M. Margala, S. R. Chalamalasetti, and S. Purohit. 2009. "Programming Model and Low-level Language for a Coarse-Grained Reconfigurable Multimedia Processor." In *Proceedings of the 2009 International Conference on Engineering of Reconfigurable Systems and Algorithms (ERSA'09)*, 195–201.

———. 2010. "A C++-Embedded Domain-Specific Language for Programming the MORA Soft Processor Array." In *Proceedings of the 21st IEEE International Conference on Application-Specific Systems Architectures and Processors (ASAP)*, 141–148. IEEE.

Varatkar, Girish V., and Radu Marculescu. 2004. "On-Chip Traffic Modeling and Synthesis for MPEG-2 Video Applications." *IEEE Transactions on Very Large Scale Integration (VLSI) Systems* 12 (1): 108–119.

Ventroux, N., and R. David. 2010. "SCMP Architecture: An Asymmetric Multiprocessor System-on-Chip for Dynamic Applications." In *Proceedings of the Second International Forum on Next-Generation Multicore/Manycore Technologies (IFMT)*, 6. Saint-Malo, France: ACM, June.

Ventroux, N., A. Guerre, T. Sassolas, L. Moutaoukil, G. Blanc, C. Bechara, and R. David. 2010. "SESAM: an MPSoC Simulation Environment for Dynamic Application Processing." In *Proceedings of the IEEE International Conference on Embedded Software and Systems (ICESS)*, 1880–1886. Bradford, UK: IEEE, July.

Ventroux, N., T. Sassolas, R. David, G. Blanc, A. Guerre, and C. Bechara. 2010. "SESAM Extension For Fast MPSoC Architectural Exploration And Dynamic Streaming Application." In *Proceedings of the IEEE/IFIP International Conference on VLSI and System-on-Chip (VLSI-SoC)*, 341–346. Madrid, Spain: IEEE, October.

Ventroux, N., T. Sassolas, A. Guerre, B. Creusillet, and R. Keryell. 2012. "SESAM/Par4All: A Tool for Joint Exploration of MPSoC Architectures and Dynamic Dataflow Code Generation." In *Proceedings of the HIPEAC Workshop on Rapid Simulation and Performance Evaluation: Methods and Tools (RAPIDO)*, 9. Paris, France, January.

Verghese, Ben, Scott Devine, Anoop Gupta, and Mendel Rosenblum. 1996. "Operating System Support for Improving Data Locality on CC-NUMA Compute Servers." *ACM SIGPLAN Notices* 31 (9): 279–289. doi:10. 1145/248209.237205.

Viaud, E., F. Pêcheux, and A. Greiner. 2006. "An Efficient TLM/T Modeling and Simulation Environment Based on Conservative Parallel Discrete Event Principles." In *Proceedings of the Conference on Design, Automation and Test in Europe (DATE)*, 94–99. Nice, France: European Design and Automation Association, April.

VMware, Inc. 2009. *vSphere Resource Management Guide: ESX 4.0, ESXi 4.0, vCenter Server 4.0.* VMware, Inc.

Waldspurger, Carl A., and William E. Weihl. 1994. "Lottery Scheduling: Flexible Proportional Share Resource Management." In *Proceedings of the 1st USENIX Conference on Operating Systems Design and Implementation (OSDI'04)*, 1–11. November.

———. 1995. *Stride Scheduling: Deterministic Proportional-Share Resource Management.* Technical report MIT/LCS/TM-528. MIT, June.

Wang, Howard, M. Petracca, A. Biberman, B. G. Lee, L. P. Carloni, and K. Bergman. 2008. "Nanophotonic Optical Interconnection Network Architecture for on-Chip and off-Chip Communications." In *Proceedings of the Conference on Optical Fiber Communication/National Fiber Optic Engineers Conference, (OFC/NFOEC'08)*, 1–3. February 24–28. doi:10. 1109/OFC.2008.4528127.

Wang, Yi, and Dan Zhao. 2007. "Design and Implementation of Routing Scheme for Wireless Network-on-Chip." In *Proceedings of the IEEE International Symposium on Circuits and Systems, 2007 (ISCAS'07)*, 1357–1360. IEEE.

Weaver, David L., and Tom Germond. 1994. *The SPARC Architecture Manual Version 9.* Sun Microsystems, Inc.

Wein, E. 2007. "Scale in Chip Interconnect Requires Network Technology." In *Proceedings of the International Conference on Computer Design, 2006 (ICCD'06)*, 180–186. IEEE.

Weldezion, Awet Yemane, Matt Grange, Dinesh Pamunuwa, Zhonghai Lu, Axel Jantsch, Roshan Weerasekera, and Hannu Tenhunen. 2009. "Scalability of Network-on-Chip Communication Architecture for 3-D Meshes." In *Proceedings of the 2009 3rd ACM/IEEE International Symposium on Networks-on-Chip*, 114–123. NOCS '09. Washington, DC, USA: IEEE Computer Society. doi:10.1109/NOCS.2009.5071459.

Wentzlaff, D., P. Griffin, H. Hoffmann, Liewei Bao, B. Edwards, C. Ramey, M. Mattina, Chyi-Chang Miao, J. F. Brown, and A. Agarwal. 2007. "On-Chip Interconnection Architecture of the Tile Processor." *IEEE Micro* 27, no. 5 (September): 15–31. doi:10.1109/MM.2007.4378780.

Wieferink, A., M. Doerper, R. Leupers, G. Ascheid, H. Meyr, T. Kogel, G. Braun, and A. Nohl. 2004. "A System Level Processor/Communication Co-Exploration Methodology for Multi-Processor System-on-Chip Platforms." In *International Conference on Design, Automation and Test in Europe (DATE)*, 2:1530–1591. Paris, France, February.

Wiggers, M., M. Bekooij, P. Jansen, and G. Smit. 2006. "Efficient Computation of Buffer Capacities for Multi-Rate Real-Time Systems with Back-Pressure." In *Proceedings of the International Conference on Hardware-Software Codesign and System Synthesis, CODES+ISSS'06*, 10–15. ACM.

Wikipedia. *Fifteen Puzzle*. http://en.wikipedia.org/wiki/-Fifteenpuzzle.

Wiklund, D., and D. Liu. 2003. "SoCBUS: Switched Network on Chip for Hard Real Time Embedded Systems." In *Proceedings of the 17th International Symposium on Parallel and Distributed Processing*. IEEE.

Wilhelm, R., J. Engblom, A. Ermedahl, N. Holsti, S. Thesing, D. Whalley, G. Bernat, C. Ferdinand, R. Heckmann, T. Mitra, et al. 2008. "The Worst-Case Execution-Time Problem – Overview of Methods and Survey of Tools." *ACM Transactions on Embedded Computing Systems* 7 (3): 1–53.

Woo, S. C., M. Ohara, E. Torrie, J. P. Singh, and A. Gupta. 1995. "The SPLASH-2 Programs: Characterization and Methodological Considerations." *ACM SIGARCH Computer Architecture News* 23 (2): 24–36.

Woo, Steven Cameron, Jaswinder Pal Singh, and John L. Hennessy. 1994. "The Performance Advantages of Integrating Block Data Transfer in Cache-Coherent Multiprocessors." *ACM SIGPLAN Notices* 29:219–229.

Xilinx Inc. *Xilinx Floating Point Operator*. http://www.xilinx.com/.

———. n.d. *Fast Simplex Link overview*. http://www.xilinx.com/products/ipcenter/FSL.htm.

Xu, Yi, Yu Du, Bo Zhao, Xiuyi Zhou, Youtao Zhang, and Jun Yang. 2009. "A Low-Radix and Low-Diameter 3D Interconnection Network Design." In *Proceedings of the 15th International Symposium on High Performance Computer Architecture, 2009 (HPCA'09)*, 30–42. IEEE, February 14–18. doi:10.1109/HPCA.2009.4798234.

Yang, F. C., C. L. Chiang, and J. Huang. 2010. "A Reverse-Encoding-Based on-Chip Bus Tracer for Efficient Circular-Buffer Utilization." *IEEE Transactions on VLSI Systems* 18 (5): 732–741. doi:10.1109/TVLSI.2009.2014872.

Yang, Z. J., A. Kumar, and Y. Ha. 2010. "An Area-Efficient Dynamically Reconfigurable Spatial Division Multiplexing Network-on-Chip with Static Throughput Guarantee." In *Proceedings of the International Conference on Field-Programmable Technology, FPT'10,* 389–392.

Yankova, Y., G. Kuzmanov, K. Bertels, G. Gaydadjiev, Y. Lu, and S. Vassiliadis. 2007. "DWARV: Delftworkbench Automated Reconfigurable VHDL Generator." In *Proceedings of the International Conference on Field Programmable Logic and Applications, (FPL'07),* 697–701. IEEE.

Yi, J. J., and D. J. Lilja. 2006. "Simulation of Computer Architectures: Simulators, Benchmarks, Methodologies, and Recommendations." *IEEE Transactions on Computers* 55, no. 3 (March): 268–280. doi:10.1109/TC.2006.44.

Ykman-Couvreur, Ch., V. Nollet, F. Catthoor, and H. Corporaal. 2006. "Fast Multidimension Multichoice Knapsack Heuristic for MP-SoC Run-Time Management." In *Proceedings of International Symposium on SoC,* 1–4. IEEE.

———. 2011. "Fast Multidimension Multichoice Knapsack Heuristic for MP-SoC Runtime Management." *ACM Transactions on Embedded Computer Systems* 10, no. 3 (May): 35:1–35:16. doi:10.1145/1952522.1952528.

Yourst, Matt T. 2007. "PTLsim: A Cycle Accurate Full System x86-64 Microarchitectural Simulator." In *Proceedings of the 2007 IEEE International Symposium on Performance Analysis of Systems & Software,* 23–34. San Jose, CA, USA: IEEE, April. doi:10.1109/ISPASS.2007.363733.

Zebchuk, Jason, Vijayalakshmi Srinivasan, Moinuddin K. Qureshi, and Andreas Moshovos. 2009. "A Tagless Coherence Directory." In *Proceedings of the 42nd Annual IEEE/ACM International Symposium on Microarchitecture (MICRO-42),* 423–434. IEEE.

Zeferino, C. Albenes, and A. Amadeu Susin. 2003. "SoCIN: A Parametric and Scalable Network-on-Chip." In *Proceedings of the 16th Symposium on Integrated Circuits and Systems Design,* 169–174.

Zhang, Hui, and Srinivasav Keshav. 1991. "Comparison of Rate-Based Service Disciplines." In *Proceedings of the Conference on Communications Architecture & Protocols,* 113–121. SIGCOMM '91. Zurich, Switzerland: ACM. doi:10.1145/115992.116004.

Zhang, Lei, Mei Yang, Yingtao Jiang, Emma Regentova, and Enyue Lu. 2006. "Generalized Wavelength Routed Optical Micronetwork In Network-on-Chip." In *Proceedings of the 18th IASTED International Conference on Parallel and Distributed Computing Systems.* Dallas, TX, USA, November 13–15. http : / / www . osti . gov / eprints / topicpages / documents/record/390/1675818.html.

Zhang, M., and K. Asanović. 2005. "Victim Replication: Maximizing Capacity while Hiding Wire Delay in Tiled Chip Multiprocessors." *ACM SIGARCH Computer Architecture News* 33 (2): 336–345.

Zhang, Xiao, Sandhya Dwarkadas, and Kai Shen. 2009. "Hardware Execution Throttling for Multi-Core Resource Management." In *Proceedings of the 2009 Conference on USENIX Annual Technical Conference,* 23–23. USENIX'09. San Diego, CA, USA: USENIX Association. http://dl.acm.org/citation.cfm?id=1855807.1855830.

Zhang, Yuting, and Richard West. 2006. "Process-Aware Interrupt Scheduling and Accounting." In *Proceedings of the 27th IEEE International Real-Time Systems Symposium, (RTSS'06),* 191–201. IEEE.

Zhao, Hongzhou, Arrvindh Shriraman, and Sandhya Dwarkadas. 2010. "SPACE: Sharing Pattern-Based Directory Coherence for Multicore Scalability." In *Proceedings of the 19th International Conference on Parallel Architectures and Compilation Techniques,* 135–146. PACT'10. Vienna, Austria: ACM. doi:10.1145/1854273.1854294.

Zhao, Li, Ravi Iyer, Ramesh Illikkal, Jaideep Moses, Don Newell, and Srihari Makineni. 2007. "CacheScouts: Fine-Grain Monitoring of Shared Caches in CMP Platforms." In *Proceedings of the Parallel Architectures and Compilation Techniques (PACT'07),* 339–352. IEEE Computer Society, September.

Ziv, J., and A. Lempel. 1977. "A Universal Algorithm for Sequential Data Compression." *IEEE Transactions on Information Theory* 23 (3): 337–343.

Index